Molecular, Cellular, and Clinical Aspects of Angiogenesis

NATO ASI Series

Advanced Science Institutes Series

A series presenting the results of activities sponsored by the NATO Science Committee, which aims at the dissemination of advanced scientific and technological knowledge, with a view to strengthening links between scientific communities.

The series is published by an international board of publishers in conjunction with the NATO Scientific Affairs Division

A	Life Sciences	Plenum Publishing Corporation
B	Physics	New York and London
C	Mathematical and Physical Sciences	Kluwer Academic Publishers
D	Behavioral and Social Sciences	Dordrecht, Boston, and London
E	Applied Sciences	
F	Computer and Systems Sciences	Springer-Verlag
G	Ecological Sciences	Berlin, Heidelberg, New York, London,
H	Cell Biology	Paris, Tokyo, Hong Kong, and Barcelona
I	Global Environmental Change	

PARTNERSHIP SUB-SERIES

1. Disarmament Technologies	Kluwer Academic Publishers
2. Environment	Springer-Verlag
3. High Technology	Kluwer Academic Publishers
4. Science and Technology Policy	Kluwer Academic Publishers
5. Computer Networking	Kluwer Academic Publishers

The Partnership Sub-Series incorporates activities undertaken in collaboration with NATO's Cooperation Partners, the countries of the CIS and Central and Eastern Europe, in Priority Areas of concern to those countries.

Recent Volumes in this Series:

Volume 283 — Eicosanoids: From Biotechnology to Therapeutic Applications
edited by Gian Carlo Folco, Bengt Samuelsson, Jacques Maclouf, and G. P. Velo

Volume 284 — Advances in Morphometrics
edited by Leslie F. Marcus, Marco Corti, Anna Loy, Gavin J. P. Naylor, and Dennis E. Slice

Volume 285 — Molecular, Cellular, and Clinical Aspects of Angiogenesis
edited by Michael E. Maragoudakis

Series A: Life Sciences

Molecular, Cellular, and Clinical Aspects of Angiogenesis

Edited by

Michael E. Maragoudakis

University of Patras Medical School
Patras, Greece

Plenum Press
New York and London
Published in cooperation with NATO Scientific Affairs Division

Proceedings of a NATO Advanced Study Institute on
Molecular, Cellular, and Clinical Aspects of Angiogenesis,
held June 16 – 27, 1995,
in Porto Carras, Halkidiki, Greece

NATO-PCO-DATA BASE

The electronic index to the NATO ASI Series provides full bibliographical references (with keywords and/or abstracts) to about 50,000 contributions from international scientists published in all sections of the NATO ASI Series. Access to the NATO-PCO-DATA BASE is possible in two ways:

—via online FILE 128 (NATO-PCO-DATA BASE) hosted by ESRIN, Via Galileo Galilei, I-00044 Frascati, Italy

—via CD-ROM "NATO Science and Technology Disk" with user-friendly retrieval software in English, French, and German (©WTV GmbH and DATAWARE Technologies, Inc. 1989). The CD-ROM also contains the AGARD Aerospace Database.

The CD-ROM can be ordered through any member of the Board of Publishers or through NATO-PCO, Overijse, Belgium.

Library of Congress Cataloging-in-Publication Data

Molecular, cellular, and clinical aspects of angiogenesis / edited by
Michael E. Maragoudakis.
        p.   cm. -- (NATO ASI series. Series A, Life sciences ; v.
285)
     "Proceedings of a NATO Advanced Study Institute on Molecular,
Cellular, and Clinical Aspects of Angiogenesis, held June 16-27,
1995, in Porto Carras, Halkidiki, Greece"--T.p. verso.
     "Published in cooperation with NATO Scientific Affairs Division."
  Includes bibliographical references and index.
  ISBN 0-306-45315-0
     1. Neovascularization--Congresses. 2. Neovascularization
inhibitors--Congresses. I. Maragoudakis, Michael E. II. North
Atlantic Treaty Organization. Scientific Affairs Division.
III. NATO Advanced Study Institute on Molecular, Cellular, and
Clinical Aspects of Angiogenesis (1995 : Porto Karras, Chalkidiki,
Greece) IV. Series.
     [DNLM: 1. Neovascularization, Pathologic--congresses.
2. Neovascularization, Physiologic--congresses. 3. Angiogenesis
Factor--physiology--congresses. WG 500 M718 1996]
QP106.6.M65 1996
612.1'3--dc20
DNLM/DLC
for Library of Congress   96-21623
                                                         CIP

ISBN 0-306-45315-0

PREFACE

There has been an explosion of research activity related to angiogenesis in recent years, and hundreds of laboratories worldwide are actively involved in many aspects of angiogenesis. The literature on angiogenesis increases exponentially every year, and more than 16,000 peer-reviewed articles have been published the past 25 years, which are scattered in basic science and clinical journals. The complexity of the cascade of events leading to new vessel formation from preexisting ones has challenged scientists in cell biology, biochemistry, physiology, pharmacology, molecular biology, developmental biology, and other fields. With their multidisciplinary approach and the powerful new techniques that have been developed, the progress in understanding angiogenesis has been impressive indeed. Only 12 years ago the mention of an angiogenic factor caused skepticism. Today we have the complete amino-acid sequence and their genes cloned for at least 9 angiogenic factors. Many laboratories are studying their role in angiogenesis, and several biotechnology firms have a keen interest in commercial developments relative to these molecules. The role of extracellular matrix components in angiogenesis and the interaction of endothelial cells with other cell types such as pericytes, smooth muscle cells, and inflammatory cells have been studied by other groups.

This rapid expansion is the result of a realization that in many disease states a common underlying pathology is a derangement in angiogenesis. Practically every medical specialty deals with disease states, some very common ones, where angiogenesis-based therapy may be possible and calls for the development of suppressors or promoters of angiogenesis. Suppressors of angiogenesis have potential clinical applications in situations where abnormal proliferation of blood vessels is related to disease progression. Such situations are: solid tumors, hemangiomas, diabetic retinopathy, inflammatory disease, etc. On the other hand, promoters of angiogenesis have therapeutic potential in conditions resulting from insufficient blood supply, such as wound healing, coronary artery disease, stroke, duodenal ulcers, organ grafts, diabetic leg ulcers, some types of hair loss, certain types of infertility, etc. There are at least 10 clinical trials in phase I and II for evaluation of promoters or inhibitors of angiogenesis. The recent findings that a positive correlation exists between vascular density and metastatic potential of many solid tumors suggests that angiogenesis can be a new and independent prognostic indicator of tumor malignancy.

Despite these impressive developments many important aspects of angiogenesis are open for investigation. We are in need of animal models to understand the role of angiogenesis in tumor progression and other angiogenic diseases. More sophisticated techniques are also needed to evaluate the phenotypic differences of endothelial cells. This may provide an explanation why in some cases angiogenesis is effective in tumor progression and not in others. It is important to understand the overall control mechanisms and the signal transduction pathways involved in physiological and pathological angiogenesis. This knowledge will undoubtedly lead to the development of new therapeutic approaches to control distinct angiogenic phenotypes.

This book contains the proceedings of the NATO Advanced Study Institute on "Molecular, Cellular and Clinical Aspects of Angiogenesis" held in Porto Carras, Halkidiki, Greece, from June 16-27, 1995. This meeting was a comprehensive review of the various aspects of angiogenesis, such as endothelial cell heterogeneity, the role of extracellular matrix, angiogenic factors and their receptors, regulation of angiogenesis, physiological and pathological angiogenesis, inhibitors of angiogenesis and biotechnological aspects of angiogenesis. The presentations and discussions of the meeting provided the opportunity for investigators from many different areas of basic science and medicine to meet, exchange ideas, information, and techniques, and evaluate the present status and future directions in the field of angiogenesis.

I wish to thank Drs. Pietro Gullino and Peter Lelkes, co-directors of this meeting, for their help in the selection of topics and speakers and also in fund raising. The International Organizing Committee that included Drs. Michael Höckel, Nicholas Kefalides, Moritz Konerding, Peter Polverini, and Marco Presta has provided an invaluable help in the organization of the meeting for which I am grateful. I thank also all the participants for their enthusiastic participation and their complementary comments on the success of the conference. Special thanks are due to the Scientific Affairs Division of NATO for providing the major portion of the grants for the organization of the meeting and also for publication of this book. The contribution of the following organizations: Becton and Dickinson/Collaborative Res. (USA), Carbomed (USA), Elpen (Greece), Galenica (Greece), Genentech (USA), Hoechst (Greece), Italfarmaco (Italy), Persetive (USA), Pharmacia (Italy), Pfizer (USA), Promega (USA), Repligen (USA), SmithKline Beecham (Greece), TAP Pharmaceuticals (USA), Vianex (Greece), which was used to support the participation of many young scientists, is gratefully acknowledged. For travel arrangements and the daily operations of the conference I am thankful to Mrs. Lydia Argyropoulou of Mondial Tours. I am particularly grateful to Mrs. Anna Marmara for her dedicated and enthusiastic work throughout the organization of the meeting and the editing of this monograph.

Michael E. Maragoudakis

CONTENTS

HETEROGENEITY OF ENDOTHELIAL CELLS

ROLE OF EXTRACELLULAR MATRIX IN ANGIOGENESIS

ANGIOGENIC FACTORS AND THEIR RECEPTORS

PHYSIOLOGICAL ANGIOGENESIS

PATHOLOGICAL ANGIOGENESIS

INHIBITORS OF ANGIOGENESIS

BIOTECHNOLOGICAL ASPECTS OF ANGIOGENESIS

ON THE POSSIBLE ROLE OF ENDOTHELIAL CELL HETEROGENEITY IN ANGIOGENESIS

Peter I. Lelkes[*], Vangelis G. Manolopoulos[*,#], Matthew Silverman[*], Shaosong Zhang[*], Soverin Karmiol[$], and Brian R. Unsworth[#]

[*]Lab. Cell Biology, Univ. Wisconsin Med School, Milwaukee Clinical Campus, Milwaukee, WI U.S.A., [$]Clonetics Corporation, San Diego, and [#]Dept. Biology, Marquette University, Milwaukee, WI

INTRODUCTION: The endothelial cell (EC[1]) lining of all blood conduits and lymphatics performs a large array of pivotal physiological functions. The generation of new blood vessels by vasculogenesis and/or angiogenesis is but one of the complex issues of the multifaceted physiology of vascular ECs (34,73,75). Functional and immunological heterogeneity, for example between venous and arterial ECs, or ECs lining large vessels and microvessels, is another manifestation of the complex biology of these pluripotent cells (22,78,93). Since the generation of new blood vessels is a highly organized, localized event which occurs in organ/tissue-specific microvasculature, we hypothesize that the local microenvironment of "angiogenic" microvessels may contain unique, site-specific cues which contribute to EC heterogeneity and which, in turn, makes these ECs especially prone to activation by angiogenic stimuli. In the wake of our increased awareness of the intricacies of both angiogenesis and endothelial cell heterogeneity, we have begun to explore possible connections between these two seemingly unrelated concepts.

In this paper we will first discuss some general aspects of EC heterogeneity, in particular differences between micro- and macro-vessel derived ECs, as they, in our view, relate to the angiogenic cascade. In the second part we will present some of our recent findings on specific aspects of EC heterogeneity (regulation of the expression of cell adhesion

---

[1] **Abbreviations**: AC: adenylyl cyclase, bFGF: basic fibroblast growth factor, BM: basement membrane, cAMP: cyclic adenosine monophosphate, EC: endothelial cell, ECM: extracellular matrix, ECs: endothelial cells, EHS: Englebreth-Holm-Swarm, ELICA: enzyme-linked cell culture assay, ELISA: enzyme-linked immunoadsorbent assay, FN: fibronectin, HMVEC: human adult dermal microvascular endothelial cell, HUVEC: human umbilical vein endothelial cell, ICAM-1: intercellular adhesion molecule-1, IL-1α: interleukin-1α, LPS: lipopolysaccharide, PECAM-1: platelet-endothelial cell adhesion molecule-1, PDE: phosphodiesterase, PKC: protein kinase C, RT-PCR: reverse transcription -polymerase chain reaction, SSRE: shear-stress responsive element, TNFα: Tumor necrosis factor-α, TF: tissue factor, tPA: tissue type plasminogen activator, VCAM-1: vascular cell adhesion molecule-1, VEGF: vascular endothelial cell growth factor

Molecular, Cellular, and Clinical Aspects of Angiogenesis
Edited by Michael E. Maragoudakis, Plenum Press, New York, 1996

molecules and tissue factor activity, extracellular matrix protein deposition, cAMP signaling) and discuss their possible relevance to one or more steps of the angiogenic cascade.

SOME GENERAL ASPECTS OF ENDOTHELIAL CELL HETEROGENEITY AS THEY MAY RELATE TO ANGIOGENESIS: All vascular and lymphatic ECs share a number of common, basic functions, most important of which is the formation of the thromboresistant, dynamic interface between the blood (lymphatics) and the underlying tissues. However, ECs residing in various anatomical locations are endowed with a unique repertoire of gene products and cellular regulatory mechanisms, which enable them to perform their basic "task" in a selective, tissue-specific manner, for instance in the brain (72,87) or in different visceral organs, such as the heart, kidney, or liver, (38,49,71).

In the past, functional dissimilarities between various EC phenotypes *in vivo* have been postulated mainly on the basis of the obvious morphological differences between ECs lining the different vascular beds (9,48,73,93). However, since the initial *in vitro* studies mainly employed large vessel ECs isolated from aortae or umbilical veins (23,35), only sporadic notion was made of an apparent EC heterogeneity (22). Indeed, much of what we know today about "EC biology" is largely based on *in vitro* studies and will need re-examination using isolated microvascular ECs. However, only recently have newer techniques become available for isolating tissue-specific microvascular ECs. These techniques include the purification of mixed primary isolates by sorting with lectin- or antibody-coated magnetic beads (33), or by flow-cytometry (12,88). It is only within last few years, that we can begin to study and compare the functional responses of distinct EC types isolated from large vessels such as vena cava, femoral artery, coronaries or pulmonary circulation and from tissue-specific microvascular beds such as brain, lung, fat, liver, bone marrow, placenta and endometrium. Thus, the concept of EC heterogeneity has truly emerged as a scientifically testable hypothesis. Indeed, the availability of tissue-specific ECs, has confirmed some general similarities between the various EC types, but at the same time has lead to the discovery of substantial differences between ECs isolated from various anatomical sites (21,82). As seen in Figure 1, a brief survey of literature listed in MEDLINE reveals a typical exponential rise in the number of annual publications that focus on *endothelial cell(s)* and *heterogeneity*.

Based on this by no means exhaustive search, we predict a significant future boom in experimental studies that will fine-tune our understanding of the tissue-specific regulation of EC physiology. A challenge for the future is to develop new approaches for studying

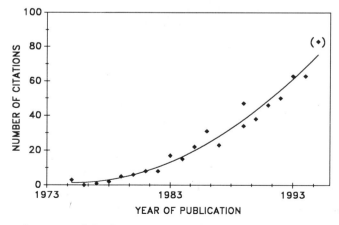

Figure 1: Current Awareness of the Concept of *Endothelial Cell Heterogeneity*. This curve, compiled from a MEDLINE search in June 1995, represents an exponential increase in the number of annual citations focussing on this topic. The last data point in parentheses is the projected value for 1995.

regional EC micro-heterogeneity: Current techniques are, as yet, only in their infancy to reliably selectively isolate and culture EC from arterioles or venules or even phenotypically diverse EC within the same (microvascular) bed (19,42).

In our view, EC heterogeneity plays a profound role in the tissue-specific regulation of angiogenesis. New (capillary) vessels are created *in vivo* from pre-existing ones mainly at the level of the smaller blood and lymph conduits. By literally watching angiogenesis occur, e.g. by time-lapse intravital video microscopy (68), one may gain a healthy respect for the complex dynamics of angiogenesis at the level of the microvasculature, and for the exquisite coordination between the various steps of the angiogenic cascade. This cascade, once initiated, is a vectorial process involving a number of highly-coordinated steps (Figure 2), including the dissolution of the subendothelial basement membrane, migration of individual ECs, EC proliferation, formation of a new vessel lumen by establishing tight inter-endothelial cell contacts, and, finally, the de-novo synthesis of a new basement membrane.

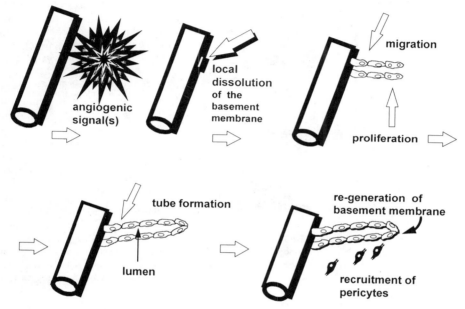

Figure 2: Principle Events During Angiogenesis. Each of the steps of this cascade may be subject of modulation by tissue-specific EC responses. For details see text.

We hypothesize that each step of the angiogenic cascade could potentially be subject to tissue-specific regulation due to the heterogeneity of EC responses in each tissue This concept is particularly evident for the very first step in angiogenesis, namely the cells' responsiveness to a particular angiogenic signal. As discussed in other chapters in this book, initiation and/or progression of angiogenesis is not a single-modal, all-or-none event. Rather, the dynamics of establishing new blood vessels is the consequence of concertedly tipping the balance of putative promoters and inhibitors of any of the above mentioned steps within the angiogenic cascade in a way that the net result will either initiate, maintain, or, alternatively, inhibit this process. Listed in Figure 3 are a number of currently endorsed promoters and inhibitors of angiogenesis. Since the local availability of any of these factors may vary in a time- and tissue-dependent manner, it is the net, cumulative action of these angiogenic compounds which determines the "angiogenic balance", that is, whether the angiogenic process is initiated, facilitated or inhibited. Under "normal" physiological conditions, this balance favors the anti-angiogenic phenotype, i.e. the EC monolayer is quiescent. Initiation of the angiogenic process requires a rise in the local concentration of these promoters. For example, ischemia-induced angiogenesis is most probably caused by hypoxia-induced elevation of the message for, and increased secretion of, vascular endothelial cell growth factor (VEGF) in (smooth, cardiac) muscle cells and/or fibroblasts (74). Inflammatory cytokines might promote angiogenesis either directly (55,91), or indirectly by triggering the local release of potent angiogenic factors, such as scatter factor (25), or other cytokines (31) from leukocytes adhering to activated ECs. Sustained angiogenesis requires the continued availability of promoters of each step (e.g., basement membrane remodelling, migration, proliferation), in the angiogenic sequence. Any deviation from this vectorial process will tip the balance from the pro-angiogenic to anti-angiogenic phenotype and result in the arrest of vessel sprouting, or initiate angio-regression. Time-lapse video microscopy suggest that these latter events are an integral part of the angiogenic process. And, yet, very little is known about (tissue-specific?) mechanisms that lead to the arrest of angiogenesis or cause the involution of already established vessels.

INHIBITORS   PROMOTERS

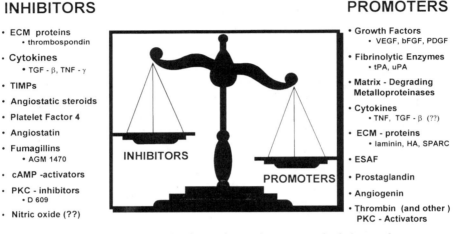

- ECM proteins
  - thrombospondin
- Cytokines
  - TGF - β, TNF - γ
- TIMPs
- Angiostatic steroids
- Platelet Factor 4
- Angiostatin
- Fumagillins
  - AGM 1470
- cAMP -activators
- PKC - inhibitors
  - D 609
- Nitric oxide (??)

- Growth Factors
  - VEGF, bFGF, PDGF
- Fibrinolytic Enzymes
  - tPA, uPA
- Matrix - Degrading Metalloproteinases
- Cytokines
  - TNF, TGF - β (??)
- ECM - proteins
  - laminin, HA, SPARC
- ESAF
- Prostaglandin
- Angiogenin
- Thrombin (and other) PKC - Activators

Figure 3: Angiogenic Balance. In order for angiogenesis to occur, the balance of promoters and inhibitors of angiogenesis must be tipped in a particular fashion which concertedly favors initiation and sequential maintenance of individual steps in the angiogenic cascade. Deviation from the cumulative pro-angiogenic balance will result in the cessation of angiogenesis and/or angioregression.

How does endothelial cell heterogeneity fit into this already complex picture? As discussed before, ECs derived from various anatomical locations differ in many aspects in their ability to perceive, transduce and respond to "angiogenic signals" (42,43). Migration and/or proliferation of ECs are tightly controlled by tissue-specific factors derived from the local microenvironment, such as humoral factors, extracellular matrix (ECM) constituents and hemodynamic forces (13,45,58). Furthermore, the local remodelling of the subendothelial basement membrane seems to be tissue specific. For example, basic fibroblast growth factor (bFGF), the "classical" promoter of angiogenesis, enhances tissue-type plasminogen activator (tPA) production in aortic ECs while suppressing it in venous ECs (92). Also, the threshold for initiating or sustaining angiogenesis by "angiogenic signals" might differ between various EC types. Specifically, microvascular ECs differ from large-vessel ECs in the expression and activation of numerous cell surface antigens, such as cell adhesion molecules (82,83). We recently found that upregulation of vascular cell adhesion molecule-1 (VCAM-1) in human umbilical vein endothelial cells (HUVEC) is by two to three orders of magnitude more sensitive to activation by tumor necrosis factor-α (TNFα) than in adult dermal microvessel derived endothelial cells (HMVEC) (Zhang *et al*, manuscript submitted for publication). Thus, based on our increased understanding of EC heterogeneity, it is prudent to use ECs isolated from microvessels in different tissues to study and fully appreciate the organ-specific regulation of angiogenesis.

The disparity in the "angiogenic potential" of HMVEC and HUVEC is exemplified in Figure 4. In this experiment, identical numbers (5 x 10⁴/well in a 12-well plate) of HUVEC and HMVEC (of the same generation/number of population doublings) were plated onto Matrigel (the basement membrane isolated from murine Englebreth-Holm-Swarm (EHS) tumors), which is widely used as a model system to study angiogenic mechanisms (26,41). At various time points we quantitated the extent of tube formation by computer- assisted video microscopy.

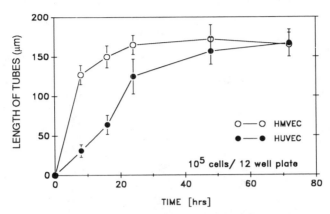

Figure 4: Time Course of Tube Formation on Matrigel by Phenotypically Diverse Human Endothelial Cells. 5 x 10⁴ ECs derived from human umbilical vein (HUVEC) and from dermal microvasculature (HMVEC) were plated onto Matrigel under identical experimental conditions into 12 well-plates. At various time-points the extent of tube formation was quantitated by computer-assisted video microscopy, as previously described (26,27). Data are presented as means ± SEM from two independent experiments carried out in triplicate wells.

As previously reported, under these conditions HUVEC required nearly 24 hours for full completion of the capillary-like tubular assembly(26). In contrast, HMVEC established a complex tubular network within approx. 8 hours. Similar results were obtained when we compared other large-vessel-derived ECs and microvascular ECs (data not shown). The reasons for the enhanced angiogenic response of microvessel-derived ECs are at present unclear. As discussed above, there are numerous steps along the route of angiogenesis in which the two ECs types might differ, e.g. in the response to angiogenic stimuli, signal transduction mechanisms, rate of migration, rate of dissolution of the basement membrane, etc.

DIFFERENTIAL EXPRESSION AND REGULATION OF CELL ADHESION MOLECULES IN HUVEC AND HMVEC: Recent studies strongly suggest that certain adhesion molecules and/or their integrin receptors may be necessary for the angiogenic cascade (5,8,10). Intercellular adhesion molecule-1 (ICAM-1) and vascular cell adhesion molecule-1 (VCAM-1) are well characterized, heterotypic cell adhesion molecules, which mediate the adhesion of circulating leukocytes to the endothelium (7). There is increasing evidence for profound differences in the degree of expression and activation of adhesion molecules and their receptors in various ECs(28). In the context of angiogenesis, regulation of ICAM-1 and VCAM-1 expression is important since the initial step of leukocyte adhesion to cytokine-activated microvascular ECs during ischemia, inflammation, infection or other pathological diseases (84) precedes the release of specific factors, which might locally trigger and/or sustain the angiogenic process.

Another cell adhesion molecule of interest, the platelet-endothelial cell adhesion molecule-1 (PECAM-1) localizes at the site of endothelial cell-cell contacts (16). In the context of angiogenesis, disruption of the endothelial monolayer integrity, e.g. following down-regulation of PECAM-1, is likely to be an essential prerequisite for the initiation of EC migration. Moroever, PECAM-1 can also function as heterotypic cell adhesion molecule: the $\alpha_v\beta_3$ integrin is a specific ligand for PECAM-1 and involved in leukocyte extravasation across the EC monolayer (65). Remarkably, the same $\alpha_v\beta_3$ integrin has been implicated as an important regulatory molecule of angiogenesis (10,11).

Using a whole cell ELISA assay (62), we found that the levels of basal expression of these three adhesion molecules are higher in HUVEC than in HMVEC (Figure 5), ranging from approx. 1.5-fold for ICAM-1, two-fold for PECAM-1, to approx. 4-fold for VCAM-1. We also observed a profound heterogeneity in the regulation of these adhesion molecules by bacterial endotoxin lipopolysaccharide (LPS).

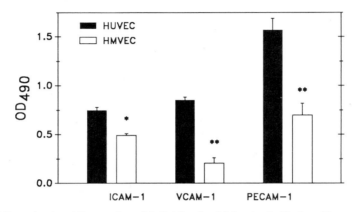

Figure 5: Differential Basal Expression of Cell Adhesion Molecules in Various Human Endothelial Cells. HUVEC and HMVEC were plated in 24 well tissue culture plates. After reaching confluence, basal cell surface expression of ICAM-1, VCAM-1, and PECAM-1 was assessed using a whole cell-based, spectrophotometric enzyme linked immunoculture assay (ELICA), as previously described (62). Primary antibodies were generous gifts of Drs. Mary E. Gerritsen (α-ICAM-1 and α-VCAM-1, monoclonal) and Peter J. Newman (α-PECAM-1, polyclonal). The data shown are from a characteristic experiment, carried out in triplicate wells and are presented as means $\pm$ SEM of the optical density (OD$_{490}$ nm) of the peroxidase substrate reaction product. Similar data were obtained in two independent experiments. **: p< 0.01, *: p < 0.05.

In line with previous reports on the action of cytokines (6,21,40), exposure of the cells for 24 hours to 1 μg/ml LPS caused a dose-dependent upregulation of ICAM-1 and VCAM-1, which was more pronounced in HUVEC than in HMVEC. By contrast, exposure to LPS induced a marked, dose-dependent downregulation of PECAM-1 in both cell types which, again, was more profound in HUVEC than in HMVEC (Figure 6). Recent data suggest that LPS is an angiogenic molecule *in vivo*, albeit probably through indirect mechanisms (53). In addition, our data suggest a possible direct function of endotoxins in angiogenesis, by perturbing endothelial barrier function and thus facilitating the migratory process of the ECs themselves.

TNFα is a well characterized inflammatory cytokine that is involved in EC activation, for example by enhancing the transcription and cell-surface expression of vascular cell adhesion molecules and of tissue factor (60,76). TNFα is also one of the more controversial putative angiogenic cytokines, which can or cannot elicit angiogenesis *in vivo*, depending on the model system used, the dosage and the site of application (18,59). The reason for these conflicting observations remain unclear. In line with our hypothesis of EC heterogeneity, we argue that the controversial effects of TNFα on angiogenesis may be due to the differential sensitivity and/or heterogeneous responses of ECs in various anatomical locations. Indeed, there are several reports in the literature attesting to an altered susceptibility of microvessel-derived ECs to activation by TNFα and cytokines in general (82,89). From our vantage point,

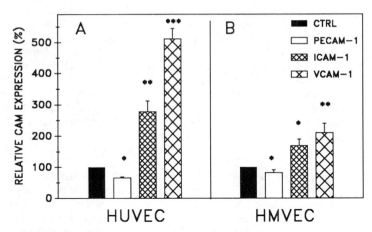

Figure 6: Differential Regulation of Cell Adhesion Molecules by Endotoxin in Various Human Endothelial Cells. The basal and LPS (10 μg/ml for 24 hours)-induced cell-surface expression of ICAM-1, VCAM-1 and PECAM-1 in HUVEC and HMVEC were assessed as described in the legends to Figure 5. After normalization for the number of cells in each well, the data are expressed relative to the basal expression of HUVEC and HMVEC, respectively. The data represent means ± SEM from 3 independent experiments carried out in triplicate wells. ***: p< 0.001, **: p<0.01, * p<0.05.

Table 1: **Heterogeneity of PECAM-1 Regulation by Endotoxin (LPS) and Cytokines (TNFα and IL-1α)**

	HUVEC	HMVEC
control, untreated cells	100	100
10 μg/ml LPS	74.6 ± 4.0[a]	78.2 ± 2.1[a]
10 ng/ml TNFα	83.0 ± 2.7[a]	110.7 ± 1.9[b]
1 ng/ml IL-1α	77.1 ±2.7[a]	110.1 ± 2.8[b]

PECAM-1 expression was determined using whole-cell based ELISA, as previously described (62) HUVEC or HMVEC (generations 3 - 8) were plated in 24 well plates at an initial density of 5 x 10⁴ cells/well. After reaching confluence, the cells were treated for 24 hours with the indicated concentrationsof the various chemicals. All experiments were carried out in triplicates and repeated at least three times. The results, normalized to the number of cells per well, represent the relative change in PECAM-1 expression as compared to untreated controls. The data are expressed as means ± SEM. a: p< 0.01, b: p< 0.05 (compared to the controls).

this observation is of importance since CAM-mediated adhesion of activated leukocytes might cause the local release of growth factors, cytokines, and proteases which might serve as angiogenic triggers.

In studying the heterogeneity of EC responses to activation by TNFα we observed several significant differences between HUVEC and HMVEC. For example, exposure of the cells for 24 hours to 10 ng/ml TNFα induced in HUVEC a down-regulation of PECAM-1 expression by approx 20 %. By contrast, in HMVEC, the same treatment caused a slight but statistically significant up-regulation of PECAM-1 expression (Table 1). Similar results were obtained following exposure of the two cell types to IL-1α.

These results indicate EC heterogeneity at two different levels: 1) between macro- and microvascular ECs, and 2) between activation by cytokines and endotoxin. We hypothesize that these differences are based on cell type-specific, differential gene expression and/or signaling pathways which may be involved in the distinct, receptor-mediated activation of cell adhesion molecules by cytokines and endotoxin, respectively.

DIFFERENTIAL MODULATION OF ENDOTHELIAL CELL ADHESION MOLECULES BY CYCLIC STRAIN: Of all the cells in the body, ECs are the ones directly exposed to the hemodynamic forces of blood flow. These pulsatile forces consist of flow-induced shear stress, cyclic strain and pulsatile pressure. For comprehensive reviews see (20,58,79). These forces appear to be instrumental in modulating a host of genes and gene products in ECs (64,66). EC responses to physical forces are frequently heterogeneous, depending in part on the anatomical location of the cells (30,50), and also on the exact nature of these forces, such as laminar vs. turbulent flow conditions (15,17).

For example, we recently described that cyclic strain differentially modulates cAMP signaling in ECs isolated from distinct sites in the vascular tree (50). Traditionally, most *in vitro* experiments have been carried out with large-vessel-derived ECs. Therefore, very little is known about the effects of hemodynamic forces on the physiology of microvessel ECs. However, the *in vivo* studies of Hudlicka and colleagues provide ample evidence for physical forces modulating angiogenesis in skeletal and cardiac muscle. For a recent review, see (29). The exact role of the various hemodynamic forces on angiogenesis remains to be explored in detail.

Table 2: Differential Regulation of Cell Adhesion Molecules in HUVEC and HMVEC by Cyclic Mechanical Strain

	HUVEC	HMVEC
Control	100	100
VCAM-1	166.7 ± 17.6^a	71.5 ± 8.9^c
ICAM-1	124.6 ± 9.4^b	79.6 ± 9.5^c
PECAM-1	$96.7 \pm 4.2^{n.s.}$	$98.3 \pm 8.1^{n.s.}$

The expression of cell adhesion molecules was assessed by a whole cell-based ELISA assay, as previously described (62). The cells were plated in 6-well flexible bottom plates and exposed to cyclic mechanical deformation in the commercially available FLEXERCELL® strain unit (20% maximal strain @ 1 Hz), as previously described. (50). After correcting for the number of cells/well (61) the data were normalized to the respective expression of VCAM-1, ICAM-1 and PECAM-1 as determined under static conditions in parallel plates not exposed to cyclic strain. All data are expressed as means $\pm$ SEM from 3 three independent experiments carried out in triplicate wells. a: $p< 0.001$, b: $p<0.01$, c: $p<0.05$, n.s. statistically not significant.

We recently compared the effects of cyclic strain on the expression of ICAM-1, VCAM-1, and PECAM-1 in HUVEC and HMVEC. Exposure for 24 hours to cyclic strain (20% maximal strain, 1 Hz) did not affect PECAM-1 expression in either cell type. (Table 2). By contrast, under the same experimental conditions, VCAM-1 and ICAM-1 expression were up-regulated in HUVEC, but down-regulated in HMVEC. Expression of VCAM-1 was more susceptible to regulation by cyclic strain than that of ICAM-1, at least in HUVEC.

This result is remarkable in that gene for ICAM-1, but not for VCAM-1, contains a "shear stress-responsive element (SSRE)" in the promoter region, which has been implicated in the regulation of gene expression by mechanical forces, including by cyclic strain (57,66). The same cis-acting SSRE has also been identified in the PECAM-1 promoter (P.Newman, personal communication) suggesting the regulability of PECAM-1 expression by hemodynamic forces. And, yet, PECAM-1 expression is not altered by cyclic strain. Thus, the differential response of various adhesion molecules to cyclic strain suggests either that the SSRE motif might play a less important role in the response to mechanical strain than to fluid shear-stress and/or that other, yet to be identified, cyclic-strain responsive elements exist, which either promote or suppress gene expression.

DIFFERENTIAL ACTIVATION OF TISSUE FACTOR (TF) BY CYTOKINES AND CYCLIC STRAIN IN HUVEC AND HMVEC:

Some of the most potent angiogenic molecules, such as thrombin, phorbol esters and cytokines also trigger the expression of a pro-thrombotic EC surface, e.g. by decreasing the expression of thrombomodulin and by increasing that of TF (2,3,80). The induction of a pro-coagulant EC phenotype might be relevant for angiogenesis, primarily for providing a local source for activated plasma proteases (e.g. thrombin) and/or for the local recruitment of leukocytes, which in turn secrete "angiogenic" compounds.

We hypothesized that the regulation of TF expression by chemical and/or mechanical activation might differ in ECs isolated from various vascular beds. HUVEC and HMVEC of the same passage (number of population doublings) expressed very similar low levels of basal TF activity. However, activation for 5 hours with 10 ng/ml TNFα lead to an approx. 6-fold increase in cellular TF activity in HMVEC and an 15-fold increase in HUVEC (Figure 7). Remarkably, exposure to cyclic mechanical strain alone induced TF activation in HMVEC, but not in HUVEC. Moreover, mechanical force and cytokines synergistically

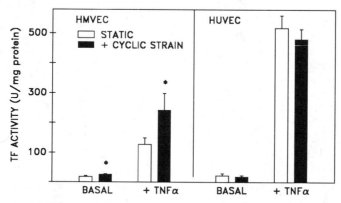

Figure 7: Differential Sensitivity of Tissue Factor Expression in Various Human Endothelial Cells to Chemical and Mechanical Stimulation. HUVEC and HMVEC were plated in 6-well flexible bottom plates. Upon reaching confluence the cells in 3 wells in each plate were stimulated with 10 ng/ml TNFα, the other three wells served as untreated controls. The plates were immediately exposed for 5 hrs to cyclic mechanical strain in the commercially available FLEXERCELL® strain unit (20% maximal strain @ 1 Hz), as previously described (50). Tissue factor activity was assessed by a two-step amidolytic assay (81) and is expressed in units of thromboplastin/mg protein per well. All data are expressed as means ± SEM from at least 3 three independent experiments carried out in triplicate wells. *:p<0.05.

induced TF activity in HMVEC, but not in HUVEC (Silverman *et al.*, manuscript submitted for publication). Previous reports indicate an upregulation of TF by fluid shear stress (1,24). which might imply the presence of a SSRE or other, similar regulatory elements in the promoter region of the recently cloned gene (54). Our data, in line with the above-mentioned results on the selective regulation of adhesion molecules, suggest the existence of additional, cell type-specific cyclic strain-sensitive mechanism(s), which regulate the functional expression of TF in EC.

EXTRACELLULAR MATRIX PROTEIN DEPOSITION IN VARIOUS ENDOTHELIAL CELLS: The composition of the subendothelial basement membrane (BM) plays a prominent role in determining the organ-specific EC phenotype (46,47). Remodelling of the extracellular matrix (ECM) occurs at least twice during the angiogenic cascade (see Figure 2). Initially, the subendothelial BM has to be locally dissolved to facilitate the migration of ECs away from the existing vasculature, while the re-formation of the BM is essential for the maturation of the nascent new capillary.

In studying the modulation of the subendothelial ECM *in vitro*, we previously reported on qualitative and quantitative temporal changes in the deposition of ECM proteins from cultured rat adrenal medullary microvascular ECs (62), In particular, we observed a biphasic appearance and disappearance of fibronectin (FN) in the subendothelial ECM which resembles that observed *in vivo* during the embryonic development of the microvasculature, e.g. in the chick chorioallantoic membrane (4) or the human kidney (56). Concomitant with the decrease in FN deposition, we also observed a decrease in the amount of immunoreactive FN inside the ECs (63). More recent studies suggest that the patterns for ECM protein synthesis and deposition from cultured microvascular cells are not necessarily typical for all EC types (Figure 8). For example, cultured aortic ECs (bovine and human) continuously deposit FN into their ECM, albeit at a much slower rate than the microvessel-derived ECs. On the other hand, cultured bovine corneal ECs, which rest *in vivo* on top of an unusually thick basement membrane (Descemet's Membrane), continuously produce large quantities of ECM proteins, including FN.

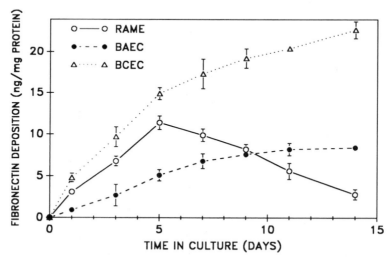

Figure 8: Heterogeneity of Fibronectin Deposition into the Subendothelial Extracellular Matrix by Various Endothelial Cell Types. Rat adrenal medullary endothelial cells (RAME), bovine aortic endothelial cells (BAEC) and bovine corneal endothelial cells (BCEC) were plated at an initial density of 10,000/well into 96 well cell culture plates. The deposition of fibronectin into the subendothelial basement membrane was assessed by ELICA, exactly as previously described (62,63). The data, expressed as ng fibronectin/mg cell protein, represent means ± SEM from three independent experiments carried out in quadruplicate wells.

These distinct patterns of ECM protein production might be due to differences in the rate of transcription and/or translation of the message for the ECM proteins. Alternatively, quantitative and qualitative differences in the amount of ECM protein deposition might also attest to the heterogeneity in the profile of matrix-degrading or fibrinolytic enzymes produced by the various EC types. We propose, that differences in the ability to initiate ECM remodelling (BM degradation, de novo synthesis and/or deposition of specific ECM proteins) might be related to the reported divergence in the angiogenic propensity of various ECs.

HETEROGENEITY OF cAMP SIGNALING IN VARIOUS EC TYPES: Cyclic AMP regulates a number of angiogenesis-related cellular functions such as growth, proliferation, differentiation, relaxation and secretion, often in a cell type-specific manner (42). The key enzyme in this ubiquitous signaling cascade is adenylyl cyclase (AC), which catalyzes the production of cAMP from ATP(44). At present, at least 8 AC isoforms are known (32,36).

Elevation of cAMP in ECs appears, in general, to exert an anti-angiogenic effect. For example, activation of the adenylyl cyclase(s) inhibits the formation of new blood vessels in the chick chorioallantoic membrane model (86) and also antagonizes *in vitro* tube formation on Matrigel (39). Also, in most cell types studied so far, including ECs, an increased in cAMP correlates with an inhibition of cell proliferation (44), rather than with its stimulation, as would be required for angiogenesis (31). Thus, activation of the cAMP signaling cascade might be an important step in the prevention of new blood vessel formation. It is also conceivable that elevation of cAMP might be pivotal for the initiation and/or for controlling endogenous, as yet unexplored EC-specific mechanism(s) that lead to the break down of already-established microvessels (angioregression).

Table 3: Adenylyl cyclase isoforms present in primary cultures of endothelial cells isolated from various rat tissues.

	II	III	IV	V	VI
Adrenal	tr	+++	tr	+	+++
Brain	++	++++	+++	++	+++
Adipose	++	++++	+++	++	+++
Lung	tr	++++	+	++	+++
Heart	+	+++	+++	+++	+++
Vena Cava	tr	+++	tr	++	++
Aorta	+	+++	tr	++	++

Total RNA was prepared from primary cultures of isolated microvascular or large vessel-derived ECs, reversed-transcribed, and used as a template for PCR amplification with primers based on unique, rat-specific sequences in the mRNA of types II-VI adenylyl cyclase. The level of expression of each isoform was calculated by comparing the intensity of the band corresponding to that isoform with the intensity of the band corresponding to a 1:100 dilution of cDNA from the same tissue and probed for the constitutively expressed house-keeping gene glyceraldehyde-3-phosphate dehydrogenase (GAPDH). Results are qualitative, relative to the intensity of the invariant GAPDH bands from + (weak, <10%) to ++++ (strong, $\geq$ 100%). tr: trace, barely visible. For details see (51).

Previous studies have yielded a number of tissue-specific, sometimes divergent, effects in ECs, which relate to cAMP signaling. For example, elevation of cAMP decreases the permeability of the EC monolayer in many large-vessel-derived EC types from many species, including bovine and porcine aortic ECs (77,94), the opposite was found to be true for rat coronary microvascular ECs (90). These divergent responses between micro- and macrovessel- derived ECs is in line with our recent findings of heterogeneous regulation of cAMP signalling in phenotypically distinct cultured ECs (50-52, 42). In particular, we observed dissimilar effects of chemical and mechanical activation on cAMP signaling in ECs isolated from different vascular beds (50,52). We hypothesized that the ostensible heterogeneity in cAMP signaling might be related to distinct, EC-specific patterns of AC isoform expression (42).

In testing our hypothesis we found by semi-quantitative RT-PCR that of the presently known eight AC isoforms at least five isoforms are concomitantly expressed in rat ECs isolated from the capillary beds of various tissues (adipose, adrenal, brain, heart, lung) and from large vessels (aorta, vena cava) (51). However there are significant differences in the amounts of each of these AC isoforms present in the various EC types (Table 3). Since some of the AC isoforms might be preferential targets in the signaling pathway of certain receptors (32) and/or are selectively activated in certain pathophysiological states (85), we postulate that selective activation or inactivation of particular AC isoforms is of importance for the initiation, maintenance and/or cessation of the angiogenic process. This testable hypothesis is of relevance in view of the known, extensive cross-talk of certain AC isoforms with other putative angiogenic signaling mechanisms, such as protein kinase C (14,37). As previously discussed (42), EC heterogeneity at the level of signal transduction is a complex issue. As of this writing, we are not aware of any relevant studies that address EC heterogeneity from the vantage point of the heterogeneity of protein kinase C (PKC) isoforms or isoforms of cyclic nucleotide phosphodiesterases (PDE). Clearly, both of these enzymes are important regulators of signaling pathways, the activation of which, respectively, facilitates (PKC) or inhibits (PDE)) angiogenesis. As such, PKC or PDE isoforms, just like the isoforms of AC, might become specific targets for selective pharmacological interventions in an attempt to control angiogenesis.

CONCLUSIONS: Heterogeneity has become a prominent issue in understanding the biology of vascular ECs. We propose that EC heterogeneity also extends to the study of angiogenesis. In particular, tissue-specific issues, such as the effects of female sex hormones on angiogenesis and angioregression during the reproductive cycle (67) or the intricacies of the development of the blood-brain barrier (69,70) can only be crudely assessed by using bovine aortic endothelial cells or the like. Just as there is no single, ubiquitous "angiogenic" effector, there is similarly no "generic" EC. Based on this premise, we firmly expect to obtain dissimilar results when testing some of the "angiogenic" agents in diverse vascular beds *in vivo* or when comparing "generic" cultured ECs vs. microvascular ECs isolated form different tissues. Thus, molecular and cellular determinants of angiogenesis (genes, gene products, signaling mechanisms) will best be studied *in vitro* using "professional, angiogenic", microvascular ECs.

Some of the specific questions on the possible connections between EC heterogeneity and angiogenesis, which we raised during the previous conferences, are currently being studied, such as the relationship between differential gene expression and the angiogenic phenotype (see chapter by D. Grant in this volume) and the regulation of the angiogenic balance in ECs through interactions between tissue-specific isoforms of pro-angiogenic and anti-angiogenic signaling pathways, such as PKC and cAMP, respectively. Accumulating evidence strongly suggests that we will have to substantially increase the levels of sophistication in our *in vitro* studies (by combining tissue-specific microvascular ECs with the appropriate microenvironmental cues, such as ECM, cytokines and mechanical forces) in order to properly model the complexity of angiogenesis *in vivo*.

The results presented here, as well as numerous studies by others, show an increasing appreciation of the distinct differences between large vessel and microvessel-derived ECs.

Often, these differences can not simply be explained by "acquired" differences in the responses to epigenetic cues. Rather, we postulate that some of the distinct functional traits of microvascular ECs are based on genetic predisposition, including their enhanced "angiogenic "potential and unique responses to biochemical and/or mechanical activation. While we don't know the details, yet, intensive studies are either already under way and/or should be launched to unravel the seemingly confounding complexity of differential gene-expression and gene-regulation in ECs. Only by transcending our current state of "higher level of confusion" (J. Folkman), will we be eventually able to arrive at the levels of understanding of the molecular and cellular mechanisms of angiogenesis which will allow the rational design of selective agonists or antagonists of the angiogenic process.

ACKNOWLEDGEMENTS: Part of our original work reported in this paper was supported by grants (to PIL) from the American Heart Association, NASA, NATO, the Mount Sinai Research Foundation and the Milwaukee Heart Research Foundation. We are grateful to Dr. Tom Hayman for critically reading the manuscript.

REFERENCES

1. Andree, H. A. M. and Y. Nemerson. 1995. Tissue factor: regulation of activity by flow and phospholipid surfaces. *Blood Coagul. Fibrinolysis* 6:189-197.
2. Archipoff, G., A. Beretz, K. Bartha, C. Brisson, C. de la Salle, C. Froget-Léon, C. Klein-Soyer, and J. P. Cazenave. 1993. Role of cyclic AMP in promoting the thromboresistance of human endothelial cells by enhancing thrombomodulin and decreasing tissue factor activities. *Br. J. Pharmacol.* 109:18-28.
3. Archipoff, G., A. Beretz, J. Freyssinet, C. Klein-Soyer, C. Brisson, and J. Cazenave. 1991. Heterogeneous regulation of constitutive thrombomodulin or inducible tissue-factor activities on the surface of human saphenous-vein endothelial cells in culture following stimulation by interleukin-1, tumor necrosis factor, thrombin or phorbol ester. *Biochem. J.* 273:679-684.
4. Ausprunk, D.H., S.M. Dethlefsen, and E.R. Higgins. 1991. Distribution of fibronectin, laminin and type IV collagen during development of blood vessels in the chick chorioallantoic membrane. In *The development of the vascular system*. R.N. Feinberg, G.K. Sherer, and R. Auerbach, editors. Karger, Basel. 93-107.
5. Bauer, J., M. Margolis, C. Schreiner, C. Edgell, J. Azizkhan, E. Lazarowski, and R. L. Juliano. 1992. In vitro model of angiogenesis using a human endothelium-derived permanent cell line: contributions of induced gene expression, G-proteins, and integrins. *J. Cell. Physiol.* 153:437-449.
6. Bender, J. R., M. M. Sadeghi, C. Watson, S. Pfau, and R. Pardi. 1994. Heterogeneous activation thresholds to cytokines in genetically distinct endothelial cells: evidence for diverse transcriptional responses. *Proc. Natl. Acad. Sci. USA* 91:3994-3998.
7. Bevilacqua, M. P., R. M. Nelson, G. Mannori, and O. Cecconi. 1994. Endothelial-leukocyte adhesion molecules in human disease. *Annu. Rev. Med.* 45:361-378.
8. Bischoff, J. 1995. Approaches to studying cell adhesion molecules in angiogenesis. *Trends Cell Biol.* 5:69-74.
9. Bottaro, D., D. Shepro, and H. B. Hechtman. 1986. Heterogeneity of intimal and microvessel endothelial cell barriers in vitro. *Microvasc. Res.* 32:389-398.
10. Brooks, P. C., R. A. F. Clark, and D. A. Cheresh. 1994. Requirement of vascular integrin $\alpha_v\beta_3$ for angiogenesis. *Science* 264:569-571.
11. Brooks, P. C., A. M. P. Montgomery, M. Rosenfeld, R. A. Reisfeld, T. Hu, G. Klier, and D. A. Cheresh. 1994. Integrin $\alpha_v\beta_3$ antagonists promote tumor regression by inducing apoptosis of angiogenic blood vessels. *Cell* 79:1157-1164.

12. Carley, W. W., M. J. Niedbala, and M. E. Gerritsen. 1992. Isolation, cultivation and partial characterization of microvascular endothelium derived from human lung. *Am. J. Respir. Cell Mol. Biol.* 7:620-630.

13. Chaudhury, A. R. and P. A. D'Amore. 1991. Endothelial cell regulation by transforming growth factor-β. *J. Cell. Biochem.* 47:224-229.

14. Chen, J. and R. Iyengar. 1993. Inhibition of cloned adenylyl cyclases by mutant-activated G_i-α and specific suppression of type 2 adenylyl cyclase inhibition by phorbol ester treatment. *J. Biol. Chem.* 268:12253-12256.

15. Davies, P. F., A. Remuzzi, E. J. Gordon, C. F. Dewey,Jr., and M. A. Gimbrone,Jr. 1986. Turbulent fluid shear stress induces vascular endothelial cell turnover in vitro. *Proc. Natl. Acad. Sci. USA* 83:2114-2117.

16. DeLisser, H. M., P. J. Newman, and S. M. Albelda. 1993. Platelet endothelial cell adhesion molecule (CD31). *Curr. Top. Microbiol. Immunol.* 184:37-45.

17. Eskin, S. G., C. L. Ives, L. V. McIntire, and L. T. Navarro. 1984. Response of cultured endothelial cells to steady flow. *Microvasc. Res.* 28:87-94.

18. Fajardo, L. F., H. H. Kwan, J. Kowalski, S. D. Prionas, and A. C. Allison. 1992. Dual role of tumor necrosis factor-α in angiogenesis. *Am. J. Pathol.* 140:539-544.

19. Fenyves, A. M., M. Saxer, and K. Spanel-Borowski. 1994. Bovine microvascular endothelial cells of separate morphology differ in growth and response to the action of interferon-γ. *Experientia* 50:99-104.

20. Frangos, J. A. 1993. *Physical forces and the mammalian cell.* Academic Press, New York.

21. Gerritsen, M. E., M. J. Niedbala, A. Szczepanski, and W. W. Carley. 1993. Cytokine activation of human macro- and microvessel-derived endothelial cells. *Blood Cells* 19:325-39; discussion 340-2.

22. Gerritsen, M. E. 1987. Functional heterogeneity of vascular endothelial cells. *Biochem. Pharmacol.* 36:2701-2711.

23. Gimbrone, M. A., R. S. Cotran, and J. Folkman. 1974. Human vascular endothelial cells in culture. *J. Cell Biol.* 60:673-684.

24. Grabowski, E. F., D. B. Zuckerman, and Y. Nemerson. 1993. The functional expression of tissue factor by fibroblasts and endothelial cells under flow conditions. *Blood* 81:3265-3270.

25. Grant, D. S., H. K. Kleinman, I. D. Goldberg, M. M. Bhargava, B. J. Nickoloff, J. L. Kinsella, P. Polverini, and E. M. Rosen. 1993. Scatter factor induces blood vessel formation in vivo. *Proc. Natl. Acad. Sci. USA* 90:1937-1941.

26. Grant, D. S., P. I. Lelkes, K. Fukuda, and H. K. Kleinman. 1991. Intracellular mechanisms involved in basement membrane induced blood vessel differentiation in vitro. *In Vitro Cell Dev. Biol.* 27A: 327-336.

27. Haralabopoulos, G. C., D. S. Grant, H. K. Kleinman, P. I. Lelkes, S. P. Papaioannou, and M. E. Maragoudakis. 1994. Inhibitors of basement membrane collagen synthesis prevent endothelial cell alignment in Matrigel in vitro and angiogenesis in vivo. *Lab. Invest.* 71:575-582.

28. Hauser, I. A., D. R. Johnson, and J. A. Madri. 1993. Differential induction of VCAM-1 on human iliac venous and arterial endothelial cells and its role in adhesion. *J. Immunol.* 151:5172-5185.

29. Hudlicka, O. and M.D. Brown. 1993. Physical forces and angiogenesis. In *Mechanoreception by the vascular wall.* G.M. Rubanyi, editor. Futura Publishing Co.,Inc. New York. pp. 197-241.

30. Iba, T. and B. E. Sumpio. 1991. Morphological response of human endothelial cells subjected to cyclic strain in vitro. *Microvasc. Res.* 42:245-254.

31. Iruela-Arispe, M. L. and E. H. Sage. 1993. Endothelial cells exhibiting angiogenesis in vitro proliferate in response to TGF-β1. *J. Cell. Biochem.* 52:414-430.

32. Iyengar, R. 1993. Molecular and functional diversity of mammalian Gs-stimulated adenylyl cyclases. *FASEB J.* 7:768-775.

33. Jackson, C. J., P. K. Garbett, B. Nissen, and L. Schrieber. 1990. Binding of human endothelium to Ulex europaeus I-coated dynabeads: application to the isolation of microvascular endothelium. *J. Cell Sci.* 96:257-262.

34. Jaffe, E. A. 1984. *Biology of endothelial cells.* Martinus Nijhoff, Boston.

35. Jaffe, E. A., R. L. Nachman, C. G. Becken, and C. R. Minick. 1973. Culture of human endothelial cells derived from umbilical veins. *J. Clin. Invest.* 52:2745

36. Johnson, B. A., G. K. Haines, L. A. Harlow, and A. E. Koch. 1993. Adhesion molecule expression in human synovial tissue. *Arthritis Rheum.* 36:137-146.

37. Kawabe, J., G. Iwami, T. Ebina, S. Ohno, T. Katada, Y. Ueda, C. J. Homcy, and Y. Ishikawa. 1994. Differential activation of adenylyl cyclase of by protein kinase C isoenzymes. *J. Biol. Chem.* 269:16554-16558.

38. Kinjo, T., M. Takashi, K. Miyake, and H. Nagura. 1989. Phenotypic heterogeneity of vascular endothelial cells in the human kidney. *Cell Tissue Res.* 256:27-34.

39. Kinsella, J. L., D. S. Grant, B. S. Weeks, and H. K. Kleinman. 1992. Protein kinase C regulates endothelial cell tube formation on basement membrane matrix, Matrigel. *Exp. Cell Res.* 199:56-62.

40. Klein, C. L., H. Köhler, F. Bittinger, M. Wagner, I. Hermanns, K. Grant, J. C. Lewis, and C. J. Kirkpatrick. 1994. Comparative studies on vascular endothelium in vitro. I. Cytokine effects on the expression of adhesion molecules by human umbilical vein, saphenous vein and femoral artery endothelial cells. *Pathobiology* 62:199-208.

41. Kubota, Y., H. K. Kleinman, G. R. Martin, and T. J. Lawley. 1988. Role of laminin and basement membrane in the morphological differentiation of human endothelial cells into capillary-like structures. *J. Cell Biol.* 107:1589-1598.

42. Lelkes, P.I., E.G. Manolopoulos, D.M. Chick, and B.R. Unsworth. 1994. Endothelial cell heterogeneity and organ-specificity. In *Angiogenesis: molecular biology, clinical aspects.* M.E. Maragoudakis, P. Guillino, and P.I. Lelkes, editors. Plenum Press, New York. 15-28.

43. Lelkes, P.I. and B.R. Unsworth. 1992. Role of heterotypic interactions between endothelial cells and parenchymal cells in organospecific differentiation: a possible trigger of vasculogenesis. In *Angiogenesis in health and disease.* M.E. Maragoudakis, P. Gullino, and P.I. Lelkes, editors. Plenum Press, New York,NY. 27-43.

44. Levitzki, A. 1988. From epinephrine to cyclic AMP. *Science* 241:800-806.

45. Madri, J. A., L. Bell, M. Marx, J. R. Merwin, C. Basson, and C. Prinz. 1991. Effects of soluble factors and extracellular matrix components on vascular cell behavior in vitro and in vivo: models of de-endothelialization and repair. *J. Cell. Biochem.* 45:123-130.

46. Madri, J. A., B. M. Pratt, and A. M. Tucker. 1988. Phenotypic modulation of endothelial cells by transforming growth factor-β depends upon the composition and organization of the extracellular matrix. *J. Cell Biol.* 106:1375-1384.

47. Madri, J. A., S. K. Williams, T. Wyatt, and C. Mezzio. 1983. Capillary endothelial cell cultures: phenotypic modulation by matrix components. *J. Cell Biol.* 97:153-165.

48. Majno, G. 1965. Ultrastructure of the vascular membrane. In *Handbook of physiology: circulation.* W.F. Hamilton and P. Dow, editors. American Physiological Society, Washington,D.C. 2293-2543.

49. Makovetskii, V. D., V. A. Kozlov, and V. D. Mishalov. 1984. Organ and tissue-specific properties of the microcirculatory bed of the heart. *Arch. Anat. Histol. Embryol.* 86:25-30.

50. Manolopoulos, V. G. and P. I. Lelkes. 1993. Cyclic strain and forskolin differentially induce cAMP production in phenotypically diverse endothelial cells. *Biochem. Biophys. Res. Commun.* 191:1379-1385.

51. Manolopoulos, V. G., J. Liu, B. R. Unsworth, and P. I. Lelkes. 1995. Adenylyl cyclase isoforms are differentially expressed in primary cultures of endothelial cells and whole tissue homogenates from various rat tissues. *Biochem. Biophys. Res. Comm.* 208:323-331.

52. Manolopoulos, V. G., M. M. Samet, and P. I. Lelkes. 1995. Regulation of the adenylyl cyclase signalling system in various types of cultured endothelial cells. *J. Cell. Biochem.* 57:590-598.

53. Mattsby-Baltzer, I., A. Jakobsson, J. Sörbo, and K. Norrby. 1994. Endotoxin is angiogenic. *Int. J. Exp. Pathol.* 75:191-196.

54. Moll, T., M. Czyz, H. Holzmüller, R. Hofer-Warbinek, E. Wagner, H. Winkler, F. H. Bach, and E. Hofer. 1995. Regulation of the tissue factor promoter in endothelial cells. *J. Biol. Chem.* 270:3849-3857.

55. Motro, B., A. Itin, L. Sachs, and E. Keshet. 1990. Pattern of interleukin 6 gene expression in vivo suggests a role for this cytokine in angiogenesis. *Proc. Natl. Acad. Sci. USA* 87:3092-3096.

56. Mounier, F., J. M. Foidart, and M. C. Gubler. 1986. Distribution of extracellular matrix glycoproteins during normal development of human kidney. *Lab. Invest.* 54 No.4:394

57. Nagel, T., N. Resnick, W. J. Atkinson, C. F. Dewey,Jr., and M. A. Gimbrone,Jr. 1994. Shear stress selectively upregulates intercellular adhesion molecule-1 expression in cultured human vascular endothelial cells. *J. Clin. Invest.* 2:885-891.

58. Nerem, R. M. 1993. Hemodynamics and the vascular endothelium. *J. Biomech. Eng.* 115:510-514.

59. Olivo, M., R. Bhardwaj, K. Schultze-Osthoff, C. Sorg, H. J. Jacob, and I. Flamme. 1992. A comparative study on the effects of tumor necrosis factor-α (TNF-α), human angiogenic factor (h-AF) and basic fibroblast growth factor (bFGF) on the chorioallantoic membrane of the chick embryo. *Anat. Rec.* 234:105-115.

60. Paleolog, E. M., S.-A. J. Delasalle, W. A. Buurman, and M. Feldman. 1994. Functional activities of receptors for tumor necrosis factor-α on human vascular endothelial cells. *Blood* 84:2578-2590.

61. Papadimitriou, E. and P. I. Lelkes. 1993. Measurement of cell numbers in microtiter culture plates using the fluorescent dye Hoechst 33258. *J. Immunol. Methods* 162:41-45.

62. Papadimitriou, E., B. R. Unsworth, M. E. Maragoudakis, and P. I. Lelkes. 1993. Time-course and quantification of extracellular matrix maturation in the chick chorioallantoic membrane and in cultured endothelial cells. *Endothelium* 1:207-219.

63. Papadimitriou, E., B.R. Unsworth, M.E. Maragoudakis, and P.I. Lelkes. 1994. Quantitative analysis of extracellular matrix formation in vivo and in vitro. In *Angiogenesis: Molecular Biology, Clinical Aspects.* M.E. Maragoudakis, P. Gullino, and P.I. Lelkes, editors. Plenum Press, New York.

64. Patrick, C. W., Jr. and L. V. McIntire. 1995. Shear stress and cyclic strain modulation of gene expression in vascular endothelial cells. *Blood Purif.* 13:112-124.

65. Piali, L., P. Hammel, C. Uherek, F. Bachmann, R. H. Gisler, D. Dunon, and B. A. Imhof. 1995. CD31/PECAM-1 is a ligand for $\alpha_v\beta_3$ integrin involved in adhesion of leukocytes to endothelium. *J. Cell Biol.* 130:451-460.

66. Resnick, N. and M. A. Gimbrone,Jr. 1995. Hemodynamic forces are complex regulators of endothelial gene expression. *FASEB J.* 9:874-882.

67. Reynolds, L. P., S. D. Killilea, and D. A. Redmer. 1992. Angiogenesis in the female reproductive system. *FASEB J.* 6:886-892.

68. Rhodin, J. A. G. and H. Fujita. 1989. Capillary growth in the mesentery of normal young rats: intravital video and electron microscope analyses. *J. Submicrosc. Cytol. Pathol.* 21:1-34.

69. Risau, W., R. Hallmann, and U. Albrecht. 1986. Differentiation-dependent expression of proteins in brain endothelium during development of the blood-brain barrier. *Dev. Biol.* 117:537-545.

70. Risau, W. and H. Wolburg. 1990. Development of the blood-brain barrier. *TINS* 13(5):174-178.

71. Scoazec, J. Y., L. Racine, A. Couvelard, J. F. Flejou, and G. Feldmann. 1994. Endothelial cell heterogeneity in the normal human liver acinus: in situ immunohistochemical demonstration. *Liver* 14:113-123.

72. Seulberger, H., F. Lottspeich, and W. Risau. 1990. The inducible blood-brain barrier specific molecule HT7 is a novel immunoglobulin-like cell surface glycoprotein. *EMBO J.* 9:2151-2158.

73. Shepro, D. and P.A. D'Amore. 1984. Physiology and biochemistry of the vascular wall endothelium. *In Handbook of physiology: section 2: the cardiovascular system: volume IV: microcirculation, part I.* E.M. Renkin and C.C. Michel, editors. American Physiological Society, Bethesda, Maryland. 103-164.

74. Shweiki, D., A. Itin, D. Soffer, and E. Keshet. 1992. Vascular endothelial growth factor induced by hypoxia may mediate hypoxia-initiated angiogenesis. *Nature* 359:843-845.

75. Simionescu, N. and M. Simionescu. 1988. *Endothelial cell biology in health and disease.* Plenum Press, New York.

76. Slowik, M. R., L. G. De Luca, W. Fiers, and J. S. Pober. 1993. Tumor necrosis factor activates human endothelial cells through the p55 tumor necrosis factor receptor but the p75 receptor contributes to activation at low tumor necrosis factor concentration. *Am. J. Pathol.* 143:1724-1730.

77. Stelzner, T. J., J. V. Weil, and R. F. O'Brien. 1989. Role of cyclic adenosine monophosphate in the induction of endothelial barrier properties. *J. Cell. Physiol.* 139:157-166.

78. Stolz, D. B. and B. S. Jacobson. 1991. Macro- and microvascular endothelial cells in vitro: maintenance of biochemical heterogeneity despite loss of ultrastructural characteristics. *In Vitro Cell. Dev. Biol.* 27A:169-182.

79. Sumpio, B. E. (editor). 1993. *Hemodynamic forces and vascular cell biology.* R. G. Landes Company, Austin.

80. Sung, C., A. J. Arleth, and E. H. Ohlstein. 1994. Involvement of protein kinase C in cytokine-induced tissue factor production in human vascular endothelial cells. *Endothelium* 2:209-216.

81. Surprenant, Y. M. and S. H. Zuckerman. 1989. A novel microtiter plate assay for the quantitation of procoagulant activity on adherent monocytes, macrophage and endothelial cells. *Thromb. Res.* 53:339-346.

82. Swerlick, R. A. and T. J. Lawley. 1993. Role of microvascular endothelial cells in inflammation. *J. Invest. Dermatol.* 100, No. 1:111S-115S.

83. Swerlick, R. A., K. H. Lee, T. M. Wick, and T. J. Lawley. 1992. Human dermal microvascular endothelial but not human umbilical vein endothelial cells express CD36 in vivo and in vitro. *J. Immunol.* 148(1):78-83.

84. Tang, S., K. C. Le-Ruppert, and V. P. Gabel. 1994. Expression of intercellular adhesion molecule-1 (ICAM-1) and vascular cell adhesion molecule-1 (VCAM-1) on proliferating vascular endothelial cells in diabetic epiretinal membranes. *Br. J. Ophthal.* 78:370-376.

85. Tobise, K., Y. Ishikawa, S. R. Holmer, M. Im, J. B. Newell, H. Yoshie, M. Fujita, E. E. Susannie, and C. J. Homcy. 1994. Changes in type VI adenylyl cyclase isoform expression correlate with a decreased capacity for cAMP generation in the aging ventricle. *Circ. Res.* 74:596-603.

86. Tsopanoglou, N. E., E. Pipili-Synetos, and M. E. Maragoudakis. 1993. Protein kinase C involvement in the regulation of angiogenesis. *J. Vasc. Res.* 30:202-208.

87. Vigne, P., G. Champigny, R. Marsault, P. Barbry, C. Frelin, and M. Lazdunski. 1989. A new type of amiloride-sensitive cationic channel in endothelial cells of brain microvessels. *J. Biol. Chem.* 264:7663-7668.

88. Voyta, J. C., D. P. Via, C. E. Butterfield, and B. R. Zetter. 1984. Identification and isolation of endothelial cells based on their increased uptake of acetylated-low density lipoprotein. *J. Cell Biol.* 99:2034-2040.

89. Warren, J. B. and R. K. Loi. 1995. Captopril increases skin microvascular blood flow secondary to bradykinin, nitric oxide, and prostaglandins. *FASEB J.* 9:411-418.

90. Watanabe, H., W. Kuhne, P. Schwartz, and H. M. Piper. 1992. A_2-adenosine receptor stimulation increases macromolecule permeability of coronary endothelial cells. *Am. J. Physiol.* 262:H1174-H1181.

91. Wojta, J., M. Gallicchio, H. Zoellner, E. L. Filonzi, J. A. Hamilton, and K. McGrath. 1993. Interleukin-4 stimulates expression of urokinase-type-plasminogen activator in cultured human foreskin microvascular endothelial cells. *Blood* 81:3285-3292.

92. Yamamoto, C., T. Kaji, M. Furuya, M. Sakamoto, H. Kozuka, and F. Koizumi. 1994. Basic fibroblast growth factor suppresses tissue plasminogen activator release from cultured human umbilical vein endothelial cells but enhances that from cultured human aortic endothelial cells. *Thromb. Res.* 73:255-263.

93. Zetter, B.R. 1988. Endothelial heterogeneity: influence of vessel size, organ localization, and species specificity on the properties of cultured endothelial cells. In *Endothelial cells: volume II.* U.S. Ryan, editor. CRC Press, Boca Raton, FL. 63-79.

94. Zink, S., P. Rösen, B. Sackmann, and H. Lemoine. 1993. Regulation of endothelial permeability by β-adrenoceptor agonists: contribution of β_1- and β_2-adrenoceptors. *Biochim. Biophys. Acta* 1178:286-298.

GENE EXPRESSION AND ENDOTHELIAL CELL DIFFERENTIATION

D. S. Grant[1], J. L. Kinsella[3], and H. K. Kleinman [2]

[1]Department of Medicine, Cardeza Foundation for Hematological Research
Thomas Jefferson University
Philadelphia, PA 19107
[2] Lab of Developmental Biology, National Institute of Dental Research
NIH, 9000 Rockville Pk, Bld 30, Bethesda MD, 20892
[3] National Institute on Aging, NIA, 4940 Eastern Ave, Baltimore, MD 21224

INTRODUCTION

Regulation of the vascular wall is an essential process that allows normal blood flow and facilitates the exchange of soluble gases, ions and vital macromolecules. Normally all vessels are composed of a nonthrombogenic layer of endothelial cells which line the intimal surface of the vessel walls. The differentiation state of this cell layer is maintained by components (factors) present in the blood the extravascular stroma and the extracellular matrix. The contribution of the matrix to vascular wall homeostasis has been unclear in the past, even though matrix comprise a significant portion of the vasculature. In fact, the endothelium is adherent to a thin, specialized extracellular layer know as a basement membrane. The basement membrane provides not only support and an adhesive surface for the endothelium but also maintains the normal differentiated phenotype of the cell layer. Vessel walls also are comprised of other vascular cells such a smooth muscle cells, pericytes and fibroblasts. The former two also have their own basement membrane, and the latter is surrounded by a collagenous insterstitium (the adventitia) and in some cases elastic fibers. Studies which examine the cells comprising the vessel walls must also evaluate the role of the matrix in the maintenance of its structure as well.

Basement membranes are composed of a meshwork of non-fibrillar collagen (type IV), the glycoproteins laminin, entactin, and fibronectin, and the proteoglycans (perlecan and chondroitin sulfate proteoglycan). Laminin (Timpl et al, 1979) and collagen type IV are the most abundant components in the matrix and have been shown to be potent regulators of angiogenesis (Grant, et al., 1991; Grant et al, 1990; Ingber and Folkman, 1989; Madri and Williams, 1983). The importance (*in vivo*) of type IV collagen in angiogenesis is well documented, as shown by experiments in which an inhibitor of collagen synthesis blocked angiogenesis in the growing chorioallantoic membrane (CAM) of the chicken embryo (Maragoudakis, et al., 1990; Maragoudakis, et al., 1988). Several distinct sites in laminin have been identified as cell binding domains. These sites promote biological activities *in vitro* such as cell spreading, migration, and cell differentiation (Kleinman et al., 1985). We have been investigating the role of these biologically active sites in laminin on endothelial cell differentiation and on angiogenesis using several angiogenic models (Grant et al., 1991; Grant et al., 1989; Grant et al., 1992; Kibbey et al., 1992; Kubota et al., 1988). We found that several domains in laminin are active with endothelial cells.

Molecular, Cellular, and Clinical Aspects of Angiogenesis
Edited by Michael E. Maragoudakis, Plenum Press, New York, 1996

Most of the information and the understanding about these basement membrane molecules has been derived from components extracted from the EHS tumor (Kleinman et al., 1987). This tumor produces a large quantity of basement membrane which can be extracted and reconstituted in vitro, this extract is termed Matrigel. Matrigel contains laminin, collagen IV, heparan sulfate proteoglycan, and several growth factors i.e., TGF-ß, EGF, insulin-like growth factor 1, basic fibroblast growth factor (bFGF), and platelet-derived growth factor (PDGF) (Vukicevic et al., 1992). Matrigel can be reconstituted *in vitro* at 37°C to form a solid gel that has biochemical and morphological features similar to common basement membrane. This extracellular matrix is biologically active and is known to promote the differentiation of a variety of cells (Kleinman et al., 1987), including inducing endothelial cells to form tube-like structures which resemble capillaries. It is also now known that several of the growth factors found in Matrigel may also be involved in angiogenesis and in endothelial cell differentiation.

Angiogenesis

Angiogenesis, the process of vessel growth or renewal, is required for development, wound healing and tumor growth (Folkman and Hanahan, 1991; Folkman 1995 and 1992 Folkman and Ingber 1992). Under some pathological conditions, such as diabetic retinopathy and psoriasis, abnormal angiogenesis can lead to tissue dysfunction. Therefore, a better understanding of the factors which regulate vessel growth is essential to treating these diseases. The specific molecular events which stabilize vessels or initiate angiogenesis to form new blood vessels are still unclear.

The molecular events which occur in stable vessels or during new blood vessel formation (angiogenesis) are still unclear. Several *in vivo* and *in vitro* models have been developed that mimic at least some aspects of the angiogenic process, and these models provide insight into the steps and initiating factors involved in stimulating endothelial cells to form and maintain structurally intact vessels (Ingber, et al., 1987; Jaffe, et al., 1973; Kramer and Fuh, 1985; Madri and Pratt, 1986; Nicosia, et al., 1984). For example, Maciag et al. (Maciag, 1990) have shown that when confluent, cultured endothelial cells are treated with phorbol esters, approximately 50% of the cells form tube networks. Using this model of endothelial cell differentiation with subtractive hybridization, a novel protein EDG-1 was identified and shown to increase during tube formation (Hla and Maciag, 1990a; Hla and Maciag, 1990b; Maciag, 1990). The morphological differentiation of endothelial cells on Matrigel has also been successfully used to identify angiogenic and anti-angiogenic factors (Grant et al., 1993).

It has been the goal of our lab to determine the role of basement membranes in endothelial cell differentition and angiogenesis. Presently, we are examining the mechanism of tube formation on Matrigel at the cellular and molecular biologic levels. In our in vivo Matrigel model, human endothelial cells (EC) are plated on Matrigel and incubated at 37°C for 20 hrs., during which the cells form a capillary-like network of tubes (Grant et al., 1989; Kubota et al., 1988). We used a subtractive hybridization model to determine what genes are expressed early in endothelial cells plated on Matrigel. This was done by differential cDNA hybridization using RNA from endothelial cells cultured for 4 hours on either plastic or Matrigel. Here we present the results of this study.

Gene expression in Matrigel assay and subtraction method

The rapid morphology differentiation of the proliferating HUVEC monolayer into capillary-like structures on Matrigel is induced in a period of time that is less than that observed in other *in vitro* models. Unlike angiogenesis *in vivo*, however, cellular proliferation on Matrigel is not observed and the cells do not invade the matrix. Therefore, it is not a complete model of angiogenesis. Nevertheless, the model provides a method of examining some aspects of angiogenesis and clearly demonstrates endothelial cell differentiation. While this model does not mimic all of the events that occur during vessel formation, it provides a powerful tool for the examination of certain molecular and biochemical events that transpire during endothelial cell morphogenesis and differentiation. The formation of capillary-like tubes from human umbilical vein endothelial cells cultured on Matrigel is a complex process which involves several steps and the biosynthesis of numerous gene products. We have shown previously (Grant, et al., 1991) that the addition of cycloheximide or actinomycin D to this assay system inhibits the differentiation of the cells into tubes. More recently, we have investigated the early genes which are induced during the endothelial cell differentiation on Matrigel (Grant et al., 1995).

As previously indicated, several studies have shown that the use of subtraction cDNA libraries or differential (expression) hybridization techniques are good approaches to study gene expression in differentiating systems. Thus, we tried to further understand the mechanism of tube formation by determining the identity of some of the early genes expressed by the cells. This was done using a differential cDNA cloning technique performed with the RNA from endothelial cells cultured 4 hrs on either plastic or Matrigel.

Identification of cDNA clones following 4hr incubation on Matrigel

HUVEC cultured on plastic display a cobblestone morphology, whereas on Matrigel the cells attach within two hours and by four hours migration and cell-cell association result in reorganization and initiation of tube morphogenesis (Figure 1). Differential hybridization of mRNA from endothelial cells on plastic vs. Matrigel at 4 hours resulted in 17 clones that were found to range in size from 500 to 3000 base pairs (Table 1). One of these gene product, encoded a small polypeptide (4.9 Kd) first identified and isolated from the calf thymus, is called thymosin β4 (Badamchian, et al., 1988; Gomez, et al., 1989; Low and Goldstein, 1984; Low and Goldstein, 1985; Low, et al., 1990). Nothing is known about the expression or the regulation of thymosin β4 synthesis in endothelial cells.

TABLE 1

PLASMID	SIZE(NB)	NORTHERN*	IDENTITY(NT)	KNOWN mRNA
Clone A	700 bp	4X increase	100 (700)"	Thymosin B-4
Clone C	600 bp	ND	74 (312)	Ribosomal protein L (frog)
Clone D	2000 bp	1.5X increase	68 (44)	Hum. LDL-rec. protein
Clone E	6000 bp	slight increase	61 (101)	HUM UPA gene
Clone F	4000 bp	3-5X increase	98 (1000)	Hum Elongation Fac. alpha 1
Clone H	2000 bp	10-20X increase	No data	Unknown
Clone I	500 bp	ND	64 (87)	Mouse pro alpha1(II)Coll
Clone N	1300 bp	Increase	95 (553)	Hum-splicing Factor-2p32
Clone Q	800 bp	2-3X inc.	99 (227)	Hum-Calmodulin

* **mRNA intensity HUVEC 4hrs on Matrigel**
"denotes number of residues sequenced

Northern analysis of thymosin β4 was determined in HUVEC on both plastic and Matrigel (Figure 2a,b). The thymosin β4 cDNA clone isolated from endothelial cells recognized a message of 800 bp, slightly larger than found in thymocytes due to an extended 3' region. Cells cultured on plastic did not significantly change their expression of thymosin β4 over a 6 hour period, whereas cells cultured on Matrigel showed a 5-fold increase at four through six hours (Figure 2a,b). This demonstrates that on Matrigel there is an increase in the synthesis of thymosin β4 mRNA.

Background information on Thymosin ß4

Thymosin ß4 is a member of a large group of thymosin proteins (both alpha and beta). These proteins were first isolated from fraction 5 (F5) of the calf thymus (Hooper et al., 1975 and Low et al., 1981). This fraction 5 preparation contained proteins ranging from 1000 to 15,000 Mr and the component peptides were named as to their migration in isoelectric focusing gels. Peptide with pI's below 5.0 we named alpha 1,2, 3 etc., and those between 5.0 and 7.0, beta 1, 2, 3 etc, and above 7.0 were identified as gamma. Thymosin ß4 has almost identical sequence to ß8, 9 and 10. The major biological properties of thymosins from fraction 5 are to induce differentiation of specific subclasses of T lymphocytes (Low and Goldstein 1984). The thymosin ß4 peptide, which was first thought to be a very specific thymic hormone, is now believed to be responsible for a much more general, though still obscure, function in the cell, due to its wide distribution in many cells and tissues (Livaniou et al., 1992). Recently it has been shown to inhibit G actin polymerization by forming a 1:1 complex with the monomer; thus, this finding leads to the hypothesis that thymosin ß4 is involved in the regulation of actin polymerization in the cell cytoplasm (Yu et al., 1993, 1994). A hint at the function of thymosin ß4 may be indicated by the change in thymosin ß4 level under certain pathological states.

At the molecular level, the thymosin ß4 gene has been described in the rat (Varghesse and Kronenberg, 1991) and was found to be intron-containing genes that have 2 kilobase pair-long transcription units containing two introns. The human thymosin ß4 gene, which is identical to the human inteferon-inducible gene 6-26, is localized to seven distinct chromosomes (1,2,4,9,11,20 and X), some of which are pseudogenes (Clauss et al., 1991). This protein is conserved. For example, a cDNA encoding human thymosin-ß4 has been isolated and found to be 84% homologous to the rat thymosin ß4 (the homology between their coding regions is 93%). The sequence of this protein is well conserved since it is over 95% homologous in human, mouse, rat and cow (Hall, 1991; Low and Goldstein, 1985) (See Table 1). At the transcription level a thymosin ß4 cDNA clone was isolated from a mouse pre-B cell line, and it was shown by Rnase protection assays that two sizes of thymosin ß4 mRNA existed, a long form containing a 98 nucleotide insertion and a short form that corresponds to the known rat and human mRNA (Rudin et.al., 1990). Southern blotting revealed that both forms are encoded by a single gene in the mouse, and that the two forms may arise by differential RNA splicing (Rudin et al.,1990). Analysis of the amino acid and cDNA nucleotide sequences of rat thymosin ß4 revealed that it is synthesized as a 44 amino acid peptide, which is processed into a 43-amino-acid peptide by removal of the first methionyl residue and acetylation of the exposed NH2-terminal seryl residue (Clauss et al., 1991). Thymosin ß4 has no signal peptide, suggesting that it is not secreted, however it is present in human serum and may be secreted by some cell types (Clauss et al., 1991).

Thymosin ß4 function and relationship to vessels

Thymosin ß4 exhibits important activities in the regulation and differentiation of thymus-dependent lymphocytes (Low and Goldstein, 1984; Low and Goldstein, 1985). For example, previous studies have demonstrated that thymosin ß4 is effective in partially or fully reconstituting immune functions in thymic-deprived or immunodeprived animals (Low and Goldstein, 1981). Thymosin ß 4 expression is also directly associated with cell cycle regulation (Schöbitz, et al., 1991b). In addition, this protein is ubiquitously expressed in many cells and tissues (Condon and Hall, 1992; Lin and Morrison-Bogorad, 1990). Thymosin ß 4 is linked to the differentiation of not only hematopoietic cells but also to cells in developing human embryonic kidney (Hall, 1991). Thymosin ß 4 also is increased in the serum of patients over 60, in newborns, and in patients suffering from AIDS, hepatitis, and inflammatory bowel disease (Naylor, et al., 1986). This small polypeptide is differentially expressed in the developing brain tissues during embryogenesis of the rat and is present in lower amounts in embryonic lung, kidney, spleen, adrenal glands, heart and liver (Lin and Morrison-Bogorad, 1990).

Distribution of thymosin β4 in capillary-like structures *in vitro* and in vessels *in vivo*

The incubation of HUVEC on Matrigel results in the elongation and reorganization of the cells into tube-like structures (Figure 1). Extensive reoganization of the cells occurs at 4-6 hrs during tube formation. Tubes which were immunostained with thymosin β4 antibody, and viewed by confocal microscopy demonstrated positive immunofluorescent staining in most of the cells comprising the tube structures (Figure 2c). Close examination of these stained networks indicated that thymosin ß4 is expressed in the tube-forming cells.

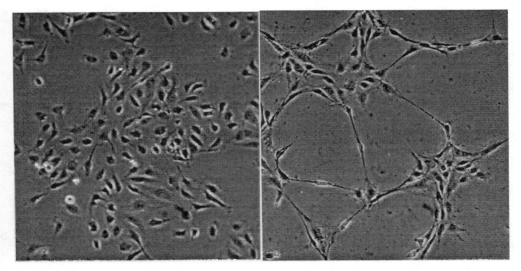

Figure 1. Morphology of HUVEC 4 hrs on Plastic and Matrigel

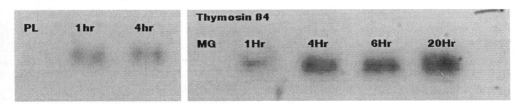

Figure 2a. Northern, RNA from HUVEC cultured on Plastic (left) or on Matrigel (right) for 1,,4, 6 and 20 hrs then probed with thymosin ß4 cDNA clone.

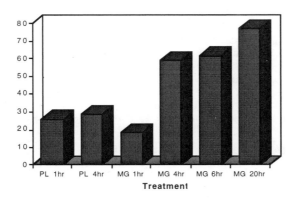

Figure 2b. Quantitation of relative densities of the Northerns above.

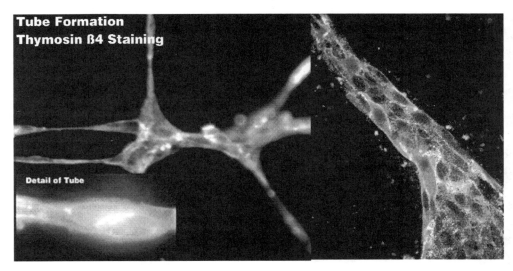

Figure 2c. Tube formation on Matrigel, immunostained for thymosin ß4. Left panel whole tube staining viewed in standard microscope, right confocal microscopic image.

When paraffin tissue sections from a Matrigel subcutaneous implant in vivo with induced angiogenesis (Passaniti et al., 1992 and Kibbey et al., 1992) were immunostained with an antibody to thymosin β4, numerous vessels forming in the tissue were positively stained (not shown). Other areas of the tissue sections were also examined and staining was found in the basal layer of the epithelium of the skin as well as in some cells of the connective tissue (not shown). Large and medium sized vessels in the dermis adjacent to the Matrigel plugs were also positively stained. Specifically, the arterial endothelium and the first layer of smooth muscle cells were stained most prominantly. In the skeletal muscle of the mouse abdomen, staining for thymosin β4 was found primarily in the capillaries situated between the skeletal muscle fibers. These data indicate that thymosin β4 is present in forming vessels in the *in vitro* and *in vivo* angiogenesis assays as well as in mature vessels.

Expression of transfected thymosin β4 in HUVEC

Thymosin β4 full length cDNA was transfected into endothelial cells and the presence of the protein was detected by FITC-immunocytochemistry. Most control cells, either untransfected or mock transfected with the vector alone, showed little or low thymosin β4 staining (Figure 3a) with approximately 10% of the cells staining strongly positively. HUVEC transfected with the thymosin β4 cDNA were strongly stained within 24 hours, with 50-60% of the cells showing strong positive staining (Figure 3b). At this time, the fluorescence was primarily diffuse and cytoplasmic, although in some cases the staining appeared brighter in perinuclear regions (Figure 3b). Staining for thymosin β4 was also present in cells undergoing mitosis (not shown). When the HUVEC were examined 48 hours after transfection, the staining was redistributed to granule-like structures located at the periphery of the cytoplasm. Foreskin fibroblasts were also transfected with the thymosin β4 clone and demonstrated a similar cytoplasmic distribution as the endothelial cells at 24 and 48 hrs (data not shown).

Behavior of thymosin β4 transfected cells

We next examined the morphology of thymosin β4-transfected HUVEC plated on plastic for 30 min. Like normal cells, the mock transfected cells were not fully adherent nor did they all spread within 30 min (cells usually take about 60 mins to attach and spread). In contrast, transfected cells became rapidly adherent and were almost completely spread within 30 min (Fig 4a, Tβ4 panel).

The effect of thymosin B4 transfection on HUVEC attachment In attachment assays normal HUVEC showed a dose-dependent increase in attachment to laminin within the short 30

Figure 3a/b. Thymosin ß4 mock and transfected HUVEC immunostained with a antibody to the thymosin B4 protein.

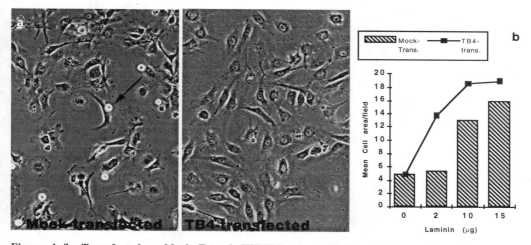

Figure 4a/b. Transfected vs Mock Transf. HUVEC. a) morphology and spread on plastic, b) dose dependent graph of attachment to laminin.

minute time span (Figure 4b). The transfected cells also showed increased cell spreading on plastic as well (Figure 4b) The mock transfected cells also demonstrated a dose-dependent attachment to laminin (Figure 4b).

The thymosin β4 transfected cells, however, demonstrated an increase in the rate of attachment to laminin (at dosages of 2 to 10 μg) over that observed with either the normal or mock transfected cells. At times greater than 30 minutes, no difference in the total number of attached cells was observed, demonstrating that thymosin β4 expression only affected the rate but not the total number of attached cells (data not shown). Thus, cells transfected with thymosin β4 are more adherent and spread more quickly than control cells, suggesting that this protein has a biological function in promoting adhesion and spreading. Similar assays were done to examine the effect on attachment to collagen IV. As seen with laminin thymosin ß4 transfected cells promoted greater attachment to collagen IV than normal or mock transfected cells (Table 2). However, the degree of attachment to collagen compared to laminin was less.

Table 2: Attachment of normal, mock- or TB4-transfected HUVEC to different substrata at 30 min.

Cell Treatment	Mean Cell Attachment to Different Substrata		
	Plastic	Laminin (10μg)	Collagen IV (10μg)
Normal HUVEC	800 +/- 90	970 +/- 26	954 +/- 58
Mock Transfected HUVEC	950 +/- 27	1180 +/- 88	1111 +/-80
TB4 Transfected HUVEC	990 +/- 60	1460 +/- 67	1380 +/- 79

We also compared normal, mock and thymosin B4 tranfected HUVEC attachment to plastic vs laminin and collagen. Transfected cells demonstrated a slight increase in attachment to laminin and collagen as compared to mock transfected or normal cells (Table 2).

The effect of thymosin B4 on tube formation

Induction of tubes by thymosin β4 was also examined in both transfected and mock transfected cells plated on Matrigel. No differences in adhesion to Matrigel were observed at any time point past 30 mins between transfected and non-transfected cells. However, morphological differences in tube formation were seen between transfected and mock transfected cells when the cells were examined at 2, and 4 hours after seeding on Matrigel (Figure 6). The mock transfected cells demonstrated the similar pattern of adhesion and alignment found in normal HUVEC, and cells remained fairly flattened (similar to HUVEC on plastic) from 2-4 hours (Figure 6, left panels).

At these same time points, the thymosin β4-transfected cells appeared more elongated, and structures resembling normal tube formation, usually observed at approximately 10 hrs, were now observable at 4 hrs (Figure 6, right panels). There was no difference between the two

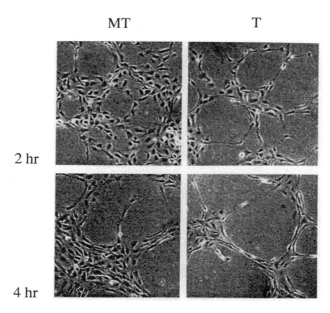

MT   T

2 hr

4 hr

Figure 5. Comparison of tube formation on Matrigel using mock- or thymosin β4 transfected cells

cell types and nontransfected cells following a 24 hour incubation on Matrigel, when complete tubes were observed (data not shown). Thus, the cells transfected with thymosin ß4 formed tubes more quickly on Matrigel than control cells and the final tube formation was similar to the normal. Therefore it appears that thymosin ß4 can help in the acceleration of tube formation. These changes may be due in part to the cells' increased ability to spread and migrate on Matrigel. We are currently examining if there are changes in matrix receptors expression in thymosin ß4 transfected cells. All these findings suggest that thymosin ß4 plays an important role in tube formation and endothelial cell differentiation. This is the first time that this protein has been shown to exhibit a biological function with endothelial cells .

Summary

We have tried to further the understanding of the process of endothelial cell differentiation by determining the identity of some of the early genes by subtractive cDNA cloning and sequencing (Sambrook et al, 1989; Pearson and Lipman, 1988). We were very sucessful and have identified several genes. Some of the genes that were found to be upregulated when examined in Northern blots were calmodulin, tubulin, a collagen-like gene, ribosomal proteins, human elongation and splicing factors, several novel genes, and thymosin ß4. **Thymosin ß4**, was increased 5-fold. Nothing is known about the expression or the regulation of thymosin ß4 in endothelial cells. Recently we reported strong evidence that this polypeptide is involved in the regulation of endothelial cell differentiation because: 1) immunostaining localized thymosin ß4 *in vivo* to growing and mature vessels, 2) endothelial cells transfected with thymosin ß4 showed increased attachment and spreading on matrix components, and an accelerated rate of tube formation on Matrigel, and 3) direct addition of purified thymosin ß4 protein to endothelial cells in culture induced greater tube formation and promoted actin cytoskeletal changes. 4) The preparation of antisense oligos to thymosin ß4 blocked thymosin protein synthesis and tube formation *in vitro*. 5) The addition of purified thymosin ß4 protein to endothelial cells in culture increased their attachment to matrix proteins, enhanced tube formation but inhibited cell proliferation. The results suggest that thymosin ß4 is induced and possibly involved in differentiating endothelial cells and is likely to play a role in vessel formation.

Our study is the first study that illustrates that the expression of thymosin ß4 is involved in endothelial cell differentiation. Further mechanistic investigations are needed to establish if thymosin ß4 is differentially expressed in vessels and if it directly induces differentiation in endothelial cells or if it is a product of the differentiation process.

REFERENCES:

Auerbach, R., Auerbach, W., and Polakowski, I. (1991) Assays for angiogenesis: A review. *Pharmacol Ther.* **1**: 1-11.

Badamchian, M., Strickler, M. P., Stone, M. J. and Goldstein, A. L. (1988). Rapid isolation of thymosin beta 4 from thymosin fraction 5 by preparative high-performance liquid chromatography. *J Chromatogr,* **459**, 291-300.

Barsigian, C. and Martinez, J. (1990) Binding and covalent processing of fibrinogen by hepatocytes and endothelial cells. *Blood Coagulation and Fibrinolysis* **1**: 551-555.

Cheng Y-F, R.H. Kramer. (1989) Human microvascular endothelial cells express integrin-related complexes that mediate adhesion to the extracellular matrix. J. Cellul. Phys., 139:275-286.

Clauss, I. M., Wathelet, M. G., Szpirer, J., Quamrul Islam, M., Levan, G., Szpirer, C. and Huez, G. A. (1991) Human thymosin ß4/6-26 gene is part of a multigene family composed of seven members located on seven different chromosomes. *Genomics* **9**: 174-180.

Condon, M. R. and Hall, A. K. (1992). Expression of thymosin beta-4 and related genes in developing human brain. *J Mol.Neurosci,* **3**, 165-70.

Cordero OJ; Sarandeses C; and Nogueira M. (1994) Prothymosin alpha receptors on peripheral blood mononuclear cells. FEBS Lett; **341**:23-7.

Fellin, M., Barsigian, C., Rich, E., and Martinez, J. (1988) Binding and cross-linking of rabbit fibronectin by rabbit hepatocytes in suspension. *J. Biol Chem* **263**(4): 1791-1797.

Folkman, J. (1985) Toward and understanding of angiogenesis: search and discovery. *Persp Biol Med* **29** 10-36.

Folkman, J. (1992). The role of angiogenesis in tumor growth. *Semin Cancer Biol,* **3**, 65-71.

Folkman, J. and Hanahan, D. (1991). Switch to the angiogenic phenotype during tumorigenesis. *Princess Takamatsu Symp,* **22**, 339-47.

Folkman, J. and Ingber, D. (1992). Inhibition of angiogenesis. *Semin Cancer Biol,* **3**, 89-96.

Folkman, J. and Shing, Y. (1992). Angiogenesis. *J Biol Chem,* **267**, 10931-4.

Giordano, T., Howard, T. H., Coleman, J., Sakamoto, K. and Howard, B. H. (1991). Isolation of a population of transiently transfected quiescent and senescent cells my magnetic affinity cell sorting. *Exp. Cell Res.*, **192**, 193-197.

Goldstein, S., Fortdis, C. M. and Howard, B. H. (1989). Enhanced transfection efficiency and improved cell survival after electroporation of G2/M-synchronized cells and treatment with sodium butyrate. *Nucleic Acids Res.*, **17**, 3959-3971.

Gomez-Marquez, J., Dosil, M., Segade, F., Bustelo, X., Pichel, J., Dominguez, F. and Freire, M. (1989). Thymosin-b4 gene: preliminary characterisation and expression in tissues, thymic cells, and lymphocytes. *J. Immuno.*, **143**, 2740-2744.

Grant, D. S., Tashiro, K. I., Segui-Real, B., Yamada, Y., Martin, G. R., and Kleinman, H. K. (1989) Two different laminin domains mediate the differentiation of human endothelial cells into capillary-like sturctures *in vitro.* Cell **58**: 933-943.

Grant, D. S., Kinsella, J. L., Fridman, R., Auerbach, R., Piasecki, B. A., Yamada, Y., Zain, M. and Kleinman, H. K. (1992). Interaction of endothelial cells with a laminin A chain peptide (SIKVAV) in vitro and induction of angiogenic behavior in vivo. *J Cell Physiol*, **153**, 614-625.

Grant, D. S., Kleinman, H. K., Goldberg, I. D., Bhargava, M. M., Nickoloff, B. J., Kinsella, J. L., Polverini, P. and Rosen, E. M. (1993). Scatter factor induces blood vessel formation in vivo. *Proc Natl Acad Sci U S A*, **90**, 1937-1941.

Grant, D. S., Kleinman, H. K. and Martin, G. R. (1990). The role of basement membranes in vascular development. *Ann N Y Acad Sci*, **588**, 61-72.

Grant, D. S., Kinsella, J. L., Kibbey, M. C., LaFlamme, S., Burbelo, P. D., Goldstein, A. L., and Kleinman, H. K. (1995) Matrigel Induces Thymosin ß4 Gene in Differentiating Endothelial Cells. Submitted, final review process.

Grant, D. S., Lelkes, P. I., Fukuda, K. and Kleinman, H. K. (1991). Intracellular mechanisms involved in basement membrane induced blood vessel differentiation in vitro. *In Vitro Cell Dev Biol.*, **27a**, 327-336.

Hall, A. (1991). Differential expression of thymosin genes in human tumors and in the developing human kidney. *Int. J. Can.*, **48**, 672-677.

Hajjar, K.,N.M. Hamel, (1990) Identification and characterization of human endothelial cell membrane binding sites for tissue plasminogen activator and urokinase. J. Biol. Chem. 265: 2908-16.

Hla, T. and Maciag, T. (1990a). An abundant transcript induced in differentiating human endothelial cells encodes a polypeptide with structural similarities to G-protein-coupled receptors. *J Biol Chem*, **265**, 9308-9313.

Hla, T. and Maciag, T. (1990b). Isolation of immediate-early differentiation mRNAs by enzymatic amplification of subtracted cDNA from human endothelial cells. *Biochem Biophys Res Commun*, **167**, 637-643.

Hooper, J. A., McDaniel, M. D., Thruman, G. B., Cohen, G. H., Schulhof, R. S., and Goldstein, A. L. (1975) Purification and properties of bovine thymosin. Ann NY Acad Sci (249): 125.

Ingber, D. E. and Folkman, J. (1989). How does extracellular matrix control capillary morphogenesis? *Cell*, **58**, 803-805.

Ingber, D. E., Madri, J. A. and Folkman, J. (1987). Endothelial growth factors and extracellular matrix regulate DNA synthesis through modulation of cell and nuclear expansion. *In Vitro*, **23**, 387-394.

Jaffe, E. A., Nachman, R. L., Becker, C. G. and Minick, C. R. (1973). Culture of human endothelial cells derived from umbilical veins-identification by morphological and immunological criteria. *J. Clin. Invest.*, **52**, 2745-2756.

Kibbey, M. C., Grant, D. S. and Kleinman, H. K. (1992). Role of the SIKVAV site of laminin in promotion of angiogenesis and tumor growth: an in vivo Matrigel model. *J Natl Cancer Inst*, **84**, 1633-1638.

Kleinman, H. K., McGarvey, M. L., Liotta, L. A., Gehron-Robbey, P., Tryggvasson, K. and Martin, G. R. (1987). Isolation and characterization of type IV procollagen, laminin and heparan sulfate proteoglycan from the EHS sarcoma. *Biochemistry*, **24**, 6188.

Kleinman, H. K., Cannon, F. B., Laurie, G. W., Hassel, J. R., Aumalley, M., Terranova, V. P., Martin, G. R., and Dalcq, M. D. B. (1985) Biological activities of laminin. *J. Cell Biol* **27**: 317-325.

Kramer, R. H. and Fuh, G. M. (1985). Type IV collagen synthesis by cultured human microvascular endothelial cells and its deposition in the subendothelial basement membrane. *Biochem.*, **24**, 7423-7430.

Kubota, Y., Kleinman, H. K., Martin, G. R. and Lawley, T. J. (1988). Role of laminin and basement membrane in the morphological differentiation of human endothelial cells into capillary-like structures. *J. Cell Biol.*, **107**, 1589-1598.

Lin, S. and Morrison-Bogorad, M. (1990). Developmental expression of mRNAs encoding thymosins b4 and b10 in rat brain and other tissues. *J. Mol. Neurosci.*, **2**, 35-44.

Livianiou, E., Mihehic, M., Evangelatos, G.P., Haritos, A., and Voelter, W. (1992) A thymosin ß4 ELISA using an antibody against the N-terminal fragment thymosin ß4 [1-14]. *J. Immun Met* **148**: 9-14.

Low, T. and Goldstein, A. (1984). Thymosins: structure, function and therapeutic application. *Thy.*, **6**, 27-42.

Low, T. L. and Goldstein, A. L. (1985). Thymosin beta 4. *Methods Enzymol*, **116**, 248-55.

Low, T. L., Lin, C. Y., Pan, T. L., Chiou, A. J. and Tsugita, A. (1990). Structure and immunological properties of thymosin beta 9 Met, a new analog of thymosin beta 4 isolated from porcine thymus. *Int J Pept Protein Res*, **36**, 481-8.

Low, T. L. K. and Goldstein, A. L. (1981). Chemical Characterization of Thymosin β4. *J. Biol. Chem.*, **257**, 1000-1006.

Maciag, T. (1990). Molecular and cellular mechanisms of angiogenesis. *Important Adv Oncol*, **1990**, 85-98.

Madri, J. A., Dryer, B., Pitlick, F. and Furthmayr, H. (1980). The collagenous components of the subendothelium: correlation of structure and function. *Lab. Invest.*, **43**, 303-315.

Madri, J. A., and Williams, S. K. (1983) Capillary endothelial cell cultures: Phenotypic modulation by matrix components. *J. Cell Biol* **97**: 153-165.

Madri, J. A. and Pratt, B. M. (1986). Endothelial cell-matrix interactions: in vitro models of angiogenesis. *J. Histochem. Cytochem.*, **34**, 85-91.

Maragoudakis, M. E., Sarmonika, M., and Panoutsacopoulou, M. (1988) Inhibition of basement membrane biosynthesis prevents angiogenesis. *J. Pharmacol Exp. Ther* **244** (2): 729-33.

Maragoudakis, M. E., Missirlis, E., Sarmonika, M., Panoutsacopoulou, M., and Karakiulakis, G. (1990) Basement membrane biosynthesis as a target to tumor therapy. *J Pharm Exp Therap* **253**: 753-757.

Montesano, R., M.S. Pepper, J.D. Vassalli, and L. Orci. (1987) Phorbol ester induces cultured endothelial cell to invade to invade a fibrin matrix in the presence of fibrinolytic inhibitors. J. Cellul Physiol. 132: 509-516.

Naylor, P. H., Friedman, K. A., Hersh, E., Erdos, M. and Goldstein, A. L. (1986). Thymosin alpha 1 and thymosin beta 4 in serum: comparison of normal, cord, homosexual and AIDS serum. *Int J Immunopharmacol*, **8**, 667-676.

Naylor, P. H., McClure, J. E., Spangelo, B. L., Low, T. L. K. and Goldstein, A. L. (1984). Immunochemical studies on thymosin: radioimmunoassay of thymosin β4. *Immunopharm.*, **7**, 9-16.

Nicosia, R. F., McCormick, J. F. and Bielunas, J. (1984). The formation of endothelial webs and channels in plasma clot culure. *Scan Elect Microsc*, **2**, 793-799.

Paku, S. and Paweletz, N. (1991). First steps of tumor-related angiogenesis. *Lab. Invest.*, **65**, 334-346.

Passaniti, A., Taylor, R. M., Pili, R., Guo, Y., Long, P. V., Haney, J. A., Pauly, R. R., Grant, D. S. and Martin, G. R. (1992). A simple, quantitative method for assessing angiogenesis and antiangiogenic agents using reconstituted basement membrane, heparin, and fibroblast growth factor. *Lab Invest*, **67**, 519-28.

Roberts, A. B., McCune, B. K., and Sporn, M. B. (1992) TGF-ß: Regulation of extracellular matrix. Kidney Int. **41**: 557-559.

Rudin, C. M., Engler, P., and Storb, U. (1990) Differential splicing of thymosin ß4 mRNA. *J. Immun* **144**(12): 4857-62.

Safer, D. (1992). The interaction of actin with thymosin beta 4 [news]. *J Muscle Res Cell Motil*, **13**, 269-71.

Safer, D., Elzinga, M. and Nachmias, V. T. (1991). Thymosin beta 4 and Fx, an actin-sequestering peptide, are indistinguishable. *J Biol Chem*, **266**, 4029-32.

Sanders, M. C., Goldstein, A. L. and Wang, Y. L. (1992). Thymosin beta 4 (Fx peptide) is a potent regulator of actin polymerization in living cells. *Proc Natl Acad Sci U S A*, **89**, 4678-82.

Schöbitz, B., Hannappel, E. and Brand, K. (1991a). The early induction of the actin-sequestering peptide thymosin beta 4 in thymocytes depends on the proliferative stimulus. *Biochim Biophys Acta*, **1095**, 230-235.

Schöbitz, B., Netzker, R., Hannappel, E. and Brand, K. (1991b). Cell-cycle-regulated expression of thymosin beta 4 in thymocytes. *Eur J Biochem*, **199**, 257-262.

Shimamura, R., Kudo, J., Kondo, H., Dohmen, K., Gondo, H., Okamura, S., Ishibashi, H. and Niho, Y. (1990). Expression of the thymosin β4 gene during differentiation of hematopoietic cells. *Blood*, **76**, 977-984.

Timpl, R., Rohde, H., Gehron Robey, P., Rennard, S. I., Foidart, J.M., and Martin, G.R. (1979). Laminin-a glycoprotein from basement membranes. *J. Biol. Chem.* **254**, 9933-9937.

Vancompernolle, K., Goethals, M., Huet, C., Louvard, D. and Vandekerckhove, J. (1992). G-to F-actin modulation by a single amino acid substitution in the actin binding site of actobindin and thymosin beta 4. *Embo J*, **11**, 4739-46.

Varghese, S., and Kroneneberg, H. M. (1991) Rat thymosin ß4 gene. *J. Biol Chem* **266**(22): 14256-61.

Vukicevic, S., Kleinman, H., Luyten, F. P., Roberts, A.B., Roche, N. S., and Reddi, A. H. (1992) Identification of multiple active growth factors in basement membrane Matrigel suggests caution in interpretation of cellular activity related to extracellular matrix components. *Exp Cell Res* **202**: 1-8

Yamamoto, M., T.Yamagishi, H. Yaginuma, K. Murakami, and N. Ueno. (1994) Localization of thymosin ß4 to the neural tissues during the development of Xenopuslaevis, as studied by insitu hybridization and immunochemistry. *Develop.Brain Res. 79,177-185.*

Yu, F. X., Lin, S. C., Morrison, B. M., Atkinson, M. A. and Yin, H. L. (1993). Thymosin beta 10 and thymosin beta 4 are both actin monomer sequestering proteins. *J Biol Chem*, **268**, 502-

Yu, F-X, S-C. lin, M. Morrison-Bogorad, and H.L. Yin. (1994) Effects of thymosin ß4 and thymosin ß10 on actin structutres in living cells. *Cell Motility and the Cytoskeleton 27: 13-25.*

CO-EXPRESSION OF THE α2-SUBUNIT OF LAMININ AND THE METASTATIC PHENOTYPE IN MELANOMA CELLS

Jing Han and Nicholas A. Kefalides

Connective Tissue Research Institute and Department of Medicine,
University of Pennsylvania, and University City Science Center,
Philadelphia, Pennsylvania 19104

INTRODUCTION

Previous studies in this laboratory demonstrated the presence of laminin M or "M" subunit of about 300 kDa, which was shown to be absent in EHS tumor laminin or in laminin synthesized by some neoplastic cell lines (Ohno et al., 1983, 1986; Jenq et al., 1993). Because of the large number of isoforms of the laminin chains isolated in different laboratories, a new nomenclature has been proposed (Burgeson, et al., 1993). Table I shows the new names given to the "classical" laminin chains isolated from the mouse EHS tumor as well as to the new isoforms of laminin isolated from a variety of normal tissues. Thus, A B1 B2 is now designated α1β1γ1. The "M" subunit first isolated by Ohno et al. (1983, 1986) in our laboratory has been given the new name α2. Merosin, the laminin which contains the α2-subunit has been designated α2β1γ1. Merosin has been identified in human placenta, striated muscle, peripheral nerve and Schwannoma cells (Leivo and Engvall, 1988; Ehrig et al., 1990; Engvall et al., 1992).

Studies by Jenq et al. (1994) demonstrated that metastatic cells (4R), (ras-transformed rat fibroblasts) and tumorigenic but not metastatic cells (RE4) (ras plus adenovirus type 2 Ela gene transformed rat fibroblasts) differ in their ability to synthesize and secrete the α2-subunit of laminin. 4R cells (highly metastatic, non-tumorigenic) produced large amounts of the α2-subunit, whereas, RE4 cells (highly tumorigenic but non-metastatic) produced little or no α2-subunit. In addition, the authors showed that the laminin secreted by the metastatic 4R cells had a higher cell adhesion promoting activity than the non-metastatic RE4 cells. In the same study, it was demonstrated that a non-metastatic melanoma cell line (W793) produced little or no α2-subunit, whereas, a metastatic cell line (W1205) that was derived from the W793 line after passage into nude mice, produced significant amounts of the α2-subunit of laminin.

In the present study we examined a number of melanoma cell lines in an attempt to find out whether a correlation existed between the expression of the α2-subunit of laminin and the metastatic phenotype. In addition, we determined the expression of several integrins by the various melanoma cell lines in an attempt to determine whether the expression of the α2-subunit and the metastatic phenotype correlated with the expression or non-expression of an integrin subunit.

Molecular, Cellular, and Clinical Aspects of Angiogenesis
Edited by Michael E. Maragoudakis, Plenum Press, New York, 1996

31

MATERIALS AND METHODS

Cell Cultures

The melanoma cell lines were kindly provided by Dr. Meenhard Herlyn, Wistar Institute, Philadelphia, PA, USA. Cells were grown in culture medium containing 800 ml MCDB-153, and 200 ml L-15 per liter, supplemented with 20 ml FBS and 5 µg/ml insulin. MCDB-153, L-15 and insulin were purchased from Sigma Chemical Co., St. Louis, MO, USA. Cultures were maintained at 37°C in a 5% CO_2 incubator.

Antibodies

Polyclonal antibody against the mouse Engelbreth-Holm-Swarm (EHS) tumor was purchased from Sigma Chemical Co., St. Louis, MO, USA, and antibodies against human integrins were purchased from Chemicon International, Inc., Temecula, CA, USA. Monoclonal antibodies were directed against integrins α2, α6 , and β1; polyclonal antibodies were directed against integrins α1, αV, β3 and β4.

Metabolic Labeling, Immunoprecipitation and SDS-Page

The immunoprecipitation procedure used was that of Aratani and Kitagawa (1988) with some modifications. Cells at 85% confluency were labeled for 18 h with 100 µCi/ml trans-labeled [^{35}S]methionine in methionine-free DMEM which contained 15 mM HEPES. The cells were extensively washed with PBS and lysed in lysis buffer (10 mM-Tris/2.5 mM-EDTA/ 1%-Triton X-100/0.4% SDS/10 µl/ml aprotinin/ 0.4 M NaCl, pH 7.4). Four hundred microliters of clarified cell lysate (from a total of 4 ml) or 600 µl conditioned medium (from a total of 10 ml) was initially incubated with 40 µl 10% (w/v) suspension of Protein A-Sepharose for 3 h with shaking to remove nonspecifically bound polypeptides. The supernatants were then incubated overnight with the appropriate laminin or integrin antibody at a dilution of 1:200, followed by 4 h incubation with 40 µl of Protein A-Sepharose with gentle agitation. All incubations were at 4° C. The packed Sepharose beads were washed three times with the lysis buffer. The washed Protein A-Sepharose pellets were then suspended in 60 µl reducing sample buffer (0.062 M Tris-HCl, pH 6.8, 2% SDS, 30% glycerol, and 0.2%

Table I - ISOFORMS OF LAMININ*

NEW NAME	NEW COMPOSITION	PREVIOUS NAME	LOCALIZATION
Laminin 1	α1β1γ1	EHS Laminin	all basement membrane except skeletal muscle
Laminin 2	α2β1γ1	Merosin	striated muscle, peripheral nerve, placenta
Laminin 3	α1β2γ1	S-Laminin	synapse, glomerus, arterial blood vessel walls
Laminin 4	α2β2γ1	S-Merosin	myotendinous junction, trophoblast
Laminin 5	α3β3γ2	Kalinin/Nicein	dermal-epidermal junction, stromal-epidermal junction
Laminin 6	α4β1γ1	K-Laminin	dermal-epidermal junction, stromal-epidermal junction
Laminin 7	α3β2γ1	KS-Laminin	amnion, fetal skin

*From Burgeson et al., 1993.

bromophenol blue containing 5% β-mercaptoethanol). Samples were boiled for 3 min in sample buffers, and gel wells were loaded with equal TCA-precipitable cpm. Gel electrophoresis of the precipitated proteins was carried out on a 5% separating gel with a 3% stacking gel according to Laemmli (1970). Finally, the labeled protein bands were visualized by fluorography.

RESULTS

Expression of the α2-Subunit of Laminin

Seven distinct melanoma cell lines were examined for their ability to express the α2-subunit of laminin (Table II). Two of the cell lines, W793 and W98-1, were from primary tumors and were non-metastatic, the other five were all able to metastasize. Although, melanoma cell lines W136-1A and A2058 were from primary tumors, the cells had a high metastatic propensity. Cell lines W373, W164 and W9 were isolated from metastatic lesions. Except for cell lines W793 and W98-1 (the two non-metastatic ones) all the other melanoma cell lines expressed significant amounts of the α2-subunit of laminin (Table II).

Expression of Integrins

Table III shows that all melanoma cell lines tested express the αV integrin subunit. However, there is no correlation between the expression of this integrin and expression of the α2-subunit of laminin or of the metastatic phenotype. On the other hand the α1 integrin subunit was uniformly absent or nearly so from all melanoma cell lines tested. The α2 and α3 integrin subunits were expressed only by the melanoma cell lines W1205 and W136-1A, both metastatic. As for the expression of the β subunits, again we noted no particular pattern attributable to either metastatic or non-metastatic phenotype.

Table II - EXPRESSION OF α2-SUBUNIT OF LAMININ BY MELANOMA CELL LINES

Cell Line	Melanoma Lesion	Metastasis	Tumor Growth in Nude Mice	LN-α2
W793	Primary	None	Yes	-
*W1205	Metastatic	Yes	Yes	++++
W98-1	Primary	None	Yes	±
W373	Metastatic	Yes	Yes	++
W164	Metastatic	Yes	ND	+++
W9	Metastatic	Yes	ND	+++
W136-1A	Primary	Yes	Yes	+++++
A2058	Primary	Yes	Yes	+++

ND: Not determined

*W1205 was derived from W793 by passage into Nude mice.

Table III - INTEGRIN EXPRESSION BY MELANOMA CELL LINES

Cell Line	Metastasis	Expression of LN-α2	Expression of Integrins								
			αV,	α1,	α2,	α3,	β1,	β2,	β3,	β4,	α6
W793	No	-	++	-	-	-	+++	ND	+++	-	+++
W1205	Yes	++++	++	-	+++	++	+++	ND	+++	ND	ND
W98-1	No	±	++	-	-	-	-	ND	++	-	-
W373	Yes	++	++	-	±	-	±	+++	+	-	-
W164	Yes	+++	+++	±	-	-	-	+	+++	-	-
W9	Yes	+++	+++	±	±	±	±	++	+++	+	+
W136-1A	Yes	+++++	++	±	+++	+++	+++	±	++	+	+

ND: Not determined

DISCUSSION

Our data show a definite correlation between the expression of the α2-subunit and the expression of the metastatic phenotype by the melanoma cell lines we tested. Recent studies from our laboratory with ras-transformed rat fibroblasts (4R) exhibiting the metastatic phenotype and with doubly transformed rat fibroblasts (RE4) exhibiting only the tumorigenic phenotype demonstrated a similar correlation. The 4R cells expressed significant amounts of the α2-subunit of laminin, whereas, the RE4 cells produced little or no α2 subunit (Jenq et al., 1994).

The expression of the αV integrin subunit by all the melanoma cell lines examined suggests that expression of this subunit does not discriminate between metastatic and non-metastatic melanoma cell line. Similarly, the expression of the β3 integrin subunit by all melanoma lines tested provides no means of distinguishing between metastatic and non-metastatic propensity.

The study of integrin expression by the melanoma cells in our laboratory was prompted by the observation by Jenq et al. (1994) that the laminin synthesized by ras-transformed metastatic fibroblasts (4R cells) had a significantly higher cell adhesion promoting activity (20-fold) than the laminin from doubly transformed non-metastatic but tumorigenic RE4 cells. When we measured the integrin production by the 4R and RE4 cells, we found a high degree of expression of the α2 and β1 integrin subunits by the former and little or no production by the latter cells (Jenq and Kefalides, unpublished data).

It was reported earlier (Jenq et al., 1993) that laminin purified from the culture media of two related mouse epithelial cell lines, a normal B82 and its tumorigenic derivative B82HT, showed quantitative differences in the subunit composition, namely the presence of higher amounts of the α2-subunit, and a higher content of α2-mRNA in B82 than in B82HT cells. In addition, laminin from B82 cells promoted cell adhesion to a greater extent than laminin from B82HT cells. A previous report by Rao and Kefalides (1991) using comparable mouse cell lines, A9 and A9HT, showed an altered ratio of the β1 and γ1 subunits in these cells. In a recent study, Monical and Kefalides (1994) have shown that although human neonatal skin fibroblasts synthesize only the β1 and γ1 subunits, keratinocytes do synthesize the α2β1γ1 isoform but not the α1 subunit.

Although the mechanism by which various activated ras oncogenes can induce the complex metastatic phenotype in suitable recipient cells (Bernstein and Weinberg, 1985, and Bradley et al., 1986) remains unclear, we speculate that the ras oncogene alone transfected into the proper recipient cells may increase the expression of the α2-subunit of laminin; on the other hand, simultaneous introduction of the viral Ela gene into the same cell suppresses its expression. The demonstration by Garbisa et al. (1987) that the c-Ha-ras oncogene alone, transfected into early passage rat embryo fibroblasts, induces these cells to secrete high levels of type IV collagenolytic metalloproteinase and to concomitantly exhibit a significant increase of spontaneous metastases in nude mice suggests that increased production of type IV collagenase is an important prerequisite for metastasis. Similarly, it would appear that increased cell adhesion correlates with the presence of high levels of the $\alpha 2$-subunit of laminin. It is tempting to speculate that tumor cells, which develop the metastatic phenotype, must regain the ability to synthesize and secrete the isoform of laminin which contains the $\alpha 2$-subunit; this in turn, may facilitate adhesion and locomotion of such cells.

SUMMARY

The present study demonstrates that the expression of the $\alpha 2$-subunit of laminin correlates with the metastatic phenotype in melanoma cell lines. The loss of this differentiated state, i.e. loss of the ability to produce the $\alpha 2$-subunit of laminin, in primary, non-metastatic, melanoma tumor cells and its recovery when these cells assume the metastatic phenotype, suggests that production of the $\alpha 2$-subunit may be a prerequisite for this phenotype to be expressed. The present data corroborate our previous findings with the ras-transformed 4R cells which are highly metastatic, produce large amounts of type IV collagenase and synthesize the $\alpha 2$-subunit of laminin.

An attempt to determine a possible correlation between the metastatic phenotype and the expression of a specific integrin, failed to reveal a direct relationship. Integrins αV and $\beta 3$ were present in both metastatic and non-metastatic melanoma cells, whereas, integrins $\alpha 1$, $\beta 4$ and $\alpha 6$ were not expressed by any of the melanoma cell lines examined. On the other hand the $\alpha 2$ and $\alpha 3$ integrins were expressed only by two of the melanoma cell lines, both of which are metastatic.

It is suggested that the synthesis and secretion of laminin containing the $\alpha 2$-subunit allows the melanoma cells to adhere onto a favorable substrate which promotes cell spreading and movement.

ACKNOWLEDGMENTS

The authors wish to express their appreciation to Dr. William G. Stetler-Stevenson for providing us 4R, RE4, and HT-1080 cells. This study was supported in part by NIH Grants AR-20553, HL-29492, and AR-07490.

REFERENCES

Aratani, Y., and Kitagawa, Y., 1988, Enhanced synthesis and secretion of type IV collagen and entactin during adipose conversion of 3T3-L1 cells and production of unorthodox laminin complex, *J. Biol. Chem.* 263:16163-16169.

Bernstein, S. C., and Weinberg, R. A., 1985, Expression of the metastatic phenotype in cells transfected with human metastatic tumor DNA, *Proc. Natl. Acad. Sci. USA* 82:1726-1730.

Bradley, M. O., Kraynak, A. R., Storer, R. D., and Gibbs, J. B., 1986, Experimental metastasis in nude mice of NIH 3T3 cells containing various ras genes, *Proc. Natl. Acad. Sci. USA* 83:5277-5281.

Burgeson, R. E., Chiqet, M., Deutzmasnn, R., Ekblom, P., Engel, J., Leeinman, H., Martin, G. R., Meneguzzi, G., Paulsson, M., Sanes, J., Timple, R., Tryggvason, K., Yamada, Y., and Yurchenco, P. D., 1994, A new nomenclature for the laminins, *Matrix Biol.* 14:209-211.

Ehrig, K., Leivo, I., Argraves, R. E., and Engvall, E., 1990, Merosin, a tissue-specific basement membrane protein, is a laminin-like protein, *Proc. Natl. Acad. Sci. USA* 87:3264-3268.

Engvall, E., Earwicker, D., Day, A., Muir, D., Manthorpe, M., and Paulsson, M., 1992, Merosin promotes cell attachment and neurite outgrowth and is a component of the neurite-promoting factor of RN22 Schwannoma cells, *Exp. Cell Res*. 198:115-123.

Garbisa, S., Pozzatti, R., Muschel, R. J., Saffiotti, U., Ballin, M., Goldfarb, R. H., Khoury, G., and Liotta, L. A., 1987, Secretion of type IV collagenolytic protease and metastatic phenotype: Induction by transfection with c-Ha-ras but not c-Ha-ras plus Ad 2-Ela, *Cancer Res.* 47:1523-1528.

Jenq, W., Wu, S. J., and Kefalides, N. A., 1993, Adhesion promoting property of laminin from normal tissue and from a tumorigenic cell line, *Connect. Tissue Res.*, 30:59-73.

Jenq, W., Wu, S. J., and Kefalides, N. A., 1994, Expression of the α2-subunit of laminin correlates with increased cell adhesion and metastatic propensity, *Differentiation* 58:29-36.

Laemmli, U. K., 1970, Cleavage of structural proteins during the assembly of the head of bacteriophage T4, *Nature* 227:680-685.

Leivo, I., and Engvall, E., 1988, Merosin: A protein specific for basement membrane of Schwann cells, striated muscle, and trophoblast, is expressed late in nerve and muscle development, *Proc. Natl. Acad. Sci. USA* 85:1544-1548.

Monical, P., Kefalides, N. A., 1994, Coculture modulates laminin synthesis and mRNA levels in epidermal keratinocytes and dermal fibroblasts, *Exp. Cell Res.* 210:154-159.

Ohno, M., Martinez-Hernandez, A., Ohno, N., and Kefalides, N. A., 1983, Isolation of laminin from human placental basement membranes: Amnion, chorion and chorionic microvessels, *Biochem. Biophys. Res. Commun.* 112:1091-1098.

Ohno, M., Martinez-Hernandez, A., Ohno, N., and Kefalides, N. A., 1986, Laminin M is found in placental basement membranes, but not in basement membranes of neoplastic origin, *Connect. Tissue Res.* 15:199-207.

Rao, N., Brinker, J. M., and Kefalides, N. A., 1991, Changes in the subunit composition of laminin during the increased tumorigenesis of mouse A9 cells, *Connect. Tissue Res.* 25:321-329.

PLASMINOGEN ACTIVATORS IN FIBRINOLYSIS AND PERICELLULAR PROTEOLYSIS. STUDIES ON HUMAN ENDOTHELIAL CELLS IN VITRO

Victor W.M. van Hinsbergh, Pieter Koolwijk and Roeland Hanemaaijer

Gaubius Laboratory TNO-PG
P.O. Box 2215
2301 CE Leiden
The Netherlands

INTRODUCTION

Fibrin is a temporary matrix, which is formed after wounding of a blood vessel and when plasma leaks from blood vessels forming a fibrous exudate, often seen in areas of inflammation and in tumors.[1] The fibrin matrix acts as a barrier preventing further blood loss, and provides a scaffolding in which new microvessels can infiltrate during wound healing. The outgrowth of new blood vessels from existing ones, angiogenesis, is an essential process during development, but normally stops when the body becomes adult. The half life time of endothelial cells in the adult body varies between 100 to 10,000 days in normal tissues, whereas it is reduced to several days in placenta an tumors.[2] With the exception of the female reproductive system, angiogenesis in the adult is associated with tissue repair after injury by wounding or inflammation. Hence, in contrast to developmental angiogenesis, which proceeds independently of inflammation or fibrin deposition, adult angiogenesis is usually accompanied by the presence of fibrin and inflammatory cells or mediators. A proper timing of the outgrowth of microvessels as well as the subsequent (partial) disappearance of these vessels is essential to ensure adequate wound healing and to prevent the formation of scar tissue. Although essential for the formation of granulation tissue and tissue repair, angiogenesis, once under control of pathological stimuli, can contribute to a number of pathological conditions, such as tumor neovascularisation, pannus formation in rheumatoid arthritis, and diabetic retinopathy. Understanding the mechanisms involved in angiogenesis may provide clues to prevent pathological angiogenesis without seriously impairing tissue repair. A number of studies and reviews have focused on angiogenic factors and the formation of capillary-like structures.[3-9] Among them, Dvorak[1,6] and Mosher[10] have pointed to the importance of fibrin in angiogenesis. Furthermore, work of Polverini and colleagues have demonstrated the involvement of monocytes and their products in the induction of angiogenesis.[9,11,12] Because of the specific roles of fibrin and inflammatory cells and mediators in pathological angiogenesis in the adult, we have focused our studies on the invasion of human endothelial cells into three-dimentional fibrin matrices

Molecular, Cellular, and Clinical Aspects of Angiogenesis
Edited by Michael E. Maragoudakis, Plenum Press, New York, 1996

and the role of endothelial plasminogen activators in this process. This chapter will survey the roles of endothelial plasminogen activators in fibrin dissolution and pericellular proteolytic events associated with the invasion of capillary-like tubular structures into fibrin.

DUAL ROLE OF ENDOTHELIAL PLASMINOGEN ACTIVATORS

It is generally believed that plasminogen activators play an important role in the migration and invasion of leukocytes and endothelial cells, and in the dissolution of the fibrin matrix.[13,14] Plasminogen activators are serine proteases, which enzymatically convert the zymogen plasminogen into the active protease plasmin, the prime protease that degrades fibrin. The production of plasminogen activators by endothelial cells not only contributes to the proteolytic events related to the formation of microvessels in a wound, but also plays a crucial role in the prevention of thrombosis. If fibrin becomes deposited within the lumen of a blood vessel, cessation of the blood flow may occur accompanied by ischemia and eventually death of the distal tissues. The endothelium contributes considerably to the maintenance of blood fluidity by exposing anticoagulant molecules, by providing factors that interfere with platelet aggregation, and by its ability to stimulate fibrinolysis. Lysis of intravascularly generated fibrin must occur rapidly. However, it should be limited to a local area, because a general elevation of fibrinolysis upon wounding would result in recurrent bleeding. Hence, endothelial cells apply plasminogen activators for initial events in angiogenesis, for the degradation of the temporary fibrin matrix during wound healing, and for the immediate dissolution of a fibrinous thrombus originating within a blood vessel[15] (Figure 1). They are able to execute all these processes adequately by orchestrating in time and space the production of both types of plasminogen activators as well as that of specific inhibitors and cellular receptors for plasminogen activators and plasminogen.

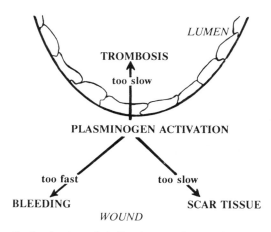

Figure 1. Plasminogen activation by the endothelium is controlled in time and space to meet various functions: prevention of thrombosis without bleeding and angiogenesis, which is necessary for adequate wound repair (from: Van Hinsbergh, *Ann. N. Y. Acad. Sci.* 667:151 (1992); with permission).

COMPONENTS OF THE FIBRINOLYTIC SYSTEM

Figure 2 summarizes the proteases and the inhibitors involved in fibrinolysis. Fibrin degradation and probably also activation of several matrix metalloproteinases can be accomplished by the serine protease plasmin, which is formed from its zymogen plasminogen by plasminogen activators (PAs). Two types of plasminogen activators are

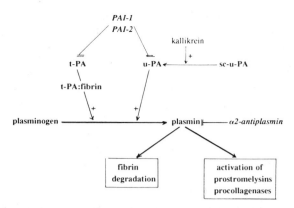

Figure 2. Schematic representation of the plasminogen activation system. +: activation; -: inhibition. Inhibitors are indicated in italics. PA: plasminogen activator; t-PA: tissue-type PA; u-PA: urokinase-type PA; sc-u-PA: single-chain u-PA; PAI: PA inhibitor.

presently known: tissue-type plasminogen activator (t-PA) and urokinase-type plasminogen activator (u-PA).[16,17] Plasminogen, t-PA and u-PA are synthesized as a single polypeptide chain, which is converted by proteolytic cleavage to a molecule with two polypeptide chains connected by a disulphide bond. The carboxy-terminal parts of the molecules (the so called B-chain) contain the proteolytically active site, whereas the amino-terminal parts of the molecule (the A-chain) are built up of domains that determine the interaction of the proteases with matrix proteins and cellular receptors. The cleavage of plasminogen and single-chain u-PA is necessary to disclose the proteolytic active site and to activate the molecule. In contrast, generation of t-PA activity does not depend on its conversion in a two-chain form, but on its interaction with a specific substrate, in particular fibrin. Once bound to this substrate, both the single-chain form and the two-chain form of t-PA are active.

The actual activity of the PAs is regulated not only by their concentration, but also by their interaction with PA inhibitors (PAIs),[18-20] cellular receptors[21,22] and, as indicated above, matrix proteins. The three proteases of the fibrinolytic system are counterregulated by potent inhibitors, which are members of the serine protease inhibitor (serpin) superfamily. Plasmin is instantaneously inhibited by α2-antiplasmin,[23] but this reaction is attenuated when plasmin is bound to fibrin. The predominant regulators of t-PA and u-PA are PAI-1 and PAI-2.[19] PAI-1 is a 50 kD glycoprotein present in blood platelets and synthesized by endothelial cells, smooth muscle cells and many other cell types in culture.[20] PAI activity in human plasma is normally exclusively PAI-1. PAI-2 is produced by monocytes/macrophages and can be found as a glycosylated secreted molecule and as an intracellular molecule.[24]

Regulation of fibrinolytic activity also occurs by cellular receptors. These receptors direct the action of PAs and plasmin to focal areas on the cell surface. High affinity binding sites for plasminogen,[25,26] t-PA[27] and u-PA[28-30] are found on various types of cells including endothelial cells. A specific u-PA receptor has been identified and cloned,[31] which binds both single-chain u-PA and two-chain u-PA via their growth factor domains.[30] The u-PA receptor is heavily glycosylated and proteolytically processed at its carboxy-terminal end; the receptor with the new carboxy-terminus is anchored in the membrane by a phosphatidyl group.[30] The N-terminal part of the u-PA receptor consists of three homologous domains, which have structural homology to the snake venom α-toxins. u-PA binds to the first (N-terminal) domain. The role of the u-PA receptor in pericellular proteolysis will be discussed in a later paragraph. In addition, the u-PA receptor may act as a cellular adhesion molecule by its ability to bind vitronectin.[32] Optimal interaction between vitronectin and u-PA receptor requires u-PA binding.

The natures of the plasmin(ogen) receptor(s) and t-PA receptor(s) on endothelial cells are less clear. A number of compounds have been indicated to act as cellular plasminogen receptors, such as α-enolase [33] (see Hajjar[34] for review). The cellular binding of plasminogen can be influenced by gangliosides [35]. The lipoprotein Lp(a), which has strong structural homology with a large part of the plasminogen molecule, can compete for plasminogen binding to endothelial cells.[26,36]. Annexin II has been identified as a t-PA receptor on endothelial cells.[37] In addition, clearance receptors exist on liver hepatocytes and liver endothelial cells, which can clear t-PA[38] and plasmin-α2-antiplasmin and PA:PAI-1 complexes[39,40] from the circulation.

REGULATION OF t-PA PRODUCTION: PREVENTION OF INTRAVASCULAR FIBRIN DEPOSITION

The fibrinolytic activity in blood is largely determined by the concentration of t-PA, which is synthesized in the endothelium.[41] The concentration of t-PA in the circulation can change rapidly. This is due to the short half life time of t-PA in the circulation, which is 5 to 10 minutes in man, and to the ability of endothelial cells to release rapidly a relatively large amount of t-PA. Clearance of t-PA occurs in the liver. Consequently, changes in the liver blood flow affect t-PA clearance and the plasma t-PA concentration. The acute release of t-PA from a storage pool in the vessel wall can be induced by vasoactive substances, such as bradykinin, platelet activating factor and thrombin.[42] This mechanism makes it possible to enhance the t-PA concentration exclusively at those places where fibrin generation occurs. Hence, it contributes to the local protection against an emerging thrombus. If a generalized stimulation of the endothelium occurs, for example by catecholamines, the acute release mechanism causes a rapid temporary increase in the blood t-PA concentration. Recently, it has become possible to mimic the acute release of t-PA in cultured endothelial cells.[43] Endothelial cells have storage granules for t-PA, which are different from the Weibel-Palade bodies, and which release their content after stimulation of the cells with vasoactive substances.[44] The size of the intracellular t-PA pool is influenced by the rate of t-PA synthesis. Hence, influencing t-PA synthesis can influence both consitutive t-PA production and acute release of t-PA. Factors that influence the synthesis of t-PA include activators of protein kinase C, retinoids and certain triazolobezodiazepines. A recent review is given by Kooistra et al.[45]

INFLAMMATORY ACTIVATION OF THE ENDOTHELIUM: EFFECT ON PAI-1 AND u-PA SYNTHESIS

In vivo, changes of the plasma levels of t-PA and its main inhibitor PAI-1 often occur in the same direction. Nevertheless, the synthesis of PAI-1 is independently regulated from that of t-PA, albeit that certain mediators, such as thrombin, can induce both t-PA and PAI-1 synthesis. In a number of diseases predominantly PAI-1 is elevated in the blood (sepsis, maturity onset diabetes, postoperative thrombosis) and/or in tissues (arteriosclerosis, sepsis). Several factors involved in inflammatory and vascular diseases, such as the cytokines tumor necrosis factor-α (TNFα) and interleukin-1 (IL-1), endotoxin (LPS), transforming growth factor-β (TGF-β), oxidized lipoproteins and thrombin, can stimulate PAI-1 production by endothelial cells in vitro. Also in vivo, administration of TNFα, IL-1, LPS or thrombin causes an increase in circulating PAI-1. After infusion of LPS in animals, PAI-1 mRNA increased in vascularized tissues and PAI-1 mRNA was elevated in the endothelium of various organs.[46,47] Administration of LPS or TNFα to patients or healthy volunteers caused after about 2 hours a large increase in circulating PAI-1, which was preceded by a rapid and sustained increase in circulating t-PA.[48-50] The mechanism underlying the stimulation of t-PA synthesis in vivo by LPS or TNFα is still unresolved, and may be the indirect result of the LPS- or TNFα-infusion by the generation of another mediator. The large increase in PAI-1 induced two hours after TNFα- or LPS-administration far exceeds the production of t-PA.[48-50] This may result - after an initial raise in fibrinolytic activity - in a prolonged attenuation of the fibrinolysis process. It is generally believed that induction of PAI-1 by inflammatory mediators may contribute to the thrombotic complications in endotoxinemia and sepsis.

However, the effect of inflammatory mediators TNFα, IL-1 and LPS on plasminogen activation is probably more complex. Simultaneous with the increase in PAI-1, these inflammatory mediators induce the synthesis of u-PA in human endothelial cells.[51] Induction of u-PA by TNFα is associated by an increased degradation of matrix proteins.[52] The enhanced secretion of u-PA occurs entirely towards the basolateral side of the cell, whereas the secretion of t-PA and PAI-1 proceed equally to the luminal and basolateral sides of the cell.[51] The polar secretion of u-PA underlines the suggestion that u-PA may be involved in local processes causing the remodelling of the basal matrix of the cell. Therefore the increase of PAI-1 induced by inflammatory mediators may represent, in addition to a role in the modulation of fibrinolysis, a protective mechanism of the cell against uncontrolled u-PA action.

THE ROLE OF THE u-PA RECEPTOR

The site, where local u-PA activity occurs, is probably directed by the u-PA receptor, a GPI-anchored glycoprotein of about 45 kD.[29,30] Human endothelial cells in vitro contain about 40,000 u-PA receptors per cell. Upon secretion, single-chain u-PA binds via its growth factor domain to the receptor, and can subsequently be converted to two-chain u-PA, by which it becomes proteolytically active. As the endothelial cell contains also plasmin(ogen) receptors, an interplay between receptor-bound u-PA and receptor-bound plasmin is likely to happen. The then generated plasmin can degrade a number of matrix proteins. On the other hand, a direct plasmin-independent action of u-PA on matrix proteins may also occur, as has been reported by Quigley et al.[53] for u-PA-dependent degradation of avian fibronectin. In this respect it is of interest to note that in various cell types the u-PA receptor has been localized in the focal attachment sites, which host integrin-matrix interactions, and in cell-cell contact areas.[54,55] The time that is allowed for u-PA to act as a protease it probably rather short. Receptor-bound two-chain u-PA is also subjected to inhibition by PAI-1. The then formed u-PA:PAI-1 complex is internalized,[56] in contrast to receptor-bound free u-PA. This occurs probably after interaction with another receptor.[57] After internalization, the u-PA:PAI-1 complex is dissociated from the receptor and degraded in the lysosomes, whereas the u-PA receptor returns to the plasma membrane.

In addition to a role in directing pericellular proteolysis to defined areas on the cell membrane, the u-PA receptor may also act as a cellular adhesion molecule, which facilitates the interaction of the cell with vitronectin.[32] Furthermore, several investigators have reported that interaction of u-PA wit its receptor evokes intracellular signals, which additionally may affect cell metabolism.[58,59]

The number of u-PA receptors is enhanced by activation of protein kinase C and elevation of the cellular cAMP concentration,[15,60] as well as by several angiogenic growth factors, including b-FGF and VEGF.[61-63] Although b-FGF enhances u-PA production in bovine endothelial cells, it is unable to induce u-PA production in various types of human

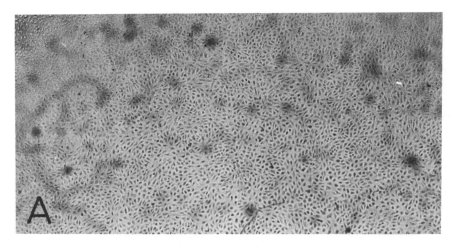

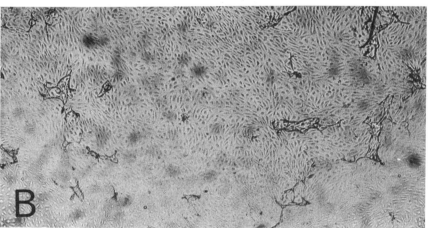

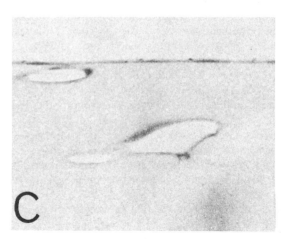

Figure 3. Formation of capillary-
like tubular structures in a three
dimensional fibrin matrix by human
endothelial cells. A. Human
microvascular endothelial cells
grown under control conditions on
top of a three-dimensional fibrin
matrix. B. Formation of tubular
structures is induced by the
simultaneous addition of the growth
factors b-FGF and VEGF and the
cytokine TNFα. C. Cross section
through the top of the fibrin layer of
b-FGF/VEGF/TNFα-stimulated
endothelial cell culture. Endothelial
cells surrounding a lumen are
visible.

endothelial cells. Therefore, the simultaneous exposure of human endothelial cells to TNFα, which induces u-PA synthesis, and to b-FGF and VEGF, which enhance the expression of u-PA receptors, are needed for a maximal increase in local u-PA activity on human endothelial cells.

PAs IN LOCAL PROTEOLYSIS: PUTATIVE ROLE IN ANGIOGENESIS

Migrating and invading cells, such as monocytes and tumor cells, express u-PA activity bound to u-PA receptors on their cellular protrusions and on focal attachment sites, which suggest that PA activity is involved in cellular migration and invasion. Such a mechanism is also likely to be involved in endothelial cell migration and in the formation of new blood vessels (angiogenesis). Proteolysis of the basement membrane of endothelial cells is a prerequisite for these processes. A direct correlation between the expression of PA activity and the migration and formation of capillary sprouts by bovine microvascular endothelial cells in vitro was demonstrated by Pepper and Montesano.[7,8,64,65] The outgrowth of tubular structures was increased by b-FGF which increases u-PA activity in bovine endothelial cells, and counteracted by TGF-β, which enhances predominantly PAI-1 and inhibits PA activity. VEGF can also stimulate bovine adrenal microvascular endothelial cells to form tubular structures, and acts cooperatively with b-FGF in this induction.[66] Studies in our laboratory[67] indicate that no or a limited number of capillary sprouts grow from a monolayer of human endothelial cells into a fibrin matrix after exposure to b-FGF and VEGF (Figure 3a). However, many tubular structures are obtained after the simultaneous exposure of the cell monolayers to TNFα, b-FGF and VEGF (Figure 3b,c). This marked outgrowth of tubular structures requires u-PA activity and is markedly reduced by inhibiting the interaction of u-PA with its receptor. Furthermore, the proteolytic activation of plasminogen by u-PA appears to be involved, because the plasmin inhibitor trasylol largely inhibits the formation of tubular structures. However, it should be stressed that these data only indicate that the formation of tubular structures by human microvascular endothelial cells into fibrin requires u-PA. However, it should be stressed that these data only indicate that the formation of tubular structures by human microvascular cells in a fibrin matrix require u-PA. Inhibition of angiogenesis by e.g inhibitors of matrix -metalloproteinases[68] suggests that in addition to u-PA/plasmin other proteases may play a role in angiogenesis. Different effects on uPA expression by various inhibitors of tube formation on collagen matrices support this hypothesis.[69] Because remodeling of the basal membrane is a prerequisite for the formation of tubular structures it is likely that matrix degrading metalloproteinases also participate in the onset of angiogenesis in a fibrin matrix.

PAs IN MATRIX REMODELLING: COOPERATION WITH MATRIX METALLOPROTEINASES

The regulation of proteolytic activation and activity is probably more complex than depicted above. It is not known whether interaction of u-PA with its receptor may affect cell signal transduction additionally. Furthermore, the local proteolytic activity is not limited to plasminogen activation. Endothelial cells in vitro can produce a number of matrix metalloproteinases (MMPs)[70,71]. They are secreted as a zymogen and depend on zinc and calcium ions for activity. Activation of these pro-enzymes can proceed by the action of other proteases, like plasmin and stromelysin (MMP-3).[71,72] In addition, activation can occur by recently cloned membrane bound metalloproteinases (MT-MMPs).[73,74] Figure 4 gives a schematic representation of the activation of these proteases and their possible

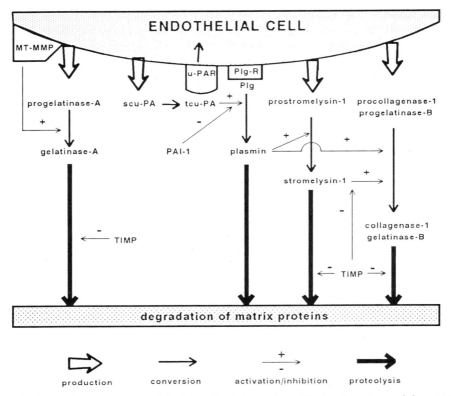

Figure 4. Schematic representation of the interaction between the u-PA-plasmin system and the presumed activation of matrix degrading metalloproteinases (MMPs). Abbreviations: PA: plasminogen activator; u-PA urokinase-type PA; sc-u-PA: single-chain u-PA; tc-u-PA: two-chain u-PA; u-PAR: u-PA receptor; Plg: plasminogen; Plg-R: Plg receptor; PAI-1: PA inhibitor-1; MT-MMP: membrane-bound MMP; TIMP: tissue inhibitor of MMP. +: stimulation; -: inhibition.

interaction.   Activation of the various MMPs results in the loss of a pro-peptide and the reduction of the molecular weight by about 8 to 10 kD. The MMPs are inhibited by tissue inhibitors of metalloproteinases, the so-called TIMPs.[75,76] Two mammalian TIMPs, TIMP-1 and TIMP-2, are well known. They have different substrate specificities. TIMP-2 inhibits preferentially gelatinase A, while TIMP-1 predomintly interacts with other MMPs, such as interstitial collagenase, stromelysin-1 and gelatinase B. Both have been demonstrated in endothelial cells in vitro. A recently identified matrix-bound TIMP (TIMP-3),[77] is also present in human EC (our unpublished results). Hence, MMP activities are regulated by (a) regulation of the synthesis of the zymogen, (b) activation of the zymogen by other proteases and (c) the presence and interaction with TIMPs. Data in our laboratory[78] have indicated that the synthesis of several MMPs, including stromelysin, is enhanced or induced by the inflammatory mediator TNFα, the same mediator that induces the synthesis of u-PA. Hence, in inflammation and in angiogenesis associated with the accompanying repair process, endothelial cells play a regulatory role by a coordinate expression and activation of locally acting matrix remodelling proteases.

44

SUMMARY

The endothelial cell uses PAs for several functions (Figure 5). The endothelium is able to respond to emerging fibrin deposits in the blood stream by regulating the production of t-PA, the main fibrinolysis regulator in blood. In addition it can fine-tune fibrinolytic activity by the simultaneous production of PAI-1, so that re-bleeding of a wound is prevented. Furthermore, the endothelium uses PAs, in particular u-PA, for proteolytically changing its interaction with its underlying matrix and for remodelling of its basement

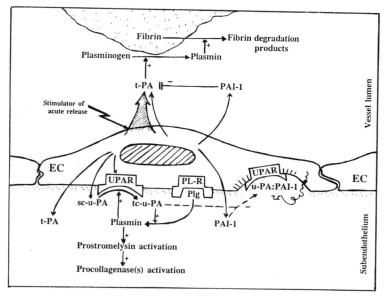

Figure 5. Schematic representation of postulated aspects of the involvement of endothelial cell plasminogen activators in fibrinolysis and local proteolysis. Abbreviations: PA: plasminogen activator; t-PA: tissue-type PA; u-PA urokinase-type PA; sc-u-PA: single-chain u-PA; tc-u-PA: two-chain u-PA; Plg: plasminogen; UPAR: u-PA receptor; PL-R: plasminogen receptor; PAI-1: PA inhibitor-1; EC: endothelial cell. +: stimulation; -: inhibition.

membrane, processes which are necessary for cell migration and angiogenesis. This process is limited in space and time by interaction of u-PA with its specific receptor and by the presence of PAI-1. As the expression of u-PA and u-PA receptor are under the control of the monocyte-derived cytokines TNFα and IL-1 and by angiogenic growth factors, respectively, monocytes may play an important role in the control of angiogenesis. In addition the induction and activation of stromelysin and other matrix metalloproteinases by the monokine TNFα and the u-PA plasmin system further contribute to the complex regulation of the local proteolytic events associated with endothelial matrix remodelling.

REFERENCES

1. H.F. Dvorak, J.A. Nagy, B. Berse, L.F. Brown, K.-T. Yeo, T.-K. Yeo, A.M. Dvorak, L. Van de Water, T.M. Sioussat, and D.R. Senger, Vascular Permeability factor, fibrin, and the pathogenesis of tumor stroma formation, *Ann. N. Y. Acad. Sci.* 667:101 (1992).

2. B. Hobson and J. Denekamp, Endothelial proliferation in tumours and normal tissues: Continuous labelling studies. *Br. J. Cancer* 49:405 (1984).

3. J. Folkman and M. Klagsbrun, Angiogenic factors. *Science* 235:442 (1987)

4. J. Folkman and Y. Shing, Angiogenesis. *J. Biol. Chem.* 267:10931 (1992).

5. W.H. Burgess and T. Maciag, The heparin-binding (fibroblast) growth factor family of proteins. *Annu. Rev. Biochem.* 58:575 (1989).

6. H.F. Dvorak, L.F. Brown, M. Detmar and A.M. Dvorak, Vascular permeability factor/vascular endothelial growth factor, microvascular hyperpermeability, and angiogenesis. *Am. J. Pathol.* 146:1029 (1995).

7. M.S. Pepper, D. Belin, R. Montesano, L. Orci, and J.-D. Vassalli, Transforming growth factor-beta 1 modulates basic fibroblast growth factor-induced proteolytic and angiogenic properties of endothelial cells in vitro, *J. Cell Biol.* 111:743 (1990).

8. R. Montesano, Regulation of angiogenesis in vitro. *Eur. J. Clin. Invest.* 22:504 (1992).

9. P.J. Polverini, Macrophage-induced angiogenesis - A review. *Macrophage-Derived Cell Regulatory Factors* 1:54 (1989).

10. R.B. Colman, Wound Healing Processes in Hemostasis and Thrombosis. *in:* "Vascular Endothelium in Hemostasis and Thrombosis," M.A. Gimbrone Jr., ed., Churchill Livingstone, Edinburgh, (1986).

11. S.J. Leibovich, P.J. Polverini, H.M. Shepard, D.M. Wiseman, V. Shively, and N. Nuseir, Macrophage-induced angiogenesis is mediated by tumour necrosis factor-α. *Nature* 329:630 (1987).

12. A.E. Koch, P.J. Polverini, S.L. Kunkel, L.A. Harlow, L.A. DiPietro, V.M. Elner, S.G. Elner and R.M. Strieter, Interleukin-8 as a macrophage-derived mediator of angiogenesis. *Science* 258:1798 (1992).

13. H.C. Kwaan, Tissue fibrinolytic activity studied by a histochemical method, *Fed. Proc.* 25:52 (1966).

14. V.W.M. Van Hinsbergh, and P. Koolwijk, Production of Plasminogen Activators and Matrix Metalloproteinases by Endothelial Cells: Their Role in Fibrinolysis and Local Proteolysis, *in:* "Angiogenesis in Health and Disease," M.E. Maragoudakis, P. Gullino, and P.I. Lelkes, eds., NATO ASI Series A Volume 227, Plenum Press, New York (1992).

15. V.W.M. Van Hinsbergh, Impact of endothelial activation on fibrinolysis and local proteolysis in tissue repair, *Ann. N. Y. Acad. Sci.* 667:151 (1992).

16. F. Bachmann, Fibrinolysis, *in:* "Thrombosis and Haemostasis 1987," M. Verstraete, J. Vermylen, R. Lijnen, and J. Arnout, eds., Leuven University Press, Leuven (1987).

17. P. Wallén, Structure and Function of Tissue Plasminogen Activator and Urokinase, *in:* "Fundamental and Clinical Fibrinolysis," P.J. Castellino, P.J. Gaffney, M.M. Samama, and A. Takada, eds., Elsevier, Amsterdam (1987).

18. E.D. Sprengers, and C. Kluft, Plasminogen activator inhibitors, *Blood* 69:381 (1987).

19. E.K.O. Kruithof, Plasminogen activator inhibitor type 1: biochemical, biological and clinical aspects, *Fibrinolysis* 2, Suppl.2:59 (1988).

20. D.J. Loskutoff, Regulation of PAI-1 gene expression, *Fibrinolysis* 5:197 (1991).

21. L.A. Miles, and E.F. Plow, Plasminogen receptors: ubiquitous sites for cellular regulation of fibrinolysis, *Fibrinolysis* 2:61 (1988).

22. E.S. Barnathan, Characterization and regulation of the urokinase receptor on human endothelial cells, *Fibrinolysis* 6, Suppl.1:1 (1992).

23. W.E. Holmes, L. Nelles, H.R. Lijnen, and D. Collen, Primary structure of human α₂-antiplasmin, a serine protease inhibitor (Serpin), *J. Biol. Chem.* 262:1659 (1987).

24. A. Wohlwend, D. Belin, and J.-D. Vassalli, Plasminogen activator-specific inhibitors produced by human monocytes/macrophages, *J. Exp. Med.* 165:320 (1987).

25. L.A. Miles, E.G. Levin, J. Plescia, D. Collen, and E.F. Plow, Plasminogen receptors, urokinase receptors, and their modulation on human endothelial cells, *Blood* 72:628 (1988).

26. R.L. Nachman, Thrombosis and atherogenesis: molecular connections. *Blood* 79:1897 (1992).

27. K.A. Hajjar, The endothelial cell tissue plasminogen activator receptor. Specific interaction with plasminogen, *J. Biol. Chem.* 266:21962 (1991).

28. J.-D. Vassalli, The urokinase receptor. *Fibrinolysis* 8 (suppl.1):172 (1994).
29. F. Blasi, M. Conese, L.B. Møller et al, The urokinase receptor: Structure, regulation and inhibitor-mediated internalisation. *Fibrinolysis* 8 (suppl.1):182 (1994).
30. K. Danø, N. Behrendt, N. Brünner, V. Ellis, M. Ploug, C. Pyke, The urokinase receptor. Protein structure and role in plasminogen activation and cancer invasion. *Fibrinolysis* 8 (suppl.1):189 (1994).
31. A.L. Roldan, M.V. Cubellis, M.T. Masucci, N. Behrendt, L.R. Lund, K. Danø, E. Appella, and F. Blasi, Cloning and expression of the receptor for human urokinase plasminogen activator, a central molecule in cell surface, plasmin dependent proteolysis. *EMBO J.* 9:467 (1990).
32. Y. Wei, D. Waltz, N. Rao, R. Drummond, S. Rosenberg and H. Chapman, Identification of the urokinase receptor as cell adhesion receptor for vitronectin. *J. Biol. Chem.* 269: 32380 (1994).
33. L.A. Miles, C.M. Dahlberg, J. Plescia, J. Felez, K. Kato and E.F. Plow, Role of cell-surface lysines in plasminogen binding to cells: Identification of alpha-enolase as a candidate plasminogen receptor. *Biochemistry* 30:1682-1691 (1991).
34. K. Hajjar, Cellular receptors in the regulation of plasmin generation. *Thromb. Haemostas.* 74:294 (1995).
35. L.A. Miles, C.M. Dahlberg, E.G. Levin and E.F. Plow, Gangliosides interact directly with plasminogen and urokinase and may mediate binding of these components to cells. *Biochemistry* 193:759 (1990).
36. L.A. Miles, G.M. Fless, E.G. Levin, A.M. Scanu, and E.F. Plow, A potential basis for the thrombotic risks associated with lipoprotein(a), *Nature* 339:301 (1989).
37. K. Hajjar, A. Jacovina and J. Chacko, An endothelial cell receptor for plasminogen tissue plasminogen activator 1 identity with annexin II. *J. Biol. Chem.* 269:21191 (1994).
38. J. Kuiper, M. Otter, D.C. Rijken, and T.J.C. Van Berkel, Characterization of the interaction in vivo of tissue-type plasminogen activator with liver cells, *J. Biol. Chem.* 263:18220 (1988).
39. K. Orth, E.L. Madison, M.-J. Gething, J.F. Sambrook, and J. Herz, Complexes of tissue-type plasminogen activator and its serpin inhibitor plasminogen-activator inhibitor type 1 are internalized by means of the low density lipoprotein receptor-related protein/α_2-macroglobulin receptor, *Proc. Natl. Acad. Sci. U.S.A.* 89:7422 (1992).
40. G. Bu, S. Williams, D.K. Strickland, and A.L. Schwartz, Low density lipoprotein receptor-related protein/α_2-macroglobulin receptor is an hepatic receptor for tissue-type plasminogen activator, *Proc. Natl. Acad. Sci. U.S.A.* 89:7427 (1992).
41. T.-C. Wun, and A. Capuano, Spontaneous fibrinolysis in whole human plasma. Identification of tissue activator-related protein as the major plasminogen activator causing spontaneous activity in vitro, *J. Biol. Chem.* 260:5061 (1985).
42. J.J. Emeis, Regulation of the acute release of tissue-type plasminogen activator from the endothelium by coagulation activation products, *Ann. N. Y. Acad. Sci.* 667:249 (1992).
43. Y. Schrauwen, J.J. Emeis, and T. Kooistra, A sensitive ELISA for human tissue-type plasminogen activator applicable to the study of acute release from cultured human endothelial cells. *Thromb. Haemostas.* 71:225 (1994).
44. Y. Schrauwen, R.E.M. de Vries, T. Kooistra and J.J. Emeis, Acute release of tissue-type plasminogen activator (t-PA) from the endothelium; regulatory mechanisms and therapeutic target. *Fibrinolysis* 8 (suppl.2):8 (1994).
45. T. Kooistra, Y. Schrauwen, J. Arts and J.J. Emeis, Regulation of endothelial cell t-PA synthesis and release. *Int. J. Hematol.* 59:233 (1994).
46. P.H.A. Quax, C.R. Van den Hoogen, J.H. Verheijen, T. Padró, R. Zeheb, T.D. Gelehrter, T.J.C. Van Berkel, J. Kuiper, and J.J. Emeis, Endotoxin induction of plasminogen activator in plasminogen activator inhibitor type 1 mRNA in rat tissues in vivo, *J. Biol. Chem.* 265:15560 (1990).
47. M. Keeton, Y. Eguchi, M. Swadey, C. Ahn, and D. Loskutoff, Cellular localization of type 1 plasminogen activator inhibitor messenger RNA and protein in murine renal tissue, *Am. J. Pathol.* 142:59 (1993).
48. A.F. Suffredini, P.C. Harpel, and J.E. Parrillo, Promotion and subsequent inhibition of plasminogen activation after administration of intravenous endotoxin to normal subjects, *N. Engl. J. Med.* 320:1165 (1989)
49. V.W.M. Van Hinsbergh, K.A. Bauer, T. Kooistra, C. Kluft, G. Dooijewaard, M.L. Sherman, and W. Nieuwenhuizen, Progess of fibrinolysis during tumor necrosis factor infusion in humans. Concomitant increase of tissue-type plasminogen activator, plasminogen activator inhibitor type-1, and fibrin(ogen) degradation products, *Blood* 76:2284 (1990).

50. S.J.H. Van Deventer, H.R. Büller, J.W. Ten Cate, L.A. Aarden, E. Hack, and A. Sturk, Experimental endotoxemia in humans: analysis of cytokine release and coagulation, fibrinolytic, and complement pathways, *Blood* 76:2520 (1990).

51. V.W.M. Van Hinsbergh, E.A. Van den Berg, W. Fiers, and G. Dooijewaard, Tumor necrosis factor induces the production urokinase-type plasminogen activator by human endothelial cells, *Blood* 75:1991 (1990).

52. M.J. Niedbala, and M. Stein Picarella, Tumor necrosis factor induction of endothelial cell urokinase-type plasminogen activator mediated proteolysis of extracellular matrix and its antagonism by γ-interferon, *Blood* 79:678 (1992).

53. J.P. Quigley, L.I. Gold, R. Schwimmer, and L.M. Sullivan, Limited cleavage of cellular fibronectin by plasminogen activator purified from transformed cells, *Proc. Natl. Acad. Sci. U.S.A.* 84:2776 (1987).

54. J. Pöllänen, K. Hedman, L.S. Nielsen, K. Danø, and A. Vaheri, Ultrastructural localization of plasma membrane-associated urokinase-type plasminogen activator at focal contacts, *J. Cell Biol.* 106:87 (1988).

55. G. Conforti, C. Dominguez-Jimenez, E. Rønne, G. Høyer-Hansen and E. Dejana, Cell-surface plasminogen activation causes a retraction of in vitro cultured human umbilical vein endothelial cell monolayer. *Blood* 83:994 (1994).

56. D. Olson, J. Pöllänen, G. Høyer-Hansen, E. Rønne, K. Sakaguchi, T.-C. Wun, E. Appella, K. Danø, and F. Blasi, Internalization of the urokinase-plasminogen activator inhibitor type-1 complex is mediated by the urokinase receptor, *J. Biol. Chem.* 267:9129 (1992).

57. A. Nykjær, C.M. Petersen, B. Møller, P.H. Jensen, S.K. Moestrup, T.L. Holtet, M. Etzerodt, H.C. Thøgersen, P. Munch, P.A. Andreasen, and J. Gliemann, Purified α_2-macroglobulin receptor/LDL receptor-related protein binds urokinase•plasminogen activator inhibitor type-1 complex. Evidence that the α_2-macroglobulin receptor mediates cellular degradation of urokinase receptor-bound complexes, *J. Biol. Chem.* 267:14543 (1992).

58. M. Del Rosso, E. Anichini, N. Pedersen, F Blasi, G. Fibbi, M. Pucci and M. Ruggiero, Urokinase-urokinase receptor interaction: non-mitogenic signal transduction in human epidermal cells. *Biochem. Biophys. Res. Comm.* 190:347 (1993).

59. I. Dumler, T. Petri and W.-D. Schleuning, Interaction of urokinase-type plasminogenactivator (u-PA) with its cellular receptor (u-PAR) induces phosphorylation on tyrosine of a 38 kDa protein, *FEBS lett.* 322:37 (1993).

60. D.J. Langer, A. Kuo, K. Kariko, M. Ahuja, B.D. Klugherz, K.M. Ivanics, J.A. Hoxie, W.V. Williams, B.T. Liang, D.B. Cines, and E.S. Barnathan, Regulation of the endothelial cell urokinase-type plasminogen activator receptor. Evidence for cyclic AMP-dependent and protein kinase C-dependent pathways, *Circ. Res.* 72:330 (1993).

61. P. Mignatti, R. Mazzieri, and D.B. Rifkin, Expression of the urokinase receptor in vascular endothelial cells is stimulated by basic fibroblast growth factor, *J. Cell Biol.* 113:1193 (1991).

62. S. Mandriota, G. Seghezzi, J-D. Vassalli, N. Ferrara, S. Wasi, R. Mazzieri, P. Mignatti and M. Pepper, Vascular endothelial growth factor increases urokinase receptor expression in vascular endothelial cells. *J. Biol. Chem.* 270: 9709 (1995).

63. V.W.M. van Hinsbergh, unpublished data

64. M.S. Pepper, J.-D. Vassalli, R. Montesano, and L. Orci, Urokinase-type plasminogen activator is induced in migrating capillary endothelial cells, *J. Cell Biol.* 105:2535 (1987).

65. M.S. Pepper, A.-P. Sappino, R. Stöcklin, R. Montesano, L. Orci and J.-D. Vassalli, Upregulation of urokinase receptor expression on migrating endothelial cells. *J. Cell Biol.* 122:673 (1995).

66. M.S. Pepper, N. Ferrara, L. Orci, and R. Montesano, Potent synergism between vascular endothelial growth factor and basic fibroblast growth factor in the induction of angiogenesis in vitro, *Biochem. Biophys. Res. Commun.* 189:824 (1992).

67. P. Koolwijk, W. de Vree, R. Hanemaaijer, M. van Erck, M. Vermeer, C. Zurcher, H. Weich and V. van Hinsbergh, Cooperative effect of VEGF, bFGF and TNFα on pericellular u-PA expression and on the formation of capillary-like tubular structures by human microvascular endothelial cells in vitro. *Fibrinolysis* 8 (suppl 1): 150 (1994).

68. E. Kohn and L. Liotta, Molecular insights into cancer invasion: strategies for prevention and intervention. *Cancer Res.* 55:1856 (1995).

69. M.S. Pepper, J.-D. Vassalli, J.W. Wilks, L. Schweigerer, L. Orci and R. Montesano, Modulation of bovine microvascular endothelial cell proteolytic properties by inhibitors of angiogenesis. *J. Cell. Biochem.* 55:419 (1994).

70. L.M. Matrisian, The matrix-degrading metalloproteinases, *BioEssays* 14:455 (1992).

71. G. Murphy, S. Atkinson, R. Ward, J. Gavrilovic, and J.J. Reynolds, The role of plasminogen activators in the regulation of connective tissue metalloproteinases, *Ann. N. Y. Acad. Sci.* 667:1 (1992).

72. H. Nagase, Y. Ogata, K. Suzuki and J.J. Enghild, Substrate specificities and activation mechanisms of matrix metalloproteinases. *Biochem. Soc. Trans.* 19: 715 (1991)..

73. H. Sato, T. Takino, Y. Okada, J. Cao, A. Shinagawa, E. Yamamoto and M. Seiki, A matrix metalloproteinase expressed on the surface of invasive tumour cells. *Nature* 370:61 (1994).

74. H. Will and B. Hinzmann, cDNA sequence and mRNA tissue distribution of a novel human matrix metalloproteinase with a potential transmembrane segment. *Eur. J. Biochem.* 231:602 (1995).

75. L.A. Liotta, P.S. Steg, and W.G. Stetler-Stevenson, Cancer metastasis and angiogenesis: an imbalance of positive and negative regulation, *Cell* 64:327 (1991).

76. A.J.P. Docherty, and G. Murphy, The tissue metalloproteinase family and the inhibitor TIMP: a study using cDNAs and recombinant proteins, *Ann. Rheumatic Diseases* 49:469 (1990).

77. M. Wick, C. Burger, S. Brusselbach, F. Lucibello and R. Muller, A novel member of human tissue inhibitor of metalloproteinases (TIMP) gene family is regulated during g(1) progression, mitogenic stimulation, differentiation, and senescence. *J. Biol. Chem.* 269: 18953 (1994).

78. R. Hanemaaijer, P. Koolwijk, L. Le Clercq, W.J.A. De Vree, and V.W.M. Van Hinsbergh, Regulation of matrix-degrading metalloproteinases (MMPs) expression in human vein and microvascular endothelial cells. Effects of TNFα, IL-1 and phorbol ester, *Biochem. J.* 296:803 (1993).

THE EFFECT OF IONIZING RADIATION ON ENDOTHELIAL CELL DIFFERENTIATION AND ANGIOGENESIS

S. K. Williamson[1], R. W. Rose[2], M. O'Hara[1], and D.S. Grant[2]

[1]Department of Radiation Oncology
[2]The Cardeza Foundation
Thomas Jefferson University
Philadelphia, PA 19107

INTRODUCTION

Angiogenesis and the structure of the vessel wall

Angiogenesis, the formation of new vascular branches from pre-existing blood vessels, is closely linked to tumor growth and metastasis. The angiogenic process is characterized by the response of endothelial cells to a variety of stimulating factors. These cells, along with other cells comprising the vessel wall, initiate the process of angiogenesis by invading the basement membrane matrix to which they are attached, proliferating, adhering to extravascular matrix to migrate to a distal site, then differentiating and forming new vascular links. This process is required for all tumors to grow beyond 1 to 2 mm in diameter, and in response to low oxygen tension and reduced nutrients, tumor cells have the ability to release several growth factors that result in rapid expansion of the local vascular as well as the tumor cell mass (Folkman, 1992).

Normally all vessels are composed of a nonthrombogenic layer of endothelial cells which line the intimal surface of the vessel walls. This endothelium is adherent to a thin, specialized extracellular layer known as the basement membrane. This basement membrane is essential and provides not only support and an adhesive surface for the endothelium, but also maintains the normal differentiated state of the cell layer. Vessel walls are also comprised of other vascular cells such as smooth muscle cells, pericytes and fibroblasts. The former two have their own basement membrane , and the latter is surrounded by a collagenous interstitium (the adventitia) and in some cases elastic fibers. Studies which examine the cells comprising the vessel walls must also evaluate the role of the matrix on the maintenance of its structure as well.

The role of the basement membrane matrix in angiogenesis

A major factor regulating endothelial cell behavior is the basement membrane, which provides many signals and maintains the endothelium in a differentiated state *in vivo* (Ingber and Folkman, 1989; Kramer and Fuh, 1985; Kubota et al., 1988; Madri et al., 1980; Vlodavsky et al., 1991). The basement membrane has been shown to exhibit biological activity and mediate cellular differentiation (Kleinman et al., 1985; Kleinman et al., 1987). During embryogenesis, the presence of basement membrane components usually correlates with the completion of the developing organ. Basement membranes may also serve as a protective meshwork which inhibits the degradation of growth factors such as TGF-ß (Roberts et al., 1992). Several reports indicate that matrix proteins may be able to modulate the activity of TGF-ß in certain cases, possibly by sequestering the peptide in specific association with the ECM and by making it available to cells at the appropriate time (Roberts et al., 1992; Vlodavsky et al, 1993).

Molecular, Cellular, and Clinical Aspects of Angiogenesis
Edited by Michael E. Maragoudakis, Plenum Press, New York, 1996

51

The structure of basement membrane

Basement membranes are primarily composed of collagen type IV, the glycoproteins laminin (Timpl et al, 1979), entactin, and fibronectin, and the proteoglycans (perlecan and chondroitin sulfate proteoglycan). Laminin and collagen type IV are the most abundant components in the matrix and have been shown to be potent regulators of angiogenesis (Grant et al., 1991; Grant et al., 1990; Ingber and Folkman, 1989; Madri and Williams, 1983; Nicosia et al., 1991). We have been investigating the role of these biologically active sites in laminin on endothelial cell differentiation and on angiogenesis using several angiogenic models (Grant et al., 1991; Grant et al., 1989; Grant et al. 1992; Kibbey et al., 1992; Kubota et al., 1988; Passaniti et al., 1992).

Most of the information and the understanding about these basement membrane molecules has been derived from components extracted from the EHS tumor (Kleinman et al., 1987). This tumor produces a large quantity of basement membrane substance from which a proteinatious substance, Matrigel, can be extracted. Matrigel contains laminin, collagen IV, heparan sulfate proteoglycan, and several growth factors (i.e., TGF-ß, EGF, insulin-like growth factor 1, bovine fibroblast growth factor (bFGF), and platelet-derived growth factor (PDGF)) (Vukicevic et al., 1992). Matrigel can be reconstituted *in vitro* at 37°C to form a solid gel that has biochemical and morphological features similar to common basement membrane. This extracted matrix is biologically active and is known to promote the differentiation of a variety of cells (Kleinman et al., 1987).

Since the matrix comprising the vessel wall provides stability and regulation, the processes of angiogenesis disrupts this organization in order to permit cell invasion and migration to form new vessels. Secondly, the matrix surrounding the vessel wall also provides a substratum that permits cell binding and interactions essential to cell mobility.

Radiation therapy and angiogenesis

Radiation therapy has been used for many years to kill tumor cells. The growth of tumor cells is directly correlated with local angiogenesis, which enables the process of metastasis and serves as a prognostic indicator of tumor response to therapy (Craft et al. 1994). Therefore it is important to study changes in the surrounding vasculature during irradiation of a tumor bed. The physical and molecular changes that occur in endothelial cells and angiogenesis in response to radiation treatment, in terms of normal tissue response, must also be understood. Many studies have been developed that use *in vitro* as well as *in vivo* models to examine the effect of radiation on endothelial cells. Some are represented by colony formation assays measuring cell survival while others examine morphology and cell behavior under various radiation doses. The role of matrix during irradiation of endothelial cells has not been fully explored. In fact the effect of radiation on the process of angiogenesis, as defined by proliferation, migration and differentiation, has not been investigated. In this manuscript we will review several *in vitro* and *in vivo* studies which examine different aspects of the effect of radiation on angiogenesis. We will also show the results of our own study which examines the effect of radiation on *in vitro* models of angiogenesis. Finally, we will propose an *in vivo* model that may be useful to examine all the processes involved in angiogenesis.

Previous studies have investigated the effect of radiation on endothelial cells by examining cell survival. Cell survival curves, which show the number of cells surviving post-irradiation, were generated. These curves use the multihit model to show radiosensitivity by plotting radiation dose versus cell survival on a logarithmic scale. Important parameters expressed on the curve include Do, Dq, and N. Do is the dose of radiation that reduces the survival to 37%, or gives one hit per critical target. Dq is the quasi-threshold dose. N is the extrapolation number of the cell survival curve, and is also a measure of the width of the shoulder of the cell survival curve (Hall 1994). In many of the previous studies, sublethal radiation damage repair (the repair that occurs between radiation doses which are split and fractionated) has been examined from cell survival curves. A second form of repair which has been examined, potential lethal damage repair, occurs when the environmental conditions of the irradiated cells are adjusted after radiation to cause repair.

In vitro models. *In vitro* models examining the effect of radiation on the survival of endothelial cells have shown that endothelial cells have a moderate radiosensitivity compared to other cell types such as fibroblasts, and that survival is dose-dependent. The first dose response survival curve was created by Degowin et al. (1976). Endothelial cells were subjected to radiation doses of 1.25 to 10 Gy, and 1.6 Gy was noted to suppress endothelial cell replication by 37%. Recovery was evident with doses of 5 Gy, but no net increase in cell number was observed above 5 Gy. In comparison, fibroblasts were slightly less sensitive under these conditions. This study defined a Do using an assay system in a culture medium on flasks although the investigators were not sure of the sensitivity of their growth assay (Degowin et al., 1976).

Other investigators have also shown cell survival to be dose-dependent and that endothelial cells exhibit a moderate repair capacity post-irradiation. Bovine aortic endothelial cells (BAEC) cultured in Dulbecco's modified Eagle's medium, when exposed to a split dose radiation of 4.5 Gy (total), showed a Do of 1.01 Gy and Dq of 0.65 Gy (Rhee et al., 1986). Response of BAEC to radiation was dose-rate dependent and BAEC were able to repair sublethal radiation damage. Radiosensitivity and sublethal damage repair by endothelial cells were also shown to be moderate (as compared to fibroblasts) when human umbilical vein endothelial cells (HUVEC) plated on 1% gelatin-coated tissue culture plastic dishes were irradiated with either single or split graded doses. The cell survival curve which was generated showed a Do of 1.65 (Hei et al., 1987).

Another more recent *in vitro* model demonstrated that the sensitivity of endothelial cells to radiation is affected by the microenvironmental conditions under which the experiment is carried out (Fuks et al., 1992). BAEC were irradiated on uncoated tissue culture plastic at doses of 2 to 6 Gy and Do of 1.07 Gy and a Dq of 0.63 Gy were observed. When the dishes were precoated with an autologous natural basement membrane-like extracellular matrix, the curves showed a similar Do of 1.06, but an increased Dq of 1.94, indicating that the components of natural ECM confer in endothelial cells an improved capacity to repair radiation lesions and to restore clonogenic capacity. The Dq was reduced to 1.56 when the cells were irradiated on foreign basement membrane. This study is the first to demonstrate the effect of radiation on the survival of bovine endothelial cells irradiated on a natural environment, the basement membrane. However, this study lacks examination of other aspects of angiogenesis, such as migration and differentiation.

Morphologically, the responses of endothelial cells to radiation are characterized by minor acute changes, which include condensation of nucleoli and mitochondrial christae. This was particularly evident in cells that had detached from the monolayer 2 hours post-irradiation. After 24 hours, dispersion of perinuclear chromatin, nucleolar aberrations, and unusual polyribosomal aggregation, particularly in association with the endoplasmic reticulum, were observed in a study done by Speidel et al. (1993). In a control group exposed to alpha particles from a Bi-212 generator, no morphological changes were observed (Speidel et al., 1993). However, phase-contrast and electron microscopy of BAEC cells cultured in RPMI-1640 medium irradiated prior to reaching confluency showed vacuolization and an increased number of lysosomes. These observations were made 24 hours post-irradiation at 15-50 Gy and at 48 hours post-irradiation at 3-5Gy. Decreases in endoplasmic reticulum and polysomes occurred late in the course of radiation injury, but no structural alteration of mitochondria was observed, suggesting a relative resistance of mitochondria to radiation (Lee et al., 1983). Irradiation of BAEC cultured in Eagle's medium showed that ionizing radiation doses of 4.0-30.0 Gy caused a dose-dependent cell loss from confluent monolayer cultures which could not immediately be compensated by cell proliferation (Rosen et al., 1989). Within 24 hrs. the remaining attached cells underwent substantial somatic hypertrophy evidenced by increased protein content, cell volume, and attachment.

In vivo models. *In vivo* models have also been used to measure cell survival and are more accurate representations of angiogenesis because cells are irradiated in their natural environments. The values of Do and Dq are higher in the *in vivo* models and range from 1.68 to 2.4, with Dq values of 3.4 compared to the *in vitro* models with Do of 1.01 and 1.65 and Dq of .63 to 1.94 as compared to non-endothelial cells with Do of 1 to 2 Gy (Reinhold et al., 1973, Reinhold et al, 1975, Van den Brenk, 1972). The *in vivo* models represent a more natural environment compared to *in vitro* models and subsequently a higher dosage is required to reduce the surviving fraction by 37%. Repair is also improved in the *in vivo* model for similar reasons. The *in vivo* models are similar to the *in vitro* models in the inverse relationship evident between radiation dose and vascular proliferation.

An example of *in vivo* studies showing the effect of radiation on vascular proliferation is a study done by Reinhold et al. (1973), who used the subcutaneous air pouch in the rat. A thin sheet of connective tissue depleted of blood vessels by freezing was irradiated, and vascular proliferation induced by uric acid and lithium lactate. The areas where blood vessels were depleted in unirradiated preparations gradually re-vascularized over a period of 12 days. However, in the irradiated preparations, the vessels were sparse and the reduction was dependent on the dose of radiation. Survival curves showed a Do value of 1.7 Gy and a Dq value of 3.4 Gy. The same author in 1975 also looked at repair occurring in capillary endothelium with the rat subcutaneous pouch. When the stimulus to proliferate after a single dose of radiation was delayed, the Do value rose from 1.68 Gy for immediate stimulation to 2.73 Gy for delayed stimulation (32 days), indicating the number of endothelial cells surviving in the tissues after radiotherapy is greater than that determined from direct proliferation assays (Reinhold et al., 1975). This delay probably allows more repair of radiation damage and thus a higher Do is observed after delay.

The Selye pouch in the rat was also used to evaluate the effect of radiation on the growth and function of granulation tissue and to measure angiogenesis (Van den Brenk, 1972). A granuloma pouch (previously described in Van den Brenk, 1972) was raised on the back of a mouse by injecting oil, and the skin was irradiated in single or split doses. Angiogenesis was stimulated in the wall of the pouch and the radiosensitivity of the vascular connective tissue was estimated by counting vascular macrocolonies which formed after irradiation. Colony counts showed Do values of 2.4 Gy for single dose and 1.8 Gy for split-dose irradiation. Increasing the interval between irradiation of intact skin and raising of the pouch was associated with rapid and marked repair of radiation damage. A delay of 2-3 weeks resulted in a dose-reduction factor of 5 to 6.25 Gy for single doses of 15 to 18 Gy given to intact skin.

Another *in vivo* model utilized a polyvinyl alcohol sponge disc containing epidermal growth factor that was implanted in the subcutis of the thorax of a mouse. The extent of growth reduction of capillaries and stromal cells post-irradiation was quantified. (Prionas et al., 1990). After graded doses of X-rays, a dose response relationship was observed when X-rays were given 11 days after implantation, with the disc removed on day 20. A dose of 15 Gy reduced the rate of incorporation of (^{3}H)TdR and decreased the total growth area, indicating that endothelial cells are important targets in the stroma especially during the active proliferation period of these cells induced by growth factors. Kowalski et al. (1992) also used the disc angiogenesis system, in which a foam disc implanted subcutaneously in experimental animals showed a decrease in angiogenesis after ionizing radiation, even when applied after the angiogenic stimulus.

Some *in vivo* studies also examined the effect of radiation dose on morphology. Narayan et al. (1982) showed local effects of radiation on 20 fully-healed rabbit ear chambers in 13 rabbits that were irradiated with single exposures of 75 Gy. This revealed an acute response of cellular infiltration of the ear chambers, loss of vasomotion in arterioles, and vasodilation. Over the next few weeks a gradual reduction in the number of blood and lymphatic vessels was noted and the endothelium contained no recognizable organelles. In certain areas the endothelial layer was completely missing, exposing the vascular basement membrane to the blood.

Although all of these studies have provided a great deal of information on the effect of ionizing radiation on endothelial cells in *in vitro* and *in vivo* models, the role that the matrix plays during irradiation of endothelial cells has not been fully explored, especially as it relates to angiogenesis. Furthermore, the effect of radiation on the process of angiogenesis in an *in vitro* model as defined by the steps of proliferation, migration and differentiation have not been fully investigated. Therefore, in this manuscript we will present some of our recent results from studies which examine the role of matrix during irradiation of endothelial cells by showing changes in the surface area of cells surviving as well as morphology.

We propose to use *in vitro* assays of angiogenesis to demonstrate the affect of ionizing radiation on human umbilical vein endothelial cells irradiated on plastic and laminin, an abundant component of basement membrane matrix using a proliferation assay to demonstrate proliferation after radiation, a wound healing assay to demonstrate migration (Morales et al., 1995) and the tube formation assay with Matrigel to demonstrate differentiation (Grant et al., 1989). We also propose an *in vivo* model using Matrigel implantation in the subcutis of a mouse and examination of angiogenesis post-irradiation.

MATERIALS AND METHODS

Culture of Human Umbilical Cord Endothelial Cells (HUVEC). (Jaffe et al., 1973) Nearly-confluent (70-85%) flasks (Nunc, T-75) of passage 4-5 HUVEC were cultured at 37°C in medium 199 (Fisher) supplemented with glutamine, penicillin/streptomycin, amphotericin, 5 units/ml heparin, 200 µg/ml endothelial cell growth suplement (ECGS, Collaborative Research, Inc.), and 20 % bovine calf serum (HyClone). The cell lines were confirmed to be endothelial by immunohistochemical detection of von Willebrand factor expression.

Irradiation of HUVEC. New T-75 flasks were either uncoated or coated with laminin (20 μg) by adding 5 ml M199 (Fisher) containing 20 μl laminin to each flask, and allowing the flasks to incubate for 30 min. at 37 °C. Two hours prior to irradiation the cells were replated on the prepared flasks and incubated at 37°C. After irradiation at 0, 1, 2, 4, 8, 12, and 20 Gy by a Cesium 137 source (Gamma Cell 40, Atomic Energy of Canada, LTD), the flasks were returned to the incubator for 72 hrs. (3 days).

The first set of assays were performed after 72 hrs. to determine the effect of radiation treament on angiogenesis by ascertaining its effect on endothelial cell proliferation, differentiation, and migration (figure 1).

Proliferation assay. The proliferation assay was set up by adding 30,000 cells from each treatment to separate wells (in duplicate) of a twenty-four well plate (Nunc) containing 500 μl of whole medium. The cells were incubated at 37°C for 3 days. After the incubation period, the wells were fixed and stained with a giemsa stain (Leukostain kit, Fisher). A Nikon Microphot SA microscope with a video camera, integrated with an Apple Power Macintosh and NIH Image software was used to quantitate cell area per well.

Wound healing assay (Morales et al, 1995). The wound healing assay (a measure of migration) was performed by adding 2 ml HUVEC medium to each well of six-well plates (Corning). 200,000 cells were then added to each well, and the plate(s) were incubated at 37°C for 2-3 days, or until each well formed a confluent monolayer. A 1 mm wide scratch was made with a cell scraper bisecting the confluent monolayer in each well. The medium was aspirated and replenished to remove all floating cells, and the plates were incubated at 37°C overnight. The cells in each well were fixed and stained. The width of the wound was quantitated using the Nikon microscope an NIH Image software as indicated above. In each well 12 linear measurements of wound width were made to ascertain the extent of healing for each treatment.

Tube formation assay. The tube assay (a measure of differentiation) was set up as follows: 270 μl of Matrigel (at 4°C) was added per well in duplicate to a 24-well plate (Nunc). Then the plate was incubated at 37°C for 20-30 minutes to polymerize the Matrigel. 500 μl medium 199 (Fisher) was added to each well on top of the Matrigel, and 500 μl whole medium containing 40,000 cells from each irradiated flask was added to corresponding wells, and incubated at 37°C overnight. The resulting tubes were fixed and stained (Leukostain kit, Fisher), and quantitation was done by measuring the total tube area in 6 random areas of each well.

Cells remaining from the assays were added to new T-75 flasks containing 10 ml whole medium and incubated at 37°C until the 7 day post-irradiation assays were performed. The same assays were then performed as illustrated above.

RESULTS

We investigated the morphological effect of irradiation on endothelial cells plated for 2 hrs either on plastic or on laminin. Cells were observed 72 hrs after irradiation and showed different morphologies compared to controls. For instance, cells irradiated at 0 and 1 Gy had very little change in their cell size or nuclear morphologies (Figure 2). Irradiated cells showed multinucleation, and in many cells nuclear fragmentation could be seen.
Some cytoplasmic granules or vacuoles were present in the cells and visible at the light microscopic level.With increasing dose of irradiation there was distinct cell cytoplasmic enlargement (at and greater than 2 Gy), all of which increased in severity with higher doses (figure 2). The morphology of the cells irradiated on plastic versus laminin was not significantly different after 72 hrs but the effect of irradiation was more apparent at 12-14 days after irradiation (data not shown).

We also examined the effect of radiation on the proliferation of HUVEC plated for 2 hrs. on plastic and laminin. Proliferation was observed to be inversely proportional to radiation dose for endothelial cells on both substrata, with an overall decrease in proliferation with an increase in radiation dose (figure 3). At three days post-irradiation, proliferation was significantly higher in the cells irradiated on laminin than those irradiated on plastic at 1 Gy and 4 Gy (figure 3). At seven days post-irradiation, proliferation was significantly higher for the cells irradiated on laminin than for those on plastic at doses of 4 Gy and 8 Gy (figure 3). The relative degree of proliferation was fairly constant for the two time points.

We investigated the effect of radiation dose on wound healing by HUVEC. The assay measured migration, and we observed that the extent of wound healing decreased with increasing radiation dose (figure 4). Wound width after 20 hrs. was proportional to radiation dose (figure 4), with wider wounds at the higher doses. Morphological differences were also observed with increasing dose, including alteration from the unirradiated HUVEC's cobblestone appearance to the more spindle-shaped irradiated cells (figure 4). Differences in wound healing for cells irradiated on plastic and laminin were not distinctly evident, but it has been observed that wound assays performed with endothelial cells irradiated on Matrigel versus lamimin show greater wound width after 20 hrs. in the laminin cells (data not shown).

Flow Chart of Methods

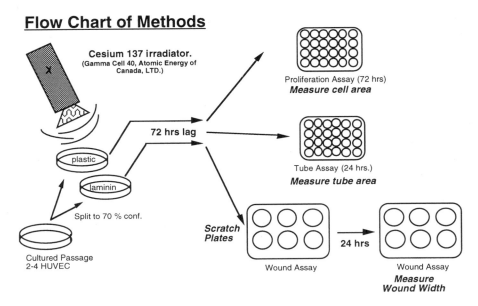

Figure 1. Flow chart of the methods.

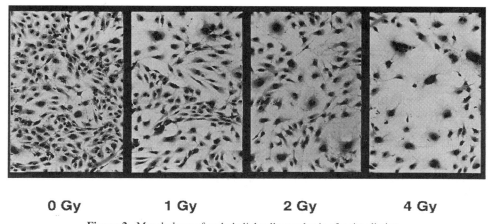

Figure 2. Morphology of endothelial cells on plastic after irradiation

The effects of radiation on tube formation (i.e. differentiation) for cells irradiated on plastic or laminin was also investigated. It was observed that total tube area decreased with increased radiation dose for both substrata at 3 and 7 days (figure 5). At three days post-irradiation, the effects of radiation were similar for both plastic and laminin at the lower doses (1 Gy), but at 4 Gy and 8 Gy tube area was higher for cells irradiated on laminin (figure 4). This difference was more evident at seven days post-irradiation; at higher doses tube area for HUVEC irradiated on laminin remained fairly high, whereas tube area for plastic decreased markedly (figure 5). At 8 Gy, tube formation by the cells irradiated on plastic had significantly decreased whereas tube formation by cells irradiated on laminin only slightly decreased.

DISCUSSION

The results demonstrate that radiation has a direct effect on proliferation, differentiation, and migration of endothelial cells *in vitro*, and that the presence of basement membrane proteins during irradiation has a significant effect on the post-radiation behavior of HUVEC *in vitro* as shown by the differences observed on cells irradiated on plastic versus laminin. The damaging effects of ionizing radiation is shown by the decrease in proliferation, tube area formation, and wound healing of HUVEC with increasing radiation dose. Since greater proliferation and tube formation at 7 days is observed at increased doses for the cells irradiated on laminin, it is evident that a protective effect is imparted to the cells by the basement membrane. This protective effect might entail a mechanism which would accomodate post-irradiation repair and would thus allow the cells to proliferate, migrate and differentiate better than cells lacking basement membrane (cells irradiated on uncoated plastic). Alternatively, this effect could impart protection to the cells during irradiation, decreasing the severity of radiation damage sustained by the cell.

The next step is to determine whether the effects of radiation treatment on endothelial cell proliferation, differentiation, and migration (angiogenesis) observed *in vitro* are also observed *in vivo*. To accomplish this we would like to perform radiation experiments using the *in vivo* Matrigel mouse assay in which Matrigel (at 4°C), containing angiogenic factors (FGF, TNF, etc), is injected subcutaneously into the anterior abdominal wall of the mouse, where it gels (Passaniti et al, 1992). After 10-14 days, the Matrigel plug is removed to ascertain the extent of angiogenesis into the plug. We propose that this assay could be coupled with radiation treatment of the injected area to determine the effect of radiation dose on angiogenesis *in vivo*. This data could then be compared with the *in vitro* observations, and an assessment of the validity of the *in vitro* model could be made.

Our study has been able to corroborate other studies on the effect of radiation on endothelial cells and angiogenesis through morphologic as well as cell proliferation analyses. Fuks et al. (1992), demonstrated the importance of the natural microenvironment on radiation repair by endothelial cells by showing that BAEC irradiated on their natural basement membrane

Proliferation, post-irradiation

Figure 3. Proliferation of HUVEC post-irradiation.

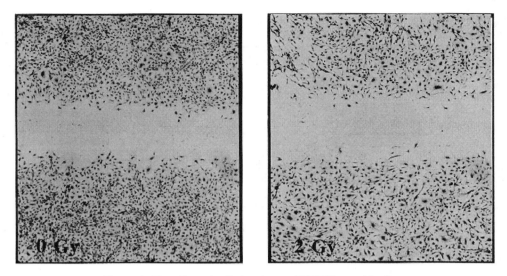

Figure 4. The effect of radiation dose on HUVEC wound healing.

Tube formation, post-irradiation

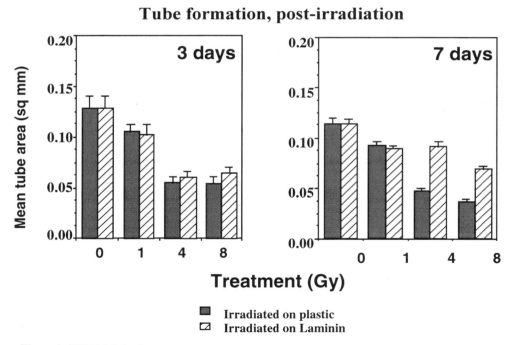

Irradiated on plastic
Irradiated on Laminin

Figure 5. HUVEC Tube formation following irradiation on plastic and laminin, early and late lag times.

had a better repair capacity compared to being irradiated on plastic or other basement membrane. This held true in all of our angiogenic assays since proliferation, wound healing and tube formation were improved in cells irradiated on laminin, a major component in the extracellular matrix, versus uncoated plastic. There was also greater improvement of repair in these three processes when the cells were irradiated on Matrigel, a natural basement membrane-like substance, as compared to laminin or plastic (data not shown). Morphologically we showed nuclear enlargement and multinucleation, as well as nuclear fragmentation and cytoplasmic enlargement with granulations that increased in a dose-dependent manner. Similar observations were also made by Lee et al. (1983) and Speidel et al. (1993) who saw an increase in chromatin condensation in the nuclei, polyribosomal aggregation in association with the endoplasmic reticulum, and an increase in the number of lysosomes. In addition, we have uniquely been able to show the effect of radiation not only on cell proliferation, but also on differentiation and migration (using the tube formation and wound healing assays), thus giving a more complete picture of the effect of radiation dose on angiogenesis.

We examined the effect of radiation dose on proliferation and not cell survival; therefore Do and Dq could not be determined. In the future, the proliferation assay and method of quantitation could be modified to measure cell survival instead of proliferation, and the Do and Dq could be obtained from the resulting survival curve. A longer post-irradiation period prior to assay might produce interesting data since endothelial cells have a low turnover rate and 7 days might not be fully adequate to see late radiation changes or a difference in the effect on repair. We would also like to use molecular biology techniques to isolate protein markers for endothelial cell damage such as von Willebrand factor.

In conclusion, radiation affects angiogenesis by causing a decrease in proliferation, migration and differentiation in a time- and dose- dependent manner above doses of 4 Gy. Extracellular matrix provides a natural environment which protects endothelial cells and potentially induces repair of radiation damage.

References

Craft, P.S., and Harris, A.L., 1994, Clinical prognostic significance of tumour angiogenesis, *Ann Oncol* 5: 305-311.

Degowin, R.L. et al, 1976, Radiation induced inhibition of human endothelial cells replicating in culture, *Radiat Res,* 68: 244-250.

Folkman, J., The role of angiogenesis in tumor growth, 1992, *Semin Cancer Biol*, 3: 65-71.

Fuks, Z, Vlodavsky, I., Andreeff, M., McLoughlin, M., and Haimovitz-Friedman, A., 1992, Effects of extracellular matrix on response of endothelial cells to radiation *in vitro*, *Eur. J. Cancer*, Vol. 28A:725-731.

Grant, D.S., Tashiro, K.-I., Segui-Real, B., Yamada, Y., Martin, G.R., and Kleinman, H.K., 1989, Two different laminin domains mediate the differentiation of human endothelial cells into capillary-like structures *in vitro*, *Cell*, 58: 933-943.

Grant, D.S., Kleinman, H.K., and Martin, G.R., 1990, The role of basement membranes in vascular development, *Ann N Y Acad Sci*, 588: 61-72.

Grant, D. S., Lelkes, P. I., Fukuda, K. and Kleinman, H. K., 1991, Intracellular mechanisms involved in basement membrane induced blood vessel differentiation *in vitro*, *In vitro Cell Dev Biol.*, 27a: 327-336.

Grant, D.S., Kinsella, J.L., Fridman, R., Auerbach, R., Piasecki, B.A., Yamada, Y., Zain, M., and Kleinman, H.K., 1992, Interaction of endothelial cells with a laminin A chain peptide (SIKVAV) *in vitro* and induction of angiogenic behavior *in vivo*, *J Cell Physiol*, 153: 614-25.

Hall, E.J., 1994, "Radiobiology for the Radiologist," 4th ed., J.B. Lippincott Co., New York.

Hei, T. K., Marchese, M.J., and Hall, E.J., 1987, Radiosensitivity and sublethal damage repair in human umbilical cord vein endothelial cells, *Int. J. Rad. Oncol. Biol. Phys*, 13:879-884.

Ingber, D. E. and Folkman, J., 1989, How does extracellular matrix control capillary morphogenesis? *Cell* 58: 803-805.

Jaffe, E.A., Nachman, R.L., Becker, C.G., and Minick, C.R., 1973, Culture of human endothelial cells derived from umbilical veins-identification by morphological and immunological criteria, *J. Clin. Invest.*, 52: 2745-2756.

Kibbey, M.C., Grant, D.S., and Kleinman, H.K., 1992, Role of the SIKVAV site of laminin in promotion of angiogenesis and tumor growth: an *in vivo* Matrigel model *J Natl Cancer Inst*, 84: 1633-8.

Kleinman, H. K., Cannon, F. B., Laurie, G. W., Hassel, J. R., Aumalley, M., Terranova, V. P., Martin, G. R., and Dalcq, M. D. B., 1985, Biological activities of laminin, *J. Cell Biol* 27: 317-325.

Kleinman, H.K., Graf, J., Iwamoto, Y., Kitten, G.T., Ogle, R.C., Sasaki, M., Yamada, Y., Martin, G.R., and Luckenbill-Edds, L., 1987, Role of basement membranes in cell differentiation, *Ann. N. Y. Acad. of Sci.*, 513: 134-145.

Kowalski, J., Kwan, H. H., Prionas, S. D., Allison, A. C., and Fajardo, L. F., 1992, Characterization and application of the disc angiogenesis system, *Exp Mol Path*, 56: 1-19.

Kramer, R. H. and Fuh, G. M., 1985, Type IV collagen synthesis by cultured human microvascular endothelial cells and its deposition in the subendothelial basement membrane, *Biochem* 24, 7423-7430.

Kubota, Y., Kleinman, H. K., Martin, G. R. and Lawley, T. J., 1988, Role of laminin and basement membrane in the morphological differentiation of human endothelial cells into capillary-like structures, *J. Cell Biol* 107, 1589-1598.

Lee, S. L., Douglas, W. H. J., Lin, P-S, and Fanburg, B. L., 1983, Ultrastructural changes of bovine pulmonary artery endothelial cells irradiated *in vitro* with a Cs-137 source, *Tissue & Cell* 15 (2): 193-204.

Madri, J. A., Dryer, B., Pitlick, F. and Furthmayr, H., 1980, The collagenous components of the subendothelium: correlation of structure and function, *Lab. Invest* 43, 303-315.

Madri, J.A. and Williams, S.K., 1983, Capillary endothelial cell cultures: phenotypic modulation by matrix components, *J. Cell Biol.*, 97: 153-165.

Morales, D.E., McGowan, K.A., Grant, D.S., Maheshwari, S., Bhartiya, D., Cid, M.C., Kleinman, H.K., and Schnaper, W., 1995, Estrogen promotes angiogenic activity in human umbilical vein endothelial cells *in vitro* and in a murine model, *Circulation*: 91 (3): 755-763.

Narayan, K and Cliff, W.J., 1982, Morphology of irradiated microvasculature: A combined *in vivo* and electron-microscopic study, *Amer. J. Pathol*, 47-62.

Nicosia, R.F., Belser, P., Bonanno, E., and Diven, J., 1991, Regulation of angiogenesis *in vitro* by collagen metabolism, *In vitro Cell Dev Biol*, 27A: 961-966.

Passaniti, A., Taylor, R.M., Pili, R., Guo, Y., Long, P.V., Haney, J.A., Pauly, R.R., Grant, D.S., and Martin, G.R., 1992, A simple, quantitative method for assessing angiogenesis and antiangiogenic agents using reconstituted basement membrane, heparin, and fibroblast growth factor, *Lab Invest*, 67: 519-28.

Prionas, S.D., Kowalski, J., Fajardo, L.F., Kaplan, I., Kwan, H.H., and Allison, A.C., 1990, Effects of X-irradiation on angiogenesis, *Radiat. Res.* 124:43-49.

Reinhold, H.S. et al, 1973, Radiosensitivity of capillary endothelium, *Br J. Radiol* 46: 54-57.

Reinhold, H.S., and Buisman, G.H., 1975, Repair of radiation damage to capillary endothelium, *Br J Radiol4* 48: 727-731.

Rhee, J.G., Lee, I., and Song, C.W., 1986, The clonogenic response of bovine aortic endothelial cells in culture to radiation, *Radiat Res* 106: 182-189.

Roberts, A.B., McCune, B.K., and Sporn, M.B., 1992, TGF-ß: regulation of extracellular matrix, *Kidney Int,* 41: 557-559.

Rosen, E. M., Vinter, D. W., and Goldberg, I. D., 1989, Hypertrophy of cultured bovine aortic endothelium following irradiation, *Rad Res* 117: 395-408.

Speidel, M. T., Holmquist, B., Kassis, A. I., Humm, J. L., Berman, R. M., Atcher, R. W., Hines, J. J. and Macklis, R. M., 1993, Morphological, biochemical, and molecular changes in endothelial cells after alpha-particle irradiation, *Rad Res* 136: 373-381.

Timpl, R., Rohde, H., Gehron Robey, P., Rennard, S. I., Foidart, J.M., and Martin, G.R., 1979, Laminin-a glycoprotein from basement membranes, *J. Biol. Chem* 254, 9933-9937.

Van den Brenk, H.A.S., 1972, Macro-colony assay for measurement of reparative angiogenesis after X-irradiation, *Int J Radiat Biol* 21 (6): 607-611.

Vlodavsky, I., Bar-Shavit, R., Ishai-Michaeli, R., Bashkin, P., and Fuks, Z., 1991, Extracellular sequestration and release of fibroblast growth factor: A regulatory mechanism? *Trends Biochem Sci* 16: 268.

Vlodavsky, I., Bar-Shavit, R., Korner, G. and Fuks, Z., 1993, Extracellular matrix-bound growth factors, enzymes, and plasma proteins, *in*: "Molecular and Cellular Aspects of Basement Membranes," P.H. Rohrbach and R. Timpl, eds., Academic Press Inc., San Diego, USA.

Vukicevic, S., Kleinman, H. K., Luyten, F.P., Roberts, A.B., Roche, N.S., and Reddi, A.H., 1992, Identification of multiple active growth factors in basement membrane Matrigel suggests caution in interpretation of cellular activity related to extracellular matrix components, *Exp. Cell Res.*, 202: 1-8.

BASIC FIBROBLAST GROWTH FACTOR EXPRESSION IN ENDOTHELIAL CELLS: AN AUTOCRINE ROLE IN ANGIOGENESIS?

Anna Gualandris, Marco Rusnati, Patrizia Dell'Era, Daniela Coltrini, Elena Tanghetti, Emanuele Nelli, and Marco Presta.

Unit of General Pathology and Immunology, Department of Biomedical Sciences and Biotechnology, University of Brescia, Brescia 25123, Italy.

INTRODUCTION

bFGF belongs to the family of the heparin-binding growth factors (Basilico and Moscatelli, 1992). The single copy human bFGF gene encodes multiple bFGF isoforms with molecular weights ranging from 24 kD to 18 kD. High molecular weight isoforms (HMW-bFGFs) are colinear NH_2-terminal extensions of the better characterized 18 kD protein (Florkiewicz and Sommer, 1989). Both low and high molecular weight bFGFs exert angiogenic activity *in vivo* and induce cell proliferation, protease production, and chemotaxis in cultured endothelial cells (Gualandris et al., 1994). Also, bFGF has been shown to stimulate endothelial cells to form capillary-like structures in collagen gels (Montesano et al., 1986) and to invade the amniotic membrane *in vitro* (Mignatti et al, 1989). The phenotype induced *in vitro* by bFGF in endothelial cells includes also modulation of integrin expression (Klein et al., 1993), gap-junctional intercellular communication (Pepper and Meda, 1992) and urokinase receptor upregulation (Mignatti et al., 1991). Experiments performed with neutralizing anti-bFGF antibodies have implicated endogenous bFGF in wound repair (Broadley et al., 1989), vascularization of the chorioallantoic membrane during chick embryo development (Ribatti et al., 1995), and tumor growth (Baird et al., 1986; Gross et al., 1993).

bFGF is thought to exert its effects on endothelial cells via a paracrine mode consequent to its release by other cells and/or mobilization from proteoglycans of ECM. On the other hand, some observations suggest that bFGF may also play an autocrine role in endothelial cells. *In vitro*, endothelial cells produce bFGF (Schweigerer et al., 1987; Vlodavsky et al., 1987; Presta et al., 1989) which modulates cell proliferation and migration, as well as the production of proteinases and their receptors (Sato and Rifkin, 1988; Itoh et al., 1992; Pepper et al., 1993). *In vivo*, expression of bFGF mRNA and/or protein occurs in endothelium adjacent to neoplastic cells of different human tumors (Schulze-Osthoff et al.,

Molecular, Cellular, and Clinical Aspects of Angiogenesis
Edited by Michael E. Maragoudakis, Plenum Press, New York, 1996

61

1990; Takahashi et al., 1990; Zagzag et al., 1990; Ohtani et al., 1993; Statuto et al., 1993). Also, tumor cells secrete angiogenic factor(s) which stimulate bFGF expression in cultured endothelial cells (Peverali et al., 1994). High levels of expression of bFGF are present in endothelial cells during the proliferating phase of human hemangiomas (Takahashi et al., 1994) and in spindle cells of endothelial origin in Kaposi's sarcoma (Ensoli et al., 1989; 1994).

To investigate the biological consequences of the expression of endogenous bFGF in vascular endothelium, a murine aortic endothelial (MAE) cell line, which do not express detectable levels of bFGF under normal culture conditions, was transfected with a retroviral expression vector harboring a human bFGF cDNA under the control of the Mo-MuLV LTR elements. Stable MAE transfectants expressing high levels of bFGF were characterized *in vitro* for their morphogenetic and invasive behavior when seeded on Matrigel, a laminin-rich gelled basement membrane matrix (Grant et al., 1989). bFGF-transfected cells were also tested *in vivo* for their capacity to induce an angiogenic response in the rabbit cornea and for their tumorigenic potential when injected in nude mice.

METHODS

bFGF cDNA transfection

The expression vector pZipbFGF, kindly provided by Dr. N. Quarto (New York University Medical Center, NY), was obtained by inserting into pZipNeoSV(X) a 1108 base pairs human bFGF cDNA (Sommer et al., 1987) encoding for 18 kD bFGF and for HMW-bFGFs under the control of the Mo-MuLV LTR elements (Quarto et al., 1989).

Balb/c mouse aortic endothelial 22106 cells (MAE cells) were obtained from Dr. R. Auerbach (University of Wisconsin) and were grown in Dulbecco's modified minimal essential medium (DMEM) added with 10% fetal calf serum (FCS). These cells were chosen as target cells since they do not express detectable levels of bFGF (see below). To obtain stable transfectants, MAE cells were plated at 8.0×10^5 cells/100-mm plate and were transfected with a calcium phosphate precipitate containing 20 μg of plasmid DNA [either pZipbFGF or pZipNeoSV(X)] and 40 μg of salmon sperm DNA. After 20 h, G418 sulfate antibiotic (Sigma) was added at 500 μg/ml to the culture medium. After 3 weeks of selective pressure, the G418-resistant clones were isolated, expanded, and tested by immunoblot and immunocytochemical analysis by using affinity purified anti-bFGF antibody as described (Gualandris et al., 1993).

Morphogenesis on Matrigel

Matrigel, an extract of the murine EHS tumor grown in C57/bl6 mice and produced as described (Kleinman et al., 1986), was provided by Dr. A. Albini (IST, Genova). 250 μl/well of Matrigel (10 mg/ml) were used to coat 24 well-plates at 4°C. After gelling at 37°C, endothelial cells were seeded onto the Matrigel layers at 75,000 cells/cm^2. Culture medium, containing the molecule to be tested, was renewed every 48 h. Matrigel cultures were fixed for 2 h with 2.5% glutaraldehyde/1% tannic acid in 0.1 M sodium cacodylate buffer, pH 7.4.

After extensive washing, gels were postfixed in 1% osmium tetroxide, dehydrated through a graded series of ethanol, stained *en bloc* with 0.3% uranyl acetate in 100% ethanol, treated with propylene oxide, and embedded in Epon 812 (Grinnell and Bennett, 1981). Semithin (1 μm) sections were cut perpendicularly to the culture plane and stained with 1% toluidine blue.

In vivo assays

The capacity of transfected MAE cells to induce an angiogenic response was tested by the rabbit cornea assay. Briefly, each rabbit received in the left cornea $2.5/3.5x10^5$ MAE cells and in the right cornea the same number of pZipbFGF2-MAE cells. The observation of the implants (12 implants per cell line) was made with a slit lamp stereomicroscope without anesthesia. It was started 48 h after surgery and then performed every other day. An angiogenic response was scored positive when budding of vessels from the limbal plexus was observed and capillaries progressed to reach the implanted pellet according to the scheme previously reported (Ziche et al., 1989).

The tumorigenic activity of bFGF transfectants was evaluated by subcutaneous inoculation of $5x10^6$ MAE cells or pZipbFGF2-MAE cells in three week-old nude mice. Animals were monitored daily and tumor size was measured with calipers. At sacrifice, tumors were removed and fixed in 4% paraformaldeyde.

RESULTS

Transfection of MAE cells with bFGF cDNA

Mouse MAE cells were transfected with pZipbFGF containing a 1.1 kb insert of the human bFGF cDNA. Control cultures were transfected with pZipNeoSV(X) containing the neo-resistance gene only. After transfection and selection, several G418-resistant clones were obtained for both vectors. bFGF-expressing clones were named pZipbFGF-MAE cells, while control clones were named pZipNeo-MAE cells.

As shown by Western blot analysis (Fig. 1), MAE cells transfected with pZipbFGF express high levels of 24 kD bFGF and of 22 kD bFGF and lower amounts of 18 kD bFGF. No bFGF immunoreactive material was observed in the extracts of control pZipNeo clones and in parental MAE cells. In keeping with the predominant expression of HMW-bFGF isoforms in respect with 18 kD bFGF, pZipbFGF clones showed a strong nuclear immunoreactivity when stained with affinity-purified anti-bFGF antibody. No specific immunoreactivity was observed in parental cells or in pZipNeo clones stained with anti-bFGF antibody or in bFGF-transfected cells stained with irrelevant IgG (data not shown).

Morphogenetic behavior of bFGF-transfected MAE cells

As already observed for NIH 3T3 cells transfected with bFGF cDNA (Quarto et al., 1989), pZipbFGF-MAE clones reach a density at confluence higher than parental cells ($5.5x10^5$ *vs* $2.0x10^5$ cells/cm^2) and show a transformed morphology when grown on tissue culture plastic.

MAEC

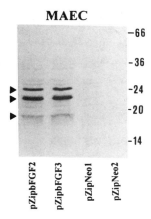

— 66

– 36

– 24

– 20

– 14

pZipbFGF2
pZipbFGF3
pZipNeo1
pZipNeo2

Figure 1. bFGF expression in transfected MAE cells. Total cell extracts from MAE cells transfected with the pZipbFGF expression vector or with the control vector pZipNeoSV(X) were probed in a Western blot with affinity-purified anti-bFGF antibody. 24 kD, 22kD, and 18 kD bFGF overexpressed isoforms are marked by arrowheads.

More interesting was the behavior of pZipbFGF-MAE cells when seeded on Matrigel, a laminin-rich gelled basement membrane matrix. Under these experimental conditions, bFGF-transfected cells originate a complex network of branching cord-like structures connecting foci of cells after 3-4 days in culture (Fig. 2). Cords, which were

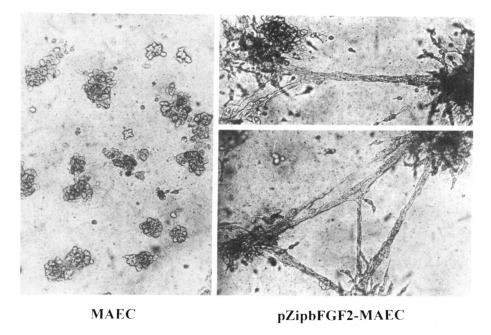

MAEC   **pZipbFGF2-MAEC**

Figure 2. Morphogenetic behavior on Matrigel of bFGF-transfected MAE cells. Parental MAE cells and pZipbFGF2-MAE cells were seeded on Matrigel at 75,000 cells/cm^2 and photographed after 4 days using an inverted phase contrast photomicroscope.

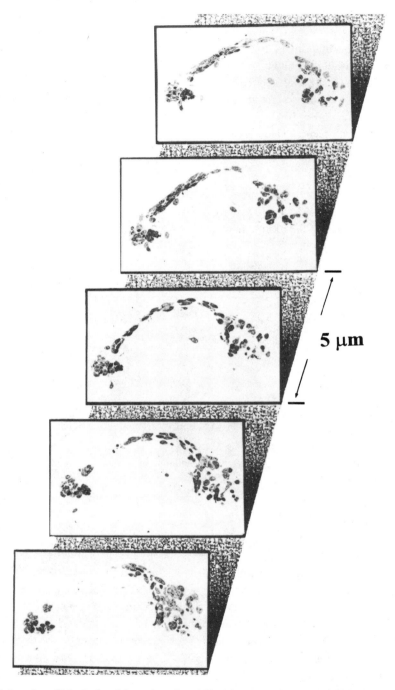

5 μm

Figure 3. Invasion of Matrigel and formation of cord-like structures by bFGF-transfected endothelial cells. Serial, longitudinal sections show a cord-like structure within Matrigel connecting two clusters of pZipbFGF cells.

devoid of lumen, were formed by a variable number of cells that organize themselves as single or multiple layers of spindle-shaped cells (Fig.3). In contrast, parental and control pZipNeo-MAE cells form small, rounded colonies scattered on the surface of the gel (Fig. 2). The organization of pZipbFGF-MAE cells on Matrigel was prevented by addition of affinity-purified anti-bFGF antibody to the culture medium during the assay, while irrelevant IgG was ineffective (Table 1). Also, the morphogenetic behavior of bFGF transfectants was abolished by addition to the culture medium of thrombospondin or heparin at concentrations that affect the interaction of bFGF with the cell surface receptors and inhibit the biological activity of the growth factor (Taraboletti et al., 1990; Rusnati et al., 1993). These findings demonstrate that the capacity of mouse endothelial cells to organize themselves in cord-like structures was dependent upon an extracellular, autocrine mode of action of overexpressed bFGF.

Table 1. Modulation of the morphogenetic behavior on Matrigel of bFGF-transfected MAE cells.

	Molecule	**Morphogenesis**
bFGF inhibitors:	Anti-bFGF Ab	Inhibition
	Heparin	Inhibition
	Thrombospondin	Inhibition
	TGF-β	Inhibition
Macromolecule synthesis inhibitors:	Cycloheximide	Inhibition
	Hydroxyurea	No effect
Cell locomotion inhibitors:	Colchicine	Inhibition
	Cytochalasin-D	Inhibition
Protease inhibitors:	1,10-Phenanthroline	Inhibition
	Trasylol	No effect

As reported in Table 1, cycloheximide, but not hydroxyurea, prevented cord formation of pZipbFGF cells, indicating that *de novo* protein synthesis, but not DNA synthesis, is required for morphogenesis. Also colchicine and cytochalasin-D prevented morphogenesis, implicating cell movement in the process of cord organization. Finally, cord formation was fully inhibited by the metallo-protease inhibitor 1,10-phenanthroline, but not by the serine-protease inhibitor trasylol. Accordingly, when parental and bFGF-transfected MAE cells were seeded on plastic and evaluated for metallo-protease production by gelatin zymography, increased levels of 92 kD gelatinase activity were observed in the conditioned medium of bFGF-expressing cells (Fig. 4). These findings are in keeping with the role played by metallo-proteases during angiogenesis *in vitro* and *in vivo* (Montesano and Orci, 1985; Klagsbrun and D'Amore, 1991; Schnaper et al., 1993).

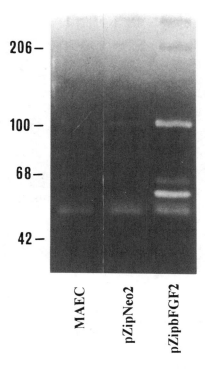

206 —

100 —

68 —

42 —

MAEC pZipNeo2 pZipbFGF2

Figure 4. Metallo-protease production in bFGF-transfected MAE cells. Gelatin-zymography of conditioned media from parental, pZipNeo, and pZipbFGF MAE cells.

Angiogenic and tumorigenic activity of bFGF-transfected MAE cells

The above data indicate that bFGF exerts an autocrine role for transfected MAE cells. We wondered whether the overexpression of bFGF in some endothelial cells could result also in the activation of neighboring quiescent endothelium *in vivo*, thus triggering an amplification of the angiogenic process. To this purpose, we compared parental and bFGF-transfected endothelial cells for the capacity to recruit quiescent host endothelial cells in the rabbit cornea angiogenesis assay. pZipbFGF-MAE cells were more potent than MAE cells in promoting the recruitment of host endothelial cells *in vivo*, the occurrence of the angiogenic response by bFGF transfectants being anticipated by 3-6 days. Indeed, 50% of pZipbFGF-MAE cell implants scored positive 3 days after surgery and 80% of implants at 12 days. In contrast, only 25% and 50% of MAE cell implants showed an angiogenic response at these times, respectively.

Then, we evaluated MAE and pZipbFGF-MAE cells for their capacity to induce tumor growth when injected in nude mice. All mice inoculated with pZipbFGF-MAE cells developed tumors within 1 week post-injection and xenografts reached a size of 3 cm^3 at 6 weeks. In contrast, lesions induced by MAE cells developed much more slowly being just palpable at 6 weeks. Histological examination of the xenografts arising from pZipbFGF-

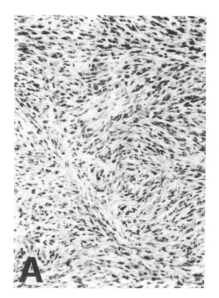
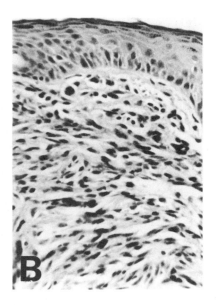

Figure 5. Histological appearance of pZipbFGF-MAE cell-induced tumors. Tumors from nude mice inoculated with pZipbFGF-MAE cells were removed 8 weeks post-injection. Hematoxylin-eoasin staining: tumors are characterized by the presence of arcs of spindled cells separated by slit-like spaces (A, x20) and by an intense neovascularization at the periphery of the tumor (B, x40).

MAE cells revealed the presence of proliferating spindle-shaped cells that form arcs intersetting one other and are separated by slit-like spaces; ectactic vascular channells within the tumor and an intense neovascularization at the periphery of the xenograft were also evident (Fig. 5). In conclusion, the histological features of pZipbFGF cell-induced tumors are characteristic of well-established lesions of Kaposi's sarcoma (Enzinger and Weiss, 1995).

CONCLUSIONS

The results demonstrate that bFGF overexpression changes the biological behavior of endothelial cells *in vitro* and *in vivo*. bFGF-transfected endothelial cells i) show an increased saturation density and a transformed morphology when grown on tissue culture plastic, ii) organize complex cord-like structures connecting foci of infiltrating cells when seeded on Matrigel, iii) are more efficient in inducing an angiogenic response in the host when implanted into the cornea of rabbit eye, iv) induce the appearance of fast-growing, highly vascularized tumors resembling Kaposi's sarcoma when injected in nude mice. Taken together, the data demonstrate that bFGF overexpression exerts an autocrine role in endothelium by affecting cell proliferation, morphogenetic potential, invasive behavior, endothelial cell recruitment, and tumorigenesis.

The capacity to organize on Matrigel is not restricted to endothelial cells (Kennedy et al., 1990; Vernon et al., 1991; Vernon et al., 1992). However, we have observed that different molecules, including TGF-β, heparin, and thrombospondin, able to affect bFGF-induced angiogenesis *in vivo* (Good et al., 1990; Passaniti et al., 1992), prevent morphogenesis *in vitro* of bFGF-transfected endothelial cells. Also, mouse NIH 3T3 fibroblasts do not organize on Matrigel after bFGF transfection (data not shown). These observations reinforce the concept that the morphogenetic organization of bFGF-transfected endothelial cells *in vitro* is triggered by the switch of an endogenous signal acting in an autocrine pattern and related to angiogenesis *in vivo*.

Some of the properties acquired by murine endothelial cells following bFGF overexpression are similar to those shown by bFGF-producing endothelial cells of different origin. For instance, the capacity to migrate and to invade the human amniotic membrane correlates with bFGF levels in different clones of cultured bovine capillary endothelial cells (Tsuboi et al., 1990). In addition, anti-bFGF antibody inhibits cell motility, plasminogen activator production, and DNA synthesis in the same cells (Sato and Rifkin, 1988). Also, antisense bFGF oligonucleotides suppress proliferation of cultured bovine aortic endothelial cells (Itoh et al., 1992). Interestingly, wounding of a bovine endothelial cell monolayer triggers an increase in gene expression of urokinase and its receptor in migrating cells which is mediated by endogenous bFGF (Pepper et al., 1993). Thus, bFGF exerts an autocrine role in cultured endothelium either when expressed under the physiological modulation of its genetic and/or epigenetic control elements or when its production is driven by a viral promoter.

It has been shown that bFGF expression occurs in the endothelium adjacent to neoplastic cells in several human tumor types, indicating that this is a common feature of vascular endothelium during tumor angiogenesis.(Takahashi et al., 1990; Schulze-Osthoff et al., 1990; Zagzag et al., 1990; Ohtani et al., 1993; Statuto et al., 1993). Our observations strongly support the hypothesis that neovascularization can be triggered by molecule(s) released by tumor cells and/or infiltrating inflammatory cells that induce bFGF upregulation in the quiescent endothelium. Indeed, Peverali et al. (1994) have shown that tumor cells release molecule(s) able to upregulate bFGF expression in endothelium. Also, bFGF itself, thrombin, and interleukin-2 stimulate bFGF production in cultured endothelial cells (Weich et al., 1991; Cozzolino et al., 1994). Newly synthesized bFGF will trigger an invasive and morphogenetic program in activated endothelial cells. Released bFGF will recruit more quiescent endothelial cells, thus amplifying the angiogenic stimulus.

Kaposi's sarcoma spindle cells of endothelial origin express bFGF *in vivo* (Xerri et al., 1991) and *in vitro* (Ensoli et al., 1989). bFGF promotes the growth of these cells in an autocrine fashion (Ensoli et a., 1989). Also, subcutaneous injection of bFGF in nude mice causes the appearance of lesions characteristic of early stage Kaposi's sarcoma (Ensoli et al., 1994). Finally, bFGF expression has been demonstrated in endothelial cells during the proliferative phase of human hemangiomas (Takahashi et al., 1994). All these observations point to an autocrine role for bFGF also in angioproliferative diseases. In keeping with this hypothesis, pZipbFGF MAE cells induce the growth of neovascularized tumors when injected in nude mice. Histological characteristics of these tumors resemble strongly those shown by well-established lesions of Kaposi's sarcoma (Enzinger and Weiss, 1995) and by experimental tumors induced by injection of Kaposi's sarcoma-derived spindle cells in nude mice (Salahuddin et al., 1988).

The pZipbFGF endothelial cell lines developed in the present study may represent an useful autocrine model of angiogenesis and angioproliferative disease and a target for the screening and evaluation of angiostatic molecules.

ACKNOWLEDGMENTS

We wish to thank Dr. M. Ziche for the rabbit cornea angiogenesis assay, Dr. L. Possati for the tumorigenic assay, and Prof. M.P. Molinari-Tosatti for histology of the xenografts. This work was supported in part by grants from C.N.R. (Progetto Finalizzato Biotecnologie e Biostrumentazioni, Sottoprogetto Biofarmaci and grant n. 94.00316.CT14) and from Associazione Italiana per la Ricerca sul Cancro (AIRC) to M. Presta.

REFERENCES

Baird, A., P. Mormède, and P. Bohlen. 1986. Immunoreactive fibroblast growth factor (FGF) in a transplantable chondrosarcoma: inhibition of tumor growth by antibodies to FGF. *J. Cell. Biochem.* 30:79-85.

Basilico, C., and D. Moscatelli. 1992. The FGF family of growth factors and oncogenes. *Adv. Cancer Res.* 59:115-165.

Broadly K.N., A.M. Aquino, S.C. Woodward, A. Buckley-Sturrock, Y. Sato, D.B. Rifkin, and J.M. Davidson. 1989. Monospecific antibodies implicate basic fibroblast growth factor in normal wound repair. *Lab. Invest.* 61:571-575.

Cozzolino, F., M. Torcia, M. Lucibello, L. Morbidelli, M. Ziche, J. Platt, S. Fabiani, J. Brett, and D. Stern. 1993. Cytokine-mediated control of endothelial cell growth: interferon-α and interleukin-2 synergistically enhance basic fibroblast growth factor synthesis and induce release promoting cell growth *in vitro* and *in vivo*. *J. Clin. Invest.* 91:2504-2512.

Ensoli, B., S. Nakamura, Z.S. Salahuddin, P. Biberfeld, L. Larsson, B. Beaver, F. Wong-Staal, and R.C. Gallo. 1989. AIDS-Kaposi's sarcoma-derived cells express cytokines wih autocrine and paracrine growth effects. *Science.* 243:223-226.

Ensoli, B., R. Gendelman, P. Markham, V. Fiorelli, S. Colombini, M. Raffeld, A. Cafaro, H.-K. Chang, J.N. Brady, and R.C. Gallo. 1994. Synergy between basic fibroblast growth factor and HIV-1 Tat protein in induction of Kaposi's sarcoma. *Nature.* 371:674-680.

Enzinger, F.M., and Weiss, S.W. 1995. Soft tissue tumors. Mosby-Year Book, Inc. St. Louis. pp.658-669.

Florkiewicz, R.Z., and A. Sommer. 1989. Human basic fibroblast growth factor gene encodes four polypeptides: three initiate translation from non-AUG codons. *Proc. Natl. Acad. Sci. U.S.A.* 86:3978-3981.

Good, D.J., P.J. Polverini, F. Rastinejad, M.M. LeBeau, R.S. Lemons, W.A. Frazier, and N.P. Bouck. 1990. A tumor suppressor-dependent inhibitor of angiogenesis is immunologically and functionally indistinguishable from a fragment of thrombospondin. *Proc. Natl. Acad. Sci. USA.* 87:6624-6628.

Grant, D.S., K.-I. Tashiro, B. Segui-Real, Y. Yamada, G.R. Martin, and H.K. Kleinman. 1989. Two different laminin domains mediate the differentiation of human endothelial cells into capillary-like structures in vitro. *Cell.* 58:933-943.

Grinnell, F., and M.H. Bennett. 1981. Fibroblast adhesion on collagen substrata in the presence and absence of plasma fibronectin. *J. Cell Sci.* 48:19-34.

Gross, J.L., W.F. Herblin, B.A. Dusak, P. Czerniak, M.D. Diamond, T. Sun, K. Eidsvoog, D.L. Dexter, and A. Yayon. 1993. Effects of modulation of basic fibroblast growth factor on tumor growth in vivo. *J. Natl. Cancer Inst.* 85:121-131.

Gualandris, A., D. Coltrini, L. Bergonzoni, A. Isacchi, S. Tenca, B. Ginelli, and M. Presta. 1993. The NH₂-terminal extension of high molecular weight forms of basic fibroblast growth factor (bFGF) is not essential for the binding of bFGF to nuclear chromatin in transfected NIH 3T3 cells. *Growth Factors.* 8:49-60.

Gualandris, A., C. Urbinati, M. Rusnati, M. Ziche, and M. Presta. 1994. Interaction of high molecular weight basic fibroblast growth factor (bFGF) with endothelium: biological activity and intracellular fate of human recombinant Mr 24,000 bFGF. *J. Cell. Physiol.* 161:149-159.

Itoh, H., M. Mukoyama, R.E. Pratt, and V.J. Dzau. 1992. Specific blockade of basic fibroblast growth factor gene expression in endothelial cells by antisense oligonucleotide. *Biochem. Biophys. Res. Commun.* 188:1205-1213.

Kennedy, A., R.N. Frank, L.B. Sotolongo, A. Das, and N.L. Zhang. 1990. Proliferative response and macromolecular synthesis by ocular cells cultured on extracellular matrix materials. *Curr. Eye Res.* 9:307-312.

Klagsbrun, M., and P.A. D'Amore. 1991. Regulators of angiogenesis. *Annu. Rev. Physiol.* 53:217-39.

Klein, S., F.G. Giancotti, M. Presta, S.M. Albelda, C.A. Buck, and D.B. Rifkin. 1993. Basic fibroblast growth factor modulates integrin expression in microvascular endothelial cells. *Mol. Biol. Cell.* 4: 973-982.

Kleinman, H.K., M.L. McGarvey, J.R. Hassell, V.L. Star, F.B. Cannon, G.W. Laurie, and G.R. Martin. 1986. Basement membrane complexes with biological activity. *Biochemistry.* 25:312-318.

Mignatti, P., R. Tauboi, E. Robbins, and D.B. Rifkin. 1989. *In vitro* angiogenesis on the human amniotic membrane: requirement for basic fibroblast growth factor. *J. Cell Biol.* 108:671-682.

Mignatti, P., R. Mazzieri, and D.B. Rifkin. 1991. Expression of the urokinase receptor in vascular endothelial cells is stimulated by basic fibroblast growth factor. *J. Cell Biol.* 113:1193-1201.

Montesano, R., and L. Orci. 1985. Tumor-promoting phorbol esters induce angiogenesis in vitro. *Cell.* 42:469-477.

Montesano, R., J.-D. Vassalli, A. Baird, R. Guillemin, and L. Orci. 1986. Basic fibroblast growth factor induces angiogenesis *in vitro. Proc. Natl. Acad. Sci. U.S.A.* 83:7297-7301.

Ohtani, H., S. Nakamura, Y. Watanabe, T. Mizoi, T. Saku, and H. Nagura. 1993. Immunocytochemical localization of basic fibroblast growth factor in carcinomas and inflammatory lesions of the human digestive tract. *Lab. Invest.* 68:520-527.

Passaniti, A., R.M. Taylor, R. Pili, Y. Guo, P.V. Long, J.A. Haney, R.R. Pauly, D.S. Grant, and G.R. Martin. 1992. A simple, quantitative method for assessing angiogenesis and antiangiogenic agents using reconstituted basement membrane, heparin, and fibroblast growth factor. *Lab. Invest.* 67:519-528.

Pepper, M.S., and P. Meda. 1992. Basic fibroblast growth factor increases junctional communication and connexin 43 expression in microvascular endothelial cells. *J. Cell. Physiol.* 153:196-205.

Pepper, M.S., A.-P. Sappino, R. Stocklin, R. Montesano, L. Orci, and J.-D. Vassalli. 1993. Upregulation of urokinase receptor expression on migrating endothelial cells. *J. Cell Biol.* 122:673-684.

Peverali, F.A., S.J. Mandriota, P. Ciana, R. Marelli, P. Quax, D.B. Rifkin, G. Della Valle, and P. Mignatti. 1994. Tumor cells secrete an angiogenic factor that stimulates basic fibroblast growth factor and urokinase expression in vascular endothelial cells. *J. Cell. Physiol.* 161:1-14.

Presta, M., J.A.M. Maier, M. Rusnati, and G. Ragnotti. 1989. Basic fibroblast growth factor: production, mitogenic response, and post-receptor signal transduction in cultured normal and transformed fetal bovine aortic endothelial cells. *J. Cell. Physiol.* 141:517-526.

Quarto, N., D. Talarico, A. Sommer, R. Florkiewicz, C. Basilico, and D.B. Rifkin. 1989. Transformation by basic fibroblast growth factor requires high levels of expression: comparison with transformation by hst/K-fgf. *Oncogene Res.* 5:101-110.

Ribatti, D., C. Urbinati, B. Nico, M. Rusnati, L. Roncali, and M. Presta. 1995. Endogenous basic fibroblast growth factor in the vascularization of the chick embryo chorioallantoic membrane. *Dev. Biol.*, in press.

Rusnati, M., C. Urbinati, and M. Presta. 1993. Internalization of basic fibroblast growth factor (bFGF) in cultured endothelial cells: role of the low affinity heparin-like bFGF receptors. *J. Cell. Physiol.* 154:152-161.

Salahuddin, S.Z., S. Nakamura, P. Biberfeld, M.H. Kaplan, P.D. Markham, L. Larsson, and R.C. Gallo. 1988. Angiogenic properties of Kaposi's sarcoma-derived cells after long-term culture in vitro. *Science.* 242:430-433.

Sato, Y., and D.B. Rifkin. 1988. Autocrine activities of basic fibroblast growth factor: regulation of endothelial cell movement, plasminogen activator synthesis, and DNA synthesis. *J. Cell Biol.* 107:1199-1205.

Schanper, H.W., D.S. Grant, W.G. Stetler-Stevenson, R. Fridman, G. D'Orazi, A.N. Murphy, R.E. Bird, M. Hoythya, T.R. Fuerst, D.L. French, J.P. Quigley, and H.K. Kleinman. 1993. Type IV collagenase(s) and TIMPs modulate endothelial cell morphogenesis in vitro. *J. Cell. Physiol.* 156:235-246

Schulze-Osthoff, K., W. Risau, E. Vollmer, and C. Sorg. 1990. In situ detection of basic fibroblast growth factor by highly specific antibodies. *Am. J. Pathol.* 137:85-92.

Schweigerer, L., G. Neufeld, J. Friedman, J.A. Abraham, J.C. Fiddes, and D. Gospodarowicz. 1987. Capillary endothelial cells express basic fibroblast growth factor, a mitogen that promotes their own growth. *Nature.* 325:257-259.

Statuto, M., M.G. Ennas, G. Zamboni, F. Bonetti, M. Pea, F. Bernardello, A. Pozzi, M. Rusnati, A. Gualandris, and M. Presta. 1993. Basic fibroblast growth factor in human pheochromocytoma: a biochemical and immunohistochemical study. *Int. J. Cancer.* 53: 5-10.

Takahashi, J.A., H. Mori, M. Fukumoto, K. Igarashi, M. Jaye, Y. Oda, H. Kikuchi, and M. Hatanaka. 1990. Gene expression of fibroblast growth factors in human gliomas and meningiomas: demonstration of cellular source of basic fibroblast growth factor mRNA and peptide in tumor tissues. *Proc. Natl. Acad. Sci. U.S.A.* 87:5710-5714.

Takahashi, K., J.B. Mulliken, H.P. Kozakewich, R.A. Rogers, J. Folkman, and R.A. Ezekowitz. 1994. Cellular markers that distinguish the phases of hemangioma during infancy and childhood. *J. Clin. Invest.* 93:2357-2364.

Taraboletti, G., D. Roberts, L.A. Liotta, and R. Giavazzi. 1990. Platelet thrombospondin modulates endothelial cell adhesion, motility, and growth: a potential angiogenesis regulatory factor. *J. Cell Biol.* 111:765-772.

Tsuboi, R., Y. Sato, and D.B. Rifkin. 1990. Correlation of cell migration, cell invasion, receptor number, proteinase production, and basic fibroblast growth factor levels in endothelial cells. *J. Cell Biol.* 110:511-517.

Vernon, R.B., T.F. Lane, J.C. Angello, and E.H. Sage. 1991. Adhesion, shape, proliferation, and gene expression of mouse Leydig cells are influenced by extracellular matrix *in vitro*. *Biol. Reprod.* 44:157-162.

Vernon, R.B., J.C. Angello, M.L. Iruela-Arispe, T.F. Lane, and E.H. Sage. 1992. Reorganization of basement membrane matrices by cellular traction promotes the formation of cellular networks *in vitro*. *Lab. Invest.* 66:536-547.

Vlodavski, I., R. Friedman, R. Sullivan, J. Sasse, and M. Klagsbrun. 1987. Aortic endothelial cells synthesize basic fibroblast growth factor which remains cell associated and platelet-derived growth factor-like protein which is secreted. *J. Cell. Physiol.* 131:402-408.

Weich, H., N. Iberg, M. Klagsbrun, and J. Folkman. 1991. Transcriptional regulation of basic fibroblast growth factor gene expression in capillary endothelial cells. *J. Cell. Biochem.* 47:158-194.

Williams, R.L., W. Risau, A.-G. Zerwes, H. Drexler, A. Aguzzi, and E.F. Wagner. 1989. Endothelioma cells expressing the polyoma middle T oncogene induce hemangiomas by host cell recruitment. *Cell.* 57:1053-1063.

Xerri, L., J. Hassoun, J. Planchet, V. Guigou, J.J. Grobb, P. Parc, D. Birnbaum, and O. deLapeyriere. 1991. Fibroblast growth factor gene expression in AIDS-Kaposi's sarcoma detected by in situ hybridization. *Am. J. Pathol.* 138:9-15.

Zagzag, D., D.C. Miller, Y. Sato, D.B. Rifkin, and D.E. Burstein. 1990. Immunohistochemical localization of basic fibroblast growth factor in astrocytomas. *Cancer Res.* 50:7393-7398.

Ziche, M., G. Alessandri, and P.M. Gullino. 1989. Gangliosides promote the angiogenic response. *Lab. Invest.* 61:629-634.

THE BIOLOGY OF VASCULAR ENDOTHELIAL GROWTH FACTOR

Napoleone Ferrara

Department of Cardiovascular Research
Genentech, Inc.
South San Francisco, CA 94080

INTRODUCTION

The establishment of a vascular supply is required for organ development and differentiation as well as for tissue repair and reproductive functions in the adult[1]. Neovascularization (angiogenesis) is also implicated in the pathogenesis of a number of disorders. These include: proliferative retinopathies, age-related macular degeneration, tumors, rheumatoid arthritis, and psoriasis[1, 2]. A strong correlation has been noted between density of microvessels in primary breast cancers and their nodal metastases and patient survival[3]. Similarly, a correlation has been reported between vascularity and invasive behavior in several other tumors[4-6].

A variety of factors have been previously identified as potential positive regulators of angiogenesis[1]. This chapter will review the molecular properties of vascular endothelial growth factor (VEGF)[7-9], an endothelial cell mitogen and angiogenesis inducer, and will discuss its role in normal and pathological angiogenesis. VEGF and its receptors appear to play a major role in the regulation of physiological angiogenesis, such as embryonic and reproductive angiogenesis[10-12]. Also, VEGF-indude angiogenesis results in development of collateral vessels in animal models of coronary or limb ischemia[13-16]. Furthermore, recent studies point to VEGF as a key mediator of neovascularization associated with a variety of disorders[17, 18].

Molecular, Cellular, and Clinical Aspects of Angiogenesis
Edited by Michael E. Maragoudakis, Plenum Press, New York, 1996

73

BIOLOGICAL ACTIVITIES OF VEGF

VEGF is a potent mitogen (ED_{50} 2-10 pM) for vascular endothelial cells derived from small or large vessels but is devoid of significant mitogenic activity for other cell types[7, 8]. VEGF is also able to promote angiogenesis in a variety of *in vivo* models[19-21]. Also, VEGF induces sprouting from rat aortic rings embedded in a collagen gel[22]. This model emphasizes the specificity of the growth factor for endothelial cells, as the proliferation induced by VEGF consisted almost exclusively of vascular endothelial cells[22]. Furthermore, VEGF induces expression of the serine proteases urokinase-type and tissue-type plasminogen activators and of PA inhibitor 1 in cultured bovine microvascular endothelial cell[23]. Also, VEGF induces expression of the metalloproteinase interstitial collagenase in human umbilical vein endothelial cells but not in dermal fibroblasts[24]. Recent studies have shown that VEGF induces expression of urokinase receptor in vascular endothelial cells[25]. Additional effects of VEGF on the vascular endothelium are the stimulation of hexose transport[26] and the induction of tissue factor expression[27].

VEGF has been independently purified and cloned as a vascular permeability factor (VPF) based on its ability to induce vascular leakage in guinea pig skin[28, 29]. It has been proposed that an increase in microvascular permeability is a crucial step in angiogenesis associated with tumors and wounds[30]. According to this hypothesis, a major function of VEGF in angiogenesis is to induce plasma protein leakage. This would result in the formation of an extravascular fibrin gel, a substrate for endothelial and tumor cell growth[30].

The mitogenic and the permeability-enhancing activity of VEGF can be potentiated[31] by placenta growth factor (PlGF), a molecule having significant structural homology with VEGF[32]. While PlGF has little or no direct mitogenic or permeability-enhancing activity, it significantly potentiates the activity of low, marginally efficacious, concentrations of VEGF[31].

Interestingly, VEGF has also been shown to induce vasodilatation *in vitro* in a dose-dependent fashion[33]. This results in a transient hypotension *in vivo* (unpublished observations). Such effects appear to be mediated primarily by endothelial cell-derived NO, as assessed by the requirement for an intact endothelium and the prevention of the effect by N-methyl-arginine[33].

THE VEGF ISOFORMS

By alternative mRNA splicing of a single gene, VEGF may exist as one of four different molecular species, having respectively 121, 165, 189 and 206 amino acids ($VEGF_{121}$, $VEGF_{165}$, $VEGF_{189}$, $VEGF_{206}$)[19, 34, 35]. VEGF purified from a variety of species and sources is a basic, heparin-binding, homodimeric glycoprotein of 45,000 daltons[8]. These properties correspond to those of $VEGF_{165}$, the predominant isoform[8]. $VEGF_{121}$ fails to bind to heparin[36]. $VEGF_{189}$ and $VEGF_{206}$ are more basic and bind to heparin with greater affinity than $VEGF_{165}$. $VEGF_{121}$ is secreted as a freely soluble protein. $VEGF_{165}$ is also secreted but a significant fraction remains bound to the cell surface or the extracellular matrix (ECM)[36, 37]. In contrast, $VEGF_{189}$ and $VEGF_{206}$ are almost completely sequestered in the ECM[36, 37]. However, they may be released in a biologically active form by plasmin cleavage[36, 37]. Plasminogen activation and generation of plasmin have been shown to play an important role in the angiogenesis cascade[1].

Interestingly, heterodimers between VEGF and PlGF have been recently identified in the medium conditioned by a rat glioma cell line[38]. The VEGF.PlGF heterodimer was ~7 fold less potent than the VEGF homodimer in promoting endothelial cell growth.

REGULATION OF VEGF GENE EXPRESSION

Oxygen tension has been shown to play a major role in the regulation of VEGF gene expression. VEGF mRNA expression is rapidly and reversibly induced by exposure to low pO_2 in a variety of cultured cells[39, 40]. Occlusion of the left anterior descending coronary artery results in a dramatic increase in VEGF RNA levels in the pig myocardium, suggesting that hypoxia-induced VEGF is a mediator of the spontaneous revascularization that follows myocardial ischemia[41].

Similarities have been noted between the mechanisms leading to hypoxic regulation of VEGF and erythropoietin (Epo) genes[42]. Also, hypoxia-inducibility appears to be conferred to both genes by homologous sequences. A 28-base sequence has been identified in the 5' promoter of the rat VEGF gene which mediated hypoxia-induced transcription in transient assays[43]. Such sequence reveals a high degree of homology and similar protein binding characteristics as the hypoxia-inducible factor 1 (HIF-1) binding site within the Epo gene, which behaves like a classic 3' transcriptional enhancer[44]. Also, activation of c-Src has been shown to participate in the hypoxic up-regulation of VEGF gene expression[45].

Several cytokines or growth factors up-regulate VEGF mRNA expression and/or induce release of VEGF protein. Exposure of quiescent human keratinocytes to serum, EGF, TGF-β or KGF resulted in a marked induction of VEGF mRNA expression[46]. Treatment of quiescent cultures of several epithelial and fibroblastic cell lines with TGF-β resulted in induction of VEGF mRNA and release of VEGF protein into the medium[47]. Furthermore, IL-1 beta has been shown to induce VEGF expression in aortic smooth muscle cells[48].

Differentiation plays an important role in the regulation of VEGF gene expression, at least in some models of cellular differentiation[49]. The VEGF mRNA was markedly up-regulated during the conversion of 3T3 preadipocytes into adipocytes. Conversely, VEGF gene expression was dramatically suppressed during the differentiation of the pheochromocytoma cell line PC12 into non-malignant, neuron-like, cells[49].

THE VEGF RECEPTORS

In agreement with the hypothesis that VEGF is an endothelial cell-specific factor, ligand autoradiography studies on fetal and adult rat tissue sections have demonstrated that high affinity VEGF binding sites are localized to the vascular endothelium of large or small vessels, but not to other cell types[50, 51].

Two tyrosine kinases have been identified as VEGF receptors[52-54]. The Flt-1 (fms-like-tyrosine kinase) and KDR (kinase domain region) proteins have been shown to bind VEGF with high affinity[52, 53]. Flk-1 (fetal liver kinase-1), the murine homologue of KDR, also binds VEGF[54].

The Flt-1 and KDR proteins have been shown to have different signal transduction properties[55, 56]. Porcine aortic endothelial cells lacking endogenous VEGF receptors display chemotaxis and mitogenesis in response to VEGF when transfected with an expression vector coding for KDR[55]. In contrast, transfected cells expressing Flt-1 lack such responses[55, 56]. While KDR/Flk-1 undergoes strong ligand-dependent tyrosine phosphorylation in intact cells[54, 55], Flt-1 reveals a very weak or undetectable response[52, 55, 56]. Transfection of Flt-1 cDNA in NIH 3T3 led to a weak VEGF-dependent tyrosine phosphorylation that did not generate any mitogenic signal[56]. These findings indicate that interaction with Flk-1/KDR is a critical requirement to elicit the full spectrum of VEGF biologic responses.

Recent studies have demonstrated that both Flt-1 and Flk-1/KDR are essential for normal development of embryonic vasculature, although their respective roles appear to be distinct[57, 58]. Mouse embryos homozygous for a targeted mutation in the flt-1 locus died in utero at day 8.5[57]. Endothelial cells developed in both embryonic and extraembryonic sites but failed to organize in normal vascular channels. Mice where the flk-1 gene had been inactivated not only lacked vasculogenesis but also failed to develop blood islands[58]. Hematopoietic precursors were severely disrupted and organized blood vessels failed to develop throughout the embryo or the yolk sac resulting in death in utero between day 8.5 and 9.5.

THE ROLE OF VEGF IN PATHOLOGIC ANGIOGENESIS

A. ANGIOGENESIS ASSOCIATED WITH TUMORS

In situ hybridization studies have shown that the VEGF mRNA is markedly up-regulated in most human tumors examined. These include: renal, bladder, breast, ovarian, and gastrointestinal tract carcinomas[59] and several intracranial tumors[60, 61]. Only sections of lobular carcinoma of the breast and papillary carcinoma of the bladder failed to reveal significant VEGF mRNA expression[59]. In all of these circumstances, VEGF mRNA is expressed by tumor cells but not by endothelial cells. A strong correlation exists between degree of vascularization of the malignancy and VEGF mRNA expression[61]. In addition, the mRNA for the VEGF receptors, Flt-1 and KDR, is up-regulated in the tumor vasculature[60, 62]. These findings suggest that VEGF-expressing tumor cells may have a growth advantage *in vivo* due to stimulation of angiogenesis[63].

More direct evidence for a role of VEGF in tumorigenesis has been made possible by specific monoclonal antibodies capable of neutralizing VEGF-induced angiogenesis[64]. Such antibodies suppress the the growth of a variety of human tumor cell lines injected subcutaneously in nude mice, including glioblastoma multiforme, rhabdomyosarcoma, leiomyosarcoma, colon and ovarian carcinoma[17, 62]. However, neither the antibodies nor VEGF itself had any effect on the *in vitro* growth of the tumor cells. In agreement with the hypothesis that inhibition of angiogenesis is the mechanism of tumor suppression, the density of microvessels was significantly lower in sections of tumors from antibody-treated animals as compared with controls[17, 62].

It has been shown that VEGF is a major mediator of the *in vivo* growth of human colon carcinoma cells in a nude mouse model of liver metastasis[62]. Treatment with anti-VEGF monoclonal antibodies resulted in a dramatic decrease in the number and size of metastases.

An independent verification of the hypothesis that VEGF action is necessary for tumorigenesis has been provided by the finding that retrovirus-mediated expression of a negative dominant Flk-1 mutant suppresses growth of glioblastoma cells *in vivo* [65].

B. ANGIOGENESIS ASSOCIATED WITH OTHER DISEASES

Diabetes mellitus, occlusion of the central retinal vein and prematurity with subsequent exposure to oxygen can all be associated with intraocular vascular proliferation[2, 66]. The new blood vessels may lead to vitreous hemorrhage, retinal detachment, neovascular glaucoma, and eventual blindness[2]. Diabetic retinopathy is the leading cause of blindness in the working population[2, 66]. All of these conditions are known to be associated with retinal ischemia[66]. As early as 1948, Michaelson proposed that the a key event in pathogenesis of such disorders is the release, by the ischemic retina, of diffusible angiogenic factor(s)[67]. Until now, the identity of such factor(s) has been unknown. VEGF, by virtue of its diffusible nature and hypoxia-inducibility, is an attractive candidate. Recently, elevations of VEGF levels in the aqueous and vitreous of eyes with proliferative retinopathy have been reported[18, 68, 69]. In agreement with these findings, *in situ* hybridization studies demonstrated up-regulation of VEGF mRNA in the retina of patients with proliferative retinopathies secondary to diabetes, central retinal vein occlusion, retinal detachment or intraocular tumors[70].

More direct evidence for the hypothesis that VEGF is a mediator of intraocular neovascularization has been provided in a primate model of iris neovascularization that closely mimics human disease and in a mouse model of retinopathy of prematurity. In the former, intraocular administration of anti-VEGF antibodies dramatically inhibited the neovascularization that follows occlusion of central retinal veins[71]. Likewise, soluble Flt-1 or Flk-1 fused to an IgG suppressed retinal angiogenesis in the mouse model[72].

It has been proposed that VEGF is involved in the angiogenesis associated with rheumatoid arthritis[73]. Levels of immunoreactive VEGF were high in the rheumatoid synovial fluid while they were very low or undetectable in the synovial fluid of patients affected by other forms of arthritis or by degenerative joint disease.

Also, elevations in VEGF expression in the skin have been recently described in three bullous disorders, bullous pemphigoid, erythema multiforme and dermatitis herpetiformis[74].

Interestingly, Lyttle et al have identified sequences having a significant homology to VEGF in the genome of *orf* virus, a parapoxvirus that affects goats, sheeps and occasionally humans[75]. Intriguingly, the lesions of goats and humans following *orf* virus infection are characterized by extensive microvascular proliferation in the skin, raising the possibility that the product of the viral VEGF-like gene is responsible for such lesions.

VEGF AS A POTENTIAL THERAPEUTIC AGENT

Growth factors able to promote the growth of new collateral vessels would be potentially of major therapeutic value for the treatment of disorders characterized by inadequate tissue perfusion. Intra-arterial or intra-muscular administration of rhVEGF$_{165}$ significantly augments perfusion and development of collateral vessels in a rabbit model where chronic hindlimb ischemia was created by removal of the femoral artery[15, 76]. Arterial gene transfer with a cDNA encoding VEGF$_{165}$ also led to revascularization of rabbit ischemic limbs[16]. In addition, the angiogenesis initiated by the administration of VEGF results in improved muscle function[77]. Similarly, it has been shown that both maximal flow velocity and maximal blood flow are significantly increased in ischemic limbs following VEGF administration[78]. Recent studies have shown that VEGF administration also leads to a recovery of normal endothelial reactivity in dysfunctional endothelium[79].

Furthermore, VEGF administration promotes increase in coronary blood flow in a dog model of coronary insufficiency[13]. In addition, extraluminal administration of as little as 2 μg of rhVEGF resulted in a significant increase in coronary blood flow in a pig model of chronic myocardial ischemia created by ameroid occlusion of the proximal circumflex artery[14]. In this model, VEGF treatment led to 2.6 fold decrease in the size of left ventricular infarct.

An additional potential therapeutic application of VEGF is the prevention of restenosis following percutaneous transluminal angioplasty (PTA). It has been proposed that damage to the endothelium is the crucial event triggering fibrocellular intimal proliferation[80]. Interestingly, VEGF administration accelerated re-endotheliazation and attenuated intimal hyperplasia in balloon-injured rat carotid artery[81].

CONCLUSIONS

The recent finding that targeted mutations inactivating the VEGF receptors genes result in a profound deficit in vasculogenesis, blood island formation, and early intrauterine death emphasizes the critical role played by the VEGF/VEGF-receptor system in the development the vascular system.

An intriguing possibility is that the VEGF protein or gene therapy with a VEGF cDNA may be used in the future to promote endothelial cell growth and collateral vessel formation. This would represent a novel therapeutic modality for conditions that frequently are refractory to conservative measures and unresponsive to pharmacological therapy.

The expression of VEGF mRNA in the vast majority of human tumors , the presence of of elevated levels of VEGF in the eyes of patients with proliferative retinopathies and in the synovial fluid of rheumatoid patients support the hypothesis that VEGF is a mediator of angiogenesis associated with various pathological conditions. The ability of anti-VEGF antibodies or soluble VEGF receptors to block tumor growth or neovascularization associated with ischemic retinal disorders provide more direct evidence for such hypothesis. Therefore, VEGF antagonists have the potential to be of therapeutic value for a variety of highly vascularized and aggressive malignancies as well as for other angiogenic disorders.

REFERENCES

1.   J. Folkman, and Y Shing. Angiogenesis. J. Biol. Chem. 267:10931 (1992)
2.   A. Garner. Vascular diseases. In: Pathobiology of ocular disease. A dynamic approach. A. Garner, and G.K. Klintworth, eds. 2nd Edition, Marcel Dekker, N.Y. pp 1625-1710 1994)
3.   N. Weidner, P. Semple, W. Welch, and J. Folkman, J. Tumor angiogenesis and metastasis. Correlation in invasive breast carcinoma. New Engl. J. Med. 324:1-(1991).
4.   S. Wakui, M. Furusato, H. Sasaki, A. Akiyama, I. Kinoshito, K. Asano, T. Tokuda, S. Aizawa, and S. Ushigome. Tumor angiogenesis in prostatic carcinoma with and without bone metastasis: a morphometric study. J. Pathol. 168:257 (1992).
5.   P. Macchiarini, G. Fontanini, M.J. Hardin, F. Squartini, and C.A. Angeletti. Relation of neovascularization to metastasis of non-small cell lung carcinoma. Lancet. 340:145 (1992).
6.   K.S. Smith-McCune, and N. Weidner, N. Demonstration and characterization of the angiogenic properties of cervical dysplasia. Cancer Res. 54:804 (1994).
7.   N. Ferrara, and W.J. Henzel. Pituitary follicular cells secrete a novel heparin-binding growth factor specific for vascular endothelial cells. Biochem. Biophys. Res. Commun. 161:851 (1989).
8.   N. Ferrara, K. Houck, L. Jakeman, and D.W. Leung. Molecular and biological properties of the vascular endothelial growth factor family of proteins. Endocr. Rev. 13:18 (1992).
9.   J. Plöuet, J. Schilling, and D. Gospodarowicz. Isolation and characterization of a newly identified endothelial cell mitogen produced by AtT20 cells. EMBO J. 8:3801 (1989).

10. G. Breier, U. Albrecht, S. Sterrer, and W. Risau. Expression of vascular endothelial growth factor during embryonic angiogenesis and endothelial cell differentiation. Development. 114:521 (1992).

11. H.S. Phillips, J. Hains, D.W. Leung, and N. Ferrara, N. Vascular endothelial growth factor is expressed in rat corpus luteum. Endocrinology. 127:965 (1990).

12. D. Shweiki, A. Itin, G. Neufeld, H. Gitay-Goren, and E. Keshet. Patterns of expression of vascular endothelial growth factor (VEGF) and VEGF receptors in mice suggest a role in hormonally regulated angiogenesis. J. Clin. Invest. 91:2235 (1993).

13. S. Banai, M.T. Jaktlish, M. Shou, D.F. Lazarous, M. Scheinowitz, S. Biro, S. Epstein, and E. Unger. Angiogenic-induced enhancement of collateral blood flow to ischemic myocardium by vascular endothelial growth factor in dogs. Circulation. 89:2183 (1994).

14. K. Harada, M. Friedman, J. Lopez, P.V. Prasad, M. Hibberd, J.D. Pearlman, F.W. Sellke, and M. Simons. Vascular endothelial growth factor improves coronary flow and myocardial function in chronically ischemic porcine hearts. Natute Medicine. In press.

15. S. Takeshita, L. Zhung, E. Brogi, M. Kearney, L.-Q. Pu, S. Bunting, N. Ferrara, J.F. Symes, and J.M. Isner. Therapeutic angiogenesis: A single intra-arterial bolus of vascular endothelial growth factor augments collateral vessel formation in a rabbit ischemic hindlimb model. J. Clin. Invest. 93:662 (1994).

16. S. Takeshita, L. Zheng, D. Cheng, R. Riessen, L. Weir, J.F. Symes, N. Ferrara, and J.M. Isner. Therapeutic angiogenesis following arterial gene transfer of vascular endothelial in a rabbit model of hindlimb ischemia. Proc. Natl. Acad. Sci. U.S.A. In press.

17. K.J. Kim, B. Li, J. Winer, M. Armanini, N. Gillett,, H.S. Phillips, and N. Ferrara, Inhibition of vascular endothelial growth factor-induced angiogenesis suppresses tumour growth in vivo. Nature. 362:841 (1993).

18. L.P. Aiello, R. Avery, R. Arrigg, B. Keyt, H. Jampel, S. Shah, L. Pasquale, H. Thieme, M. Iwamoto, J.E. Park, H. Nguyen, L.M. Aiello, N. Ferrara, and G.L.King. Vascular endothelial growth factor in ocular fluid of patients with diabetic retinopathy and other retinal disorders. N. Engl. J. Med. 331:1480 (1994).

19. D.W. Leung, G. Cachianes,W.-J. Kuang, D.V. Goeddel, and N. Ferrara. Vascular endothelial growth factor is a secreted angiogenic mitogen. Science. 246:1306 (1989).

20. G.D. Phillips, A.M. Stone, B.D. Jones, J.C. Schultz, R.A. Whitehead, and D.R. Knighton. Vascular endothelial growth factor (rhVEGF165) stimulates direct angiogenesis in the rabbit cornea. In Vivo. 8:961 (1995).

21. D.T. Connolly, D.M. Heuvelman, R. Nelson, J.V. Olander, B.L. Eppley, J.J. Delfino, N.R. Siegel, R.M. Leimgruber, and J. Feder. Tumor vascular permeability factor stimulates endothelial cell growth and angiogenesis. J. Clin. Invest. 84:1470 (1989).

22. R.F. Nicosia, S.V. Nicosia, and M. Smith. Vascular endothelial growth factor, platelet-derived growth factor and insulin-like growth factor-1 promote rat aortic angiogenesis in vitro. Am. J. Pathol. 145:1023 (1995).

23. M.S. Pepper, N. Ferrara, L. Orci, and R. Montesano. Vascular endothelial growth factor (VEGF) induces plasminogen activators and plasminogen activator inhibitor type 1 in microvascular endothelial cells. Biochem. Biophys. Res. Commun. 181:902 (1991).

24. E. Unemori, N. Ferrara, E.A. Bauer, and E.P Amento. Vascular endothelial growth factor induces interstitial collagenase expression in human endothelial cells. J.Cell. Physiol. 153:557 (1992).

25. S. Mandriota, R. Montesano, L. Orci, G. Seghezzi, J.-D. Vassalli, N. Ferrara, P. Mignatti, and M.S. Pepper. Vascular endothelial growth factor increases urokinase receptor expression in vascular endothelial cells. J. Biol. Chem. 270:9709 (1995).

26. P. Pekala, M. Marlow, D. Heuvelman, and D. Connolly. Regulation of hexose transport in aortic endothelial cells by vascular permeability factor and tumor necrosis factor-alpha, but not by insulin. J. Biol. Chem.265:18051 (1990).

27. M. Clauss, M. Gerlach, H. Gerlach, F. Brett, F. Wang, P.C. Familletti, Y.-C. Pan, J.V. Olander, D.T. Connolly, and D.T. Stern. D. Vascular permeability factor: a tumor-derived polypeptide that induces endothelial cell and monocyte procoagulant activity, and promotes monocyte migration. J. Exp. Med. 172:1535 (1990).

28. D.T. Connolly, J.V. Olander, D. Heuvelman, R. Nelson, R. Monsell, N. Siegel, B.L.Haymore, R. Leingruber, and J. Feder. Human vascular permeability factor. Isolation from U937 cells. J. Biol. Chem. 254:20017 (1989).

29. P.J. Keck, S.D. Hauser, G. Krivi, K. Sanzo, T. Warren, J. Feder, and D.T. Connolly. Vascular permeability factor, an endothelial cell mitogen related to platelet derived growth factor. Science. 246:1309 (1989).

30. H.F. Dvorak, V.S. Harvey, P. Estrella, L.F. Brown, J. McDonagh, and A.M. Dvorak. Fibrin containing gels induce angiogenesis: implications for tumor stroma generation and wound healing. Lab. Invest. 57:673 (1987).

31. J.E. Park, H. Chen, J. Winer, K. Houck, and N. Ferrara. Placenta growth factor. Potentiation of vascular endothelial growth factor bioactivity, in vitro and in vivo, and high affinity binding to Flt-1 but not to Flk-1/KDR. J. Biol. Chem. 269:25646 (1994).

32   D. Maglione, V. Guerriero, G. Viglietto, P. Delli-Bovi, and M.G. Persico. Isolation of a human placenta cDNA coding for a protein related to the vascular permeability factor. Proc. Natl. Acad. Sci. U.S.A. 88:9267 (1991).

33. D.D. Ku, J.K. Zaleski, S. Liu, and T. Brock,. Vascular endothelial growth factor induces EDRF-dependent relaxation of coronary arteries. Am. J. Physiol. 265:H586 (1993).

34. E. Tisher, R. Mitchell, T. Hartmann, M. Silva, D. Gospodarowicz, J. Fiddes, and J. Abraham. The human gene for vascular endothelial growth factor. J. Biol. Chem. 266:11947 (1991).

35. K.A. Houck,, N. Ferrara, J. Winer, G. Cachianes, B. Li, and D.W. Leung. The vascular endothelial growth factor family: Identification of a fourth molecular species and characterization of alternative splicing of RNA. Mol. Endocrinol. 5:1806 (1991).

36. K.A. Houck, D.W. Leung, A.M. Rowland, J. Winer, and N. Ferrara. Dual regulation of vascular endothelial growth factor bioavailability by genetic and proteolytic mechanisms. J. Biol. Chem. 267:26031 (1992).

37. J.E. Park, G.-A. Keller, and N. Ferrara. The vascular endothelial growth factor (VEGF) isoforms: Differential deposition into the subepithelial extracellular matrix and bioactivity of ECM-bound VEGF. Mol. Biol. Cell. 4:1317 (1993).

38. J. DiSalvo, M.L. Bayne, G. Conn, P.W. Kwok, P.G. Trivedi, D.D. Soderman, T.M. Palisi, K. Sullivan, and K.A. Thomas. Purification and characterization of a naturally occurring vascular endothelial growth factor·placenta growth factor heterodimer. J. Biol. Chem. 270:7717 (1995).

39. D. Shweiki, A. Itin, D. Soffer, and E. Keshet. Vascular endothelial growth factor induced by hypoxia may mediate hypoxia-initiated angiogenesis. Nature. 359:843 (1992).

40. D.T. Shima, A.P. Adamis, N. Ferrara, K.-T. Yeo, T.-K. Yeo, A. Allende, J. Folkman, and P.A. D'Amore. Hypoxic induction of vascular endothelial cell growth factors in the retina: Identification and characterization of vascular endothelial growth factor (VEGF) as the sole mitogen. Molec. Med. 2:64 (1995).

41. S. Banai, D. Shweiki, A. Pinson, M. Chandra, G. Lazarovici, and E. Keshet. Upregulation of vascular endothelial growth factor expression induced by myocardial ischemia: implications for coronary angiogenesis. Cardiovasc. Res. 28:1176 (1994).

42. M.A. Goldberg, and T.J. Schneider. Similarities between the oxygen-sensing mechanisms regulating the expression of vascular endothelial growth factor and erythropoietin. J. Biol. Chem. 269:4355 (1994).

43. A.P. Levy, N.S. Levy, S. Wegner, and M.A. Goldberg. Trancriptional regulation of the rat vascular endothelial growth factor gene by hypoxia. J. Biol. Chem. 1270:13333 (1995).

44. A. Madan, and P.T. Curtin. A 24-base pair sequence 3' to the human erythropoietin contains a hypoxia-responsive transcriptional enhancer. Proc. Natl. Acad. Sci. U.S.A. 90:3928 (1993).

45. D. Mukhopadhyay, L. Tsilokas, X.-M. Zhou, D. Foster, J.S. Brugge, and V.P. Sukhatme. Hypoxic induction of human vascular endothelial growth factor expression through c-Src activation. Nature. 375:577 (1995).

46. S. Frank, G. Hubner, G. Breier, M.T. Longaker, D.G. Greenhalgh, and S. Werner. Regulation of VEGF expression in cultured keratinocytes. Implications for normal and impaired wound healing. J. Biol. Chem. 270:12607 (1995).

47. L. Pertovaara, A. Kaipainen, T. Mustonen, A. Orpana, N. Ferrara, O. Saksela, O, and K. Alitalo. Vascular endothelial growth factor is induced in response to transforming growth factor-β in fibroblastic and epithelial cells. J. Biol. Chem. 269:6271 (1994).

48. J. Li, M.A. Perrella, J.C. Tsai, S.F. Yet, C.M. Hsieh, M. Yoshizumi, C. Patterson, W.O. Endego, F. Zhou, and M. Lee. Induction of vascular endothelial growth factor gene expression by interleukin-1 beta in rat aortic smooth muscle cells. J. Biol. Chem. 270:308 (1995).

49. K.P. Claffey, W.O. Wilkinson, and B.M. Spiegelman. Vascular endothelial growth factor. Regulation by cell differentiation and activated second messenger pathways. J. Biol. Chem. 267:16317 (1992).

50. L.B. Jakeman, J. Winer, G.L. Bennett, C.A. Altar, and N. Ferrara,. Binding sites for vascular endothelial growth factor are localized on endothelial cells in adult rat tissues. J. Clin. Invest. 89:244 (1992).

51. L.B. Jakeman, M. Armanini, H.S. Phillips, and N. Ferrara. Developmental expression of binding sites and mRNA for vascular endothelial growth factor suggests a role or this protein in vasculogenesis and angiogenesis. Endocrinology. 133:848 (1993).

52. C. deVries, J.A. Escobedo, H. Ueno, K. Houck, N. Ferrara, and L.T. Williams. The fms-like tyrosine kinase, a receptor for vascular endothelial growth factor. Science. 255:989 (1992).

53. B.I. Terman, M.D. Vermazen, M.E. Carrion, D. Dimitrov, D.C. Armellino, D. Gospodarowicz, and P. Bohlen. Identification of the KDR tyrosine kinase as a receptor for vascular endothelial growth factor. Biochem. Biophys. Res. Commun. 34:1578 (1992).

54. T. Quinn, K.G. Peters, C. deVries, N. Ferrara, and L.T. Williams. Fetal liver kinase 1 is a receptor for vascular endothelial growth factor and is selectively expressed in vascular endothelium. Proc. Natl. Acad. Sci. U.S.A. 90:7533 (1993).

55. J. Waltenberger, L. Claesson-Welsh, A. Siegbahn, M. Shibuya, and C.-H. Heldin. Different signal transduction properties of KDR and Flt1, two receptors for vascular endothelial growth factor. J. Biol. Chem. 269:26988 (1994).

56. L. Seetharam, N. Gotoh, Y. Maru, G. Neufeld, S. Yamaguchi, and M. Shibuya. A unique signal transduction pathway for the FLT tyrosine kinase, a receptor for vascular endothelial growth factor. Oncogene.10:135 (1995).

57. G.-H. Fong, J. Rassant, M. Gertenstein, and M. Breitman. Role of Flt-1 receptor tyrosine kinase in regulation of assembly of vascular endothelium. Nature. 376:66 (1995).

58. F. Shalabi, J. Rossant, T.P. Yamaguchi, M. Gertenstein, X.-F. Wu, M. Breitman, and A.C. Schuh. Failure of blood island formation and vasculogenesis in Flk-1 deficient mice. Nature. 376:62 (1995).

59. H.F. Dvorak, L.F. Brown, M. Detmar, and A.M. Dvorak. Vascular permeability factor/ vascular endothelial growth factor, microvascular permeability and angiogenesis. Am. J. Pathol. 146:1029 (1995).

60. K.H. Plate, G. Breier, H.A. Weich, and W. Risau, W. Vascular endothelial growth is a potential tumour angiogenesis factor in vivo. Nature. 359:845 (1992).

61. R.A. Berkman, M.J. Merrill, W.C. Reinhold, W.T. Monacci, A. Saxena, W.C. Clark, J.T. Robertson, I.U. Ali, and E.H. Oldfield. Expression of the vascular permeability/vascular endothelial growth factor gene in central nervous system neoplasms. J. Clin. Invest. 91:153 (1993).

62. R.S. Warren, H. Yuan, M.R. Matli, N. Gillett, and N. Ferrara. Regulation by vascular endothelial growth factor of human colon cancer tumorigenesis in a mouse model of experimental liver metastasis. J. Clin. Invest. 95:1789 (1995).

63. N. Ferrara, J. Winer, T. Burton, A. Rowland, M. Siegel, H.S. Phillips, T. Terrell, G.-A. Keller, and A.D. Levinson. Expression of vascular endothelial growth

factor does not promote transformation but confers a growth advantage in vivo to chinese hamster ovary cells. J. Clin. Invest. 91:160 (1993).

64. K.J. Kim, B. Li, K. Houck, J. Winer, and N. Ferrara. The vascular endothelial growth factor proteins: Identification of biologically relevant regions by neutralizing monoclonal antibodies. Growth Factors. 7:53 (1992).

65. B. Millauer, L.K. Shawver, K. Plate, W. Risau, and A. Ullrich. Glioblastoma growth is inhibited in vivo by a negative dominant Flk-1 mutant. Nature. 367:576 (1994).

66. A. Patz. Studies on retinal neovascularization. Invest. Ophthalmol. Vis. Sci. 19:1133 (1980).

67. I.C. Michaelson. The mode of development of the vascular system of the retina with some observations on its significance for certain retinal disorders. Trans. Ophthalmol. Soc. U.K.68:137 (1948).

68. A.P. Adamis, J.W. Miller, M.-T. Bernal, D. D'Amico, J. Folkman, T.-K. Yeo, and K.-T. Yeo. Increased vascular endothelial growth factor in the vitreous of eyes with proliferative diabetic retinopathy. Am. J. Ophthalmol. 118:445 (1994).

69. F. Malecaze, S. Clemens, V. Simorer-Pinotel, A. Mathis, P. Chollet, P. Favard, F. Bayard, and J. Ploüet. Detection of vascular endothelial growth factor mRNA and vascular endothelial growth factor-like activity in proliferative diabetic retinopathy. Arch. Ophthalmol. 112:1476 (1994).

70. J. Pe'er, D. Shweiki, A. Itin, I. Hemo, H. Gnessin, and E. Keshet. Hypoxia-inducedexpression of vascular endothelial growth factor (VEGF) by retinal cells is a common factor in neovascularization. Lab. Invest. 72:638 (1995).

71. A.P. Adamis, D.T. Shima, M. Tolentino, E. Gragoudas, N. Ferrara, J. Folkman, P.A. D'Amore, and J.W. Miller JW. Inhibition of VEGF prevents ocular neovascularization in a primate. Arch. Ophthalmol. In press.

72. L.P.Aiello, E.A. Pierce, E.D. Foley, H. Takagi, L. Riddle, H. Chen, N. Ferrara, G.L. King, and L.E. Smith. Suppression of retinal neovascularization in vivo by inhibition of vascular endothelial growth factor (VEGF) using soluble VEGF-receptor chimeric proteins. Proc. Natl. Acad. Sci. U.S.A. In press.

73. A.E. Koch, L. Harlow, G.K. Haines, E.P. Amento, E.N. Unemori, W.-L. Wong, R.M. Pope, and N. Ferrara. Vascular endothelial growth factor: a cytokine modulating endothelial function in rheumatoid arthritis. J. Immunol. 152:4149 (1994).

74. L.F. Brown, T.J. Harris, K.-T. Yeo, M. Stahle-Backdahl, R.W. Jackman, B. Berse, K. Tognazzi, H.F. Dvorak, and M. Detmar. Increased expression of vascular permeability factor (vascular endothelial growth factor) in bullous pemphigoid, dermatitis herpetiformis and erythema multiforme. J. Invest. Dermatol. 104:744 (1995).

75. D.J. Lyttle, K.M. Fraser, S.B. Flemings, A.A. Mercer, and A.J. Robinson. Homologs of vascular endothelial growth factor are encoded by the poxvirus orf virus. J. Virol. 68:84 (1994).

76. S. Takeshita, l.-Q. Pu, L.A. Stein, A.D. Sniderman, S. Bunting, N. Ferrara, J.M. Isner, and J.F. Symes, JF. Intramuscular administration of vascular endothelial growth factor induces dose-dependent collateral artery augmentation in a rabbit model of chronic limb ischemia. Circulation. 90:II228 (1994).

77. C.E. Walder, C.J. Errett, J. Ogez, H. Heinshon, S. Bunting, P. Lindquist, N. Ferrara, and G.R. Thomas. Vascular endothelial growth factor (VEGF) improves blood flow and function in a chronic ischemic hind-limb model. J. Cardiovasc. Pharmacol. In press.

78. C. Bauters, T. Asahara, L.P. Zheng, S. Takeshita, S. Bunting, N. Ferrara, J.F. Symes, and J.M. Isner. Physiologic assessment of augmented vascularity induced by VEGF in a rabbit ischemic hindlimb model. Am. J. Physiol. 267:H1263 (1994).

79. C. Bauters, T. Asahara, L.P. Zheng, S. Takeshita, S. Bunting, N. Ferrara, J.F. Symes, and J.M. Isner. Recovery of disturbed endothelium-dependent flow in collateral-perfused rabbit ischemic hindlimb following administration of VEGF . Circulation. 91:2793 (1995).

80. R.A. Graor, and B.H. Gray. Interventional treatment of peripheral vascular disease. In: Peripheral Vascular DiseasesJ.R. Young, R.A. Graor, J.W. Olin, and J.R. Bartholomew, eds. Mosby, St. Louis, MO. pp 111-33 (1991).

81. T. Asahara, C. Bauters, C. Pastore, S. Bunting, N. Ferrara, J.F. Symes, and J.M. Isner, JM. Local delivery of vascular endothelial growth factor accelerates re-endothelialization and attenuates intimal hyperplasia in balloon-injured rat carotid artery. Circulation. 91:2802 (1995).

SCATTER FACTOR AS A POTENTIAL TUMOR ANGIOGENESIS FACTOR

Eliot M. Rosen, Itzhak D. Goldberg

Department of Radiation Oncology
Long Island Jewish Medical Center
The Long Island Campus for Campus for Albert Einstein College of Medicine
270-05 76th Avenue
New Hyde Park, New York 11040

INTRODUCTION

SF is a stromal cell (fibroblast, smooth muscle)-derived protein that dissociates ("scatters") sheets of epithelium (Stoker et. al., 1987; Rosen et al., 1989). SF is now known to be identical to hepatocyte growth factor (HGF) (Weidner et al., 1991a; Bhargava et al., 1992), a serum mitogen that thought to function as an hepatotrophic factor (Miyazawa et al., 1989; Nakamura et al., 1989). SF is a heparin-binding glycoprotein consisting of a 60 kDa α-chain and a 30 kDa ß-chain (Gherardi et al., 1989; Rosen et al., 1990; Weidner et al., 1990). The α-chain contains an N-terminal hairpin loop and four "kringles" (disulfide looped structures that mediate protein:protein interactions). The ß-chain is homologous to serine proteases. SF has 38% amino acid sequence identity to the pro-enzyme plasminogen (Nakamura et al., 1989), but lacks protease activity (Rosen et al., 1990b) due to replacement of two essential amino acids at the catalytic triad of the ß-chain. SF is synthesized as a 728 amino acid precursor (preproSF); intracellular cleavage of a 31 amino acid signal peptide results in its secreted single-chain form (proSF), which is biologically inactive (Lokker et al., 1992). Extracellular cleavage of proSF at [494]arg-[495]val yields active two-chain SF. HGF activator, a novel serine protease homologous to coagulation factor XII (Hagemann factor), may be a physiologic cleavage enzyme for SF (Miyazawa et al., 1993). Plasminogen activators (uPA and tPA) can also cleave and activate proSF, but only at supraphysiologic concentrations (Naldini et al., 1992; Mars et al., 1993).

The SF receptor is the product of the *c-met* proto-oncogene (Bottaro et al., 1991), a tyrosine kinase (TK) expressed predominantly by epithelia (Gonzatti-Haces et al., 1988). The *c-met* receptor is a 190 kDa glycoprotein consisting of a 145 kDa membrane-spanning ß-chain and a 50 kDa α-chain that is expressed on the cell surface. The extracellular binding, transmembrane, and intracellular kinase, and non-catalytic phosphate acceptor domains are located on the ß-chain. Recent studies suggest that much of the signal transduction from the SF-activated *c-met* receptor occurs through the interaction of a novel tandem YV(H/N)V motif with the *src* homology-2 (SH2) domains of various intracellular signalling molecules (Ponzetto et al., 1994). Tyrosine phosphorylation at this site mediates the binding of *c-met* to phosphatidylinositol-3'-kinase, protein tyrosine phosphatase 2, phospholipase C-gamma, pp60[c-src], and the *grb2/hSos1* complex. Two receptor TKs related to *c-met*, *c-sea* and *Ron*, have been

Molecular, Cellular, and Clinical Aspects of Angiogenesis
Edited by Michael E. Maragoudakis, Plenum Press, New York, 1996

described (Ronsin et al., 1993; Huff et al., 1994). The ligand for the *Ron* receptor was recently identified as macrophage-stimulating protein (Wang et al., 1994), a kringle protein with 50% sequence identity to SF (Yoshimura et al., 1993).

SF stimulates three classes of biologic actions *in vitro*: motility, proliferation, and morphogenesis. Studies utilizing chimeric receptors indicate that all of these actions are mediated by the *c-met* TK (Weidner et al., 1993). In addition to cell dissociation, SF induces random motility of isolated epithelial cells, chemotactic migration, migration from carrier beads to flat surfaces, and invasion through extracellular matrix proteins (Rosen et al., 1990b,c, 1991a; Weidner et al., 1990; Bhargava et al., 1992; Li et al., 1994). SF stimulates the expression of both uPA and uPA receptor (uPAR) (Pepper et al., 1992; Grant et al., 1993; Rosen et al., 1994b). The net effect is to place uPA bound to uPAR on the cell surface. Receptor bound uPA on the cell surface is thought to mediate focal degradation of the extracellular matrix to clear a path for invading cells (Saksela and Rifkin, 1988). Thus, SF may "switch on" a program of cell activities for invasion. SF is mitogenic for various normal cell types, including epithelial cells, vascular endothelial cells, and melanocytes (Kan et al., 1991; Rubin et al., 1991; Halaban et al., 1992). SF is also a potent morphogen. It induces MDCK epithelial cells incubated in collagen I gels to organize into a network of branching tubules (Montesano et al., 1991). SF also induces mammary epithelial cells to form duct-like structures (Tsarfaty et al., 1992). Thus, SF activates specific programs of cell differentiation depending on the cell type and environment.

SF INDUCES AN ANGIOGENIC PHENOTYPE IN CULTURED ENDOTHELIUM

During the early stages of angiogenesis *in vivo*, endothelial cells (ECs) from pre-existing microvessels [eg., venules that lack a smooth muscle cell (SMC) layer] focally degrade the subendothelial basement membrane, migrate into the interstitium toward an angiogenic stimulus, and form capillary sprouts (Folkman, 1985). Sprouting ECs proximal to the migrating tip proliferate; subsequently, the EC sprouts organize into an anastamosing network of capillary tubes. Finally, these ECs synthesize new basement membrane. Adhesion of pericytes and formation of new basement membrane occur at the end of the angiogenic response (Folkman, 1985; Antonelli-Orlidge et al., 1989). Stimulation of EC motility, proliferation, and capillary-like tube formation *in vitro* are believed to correlate with the ability to induce angiogenesis *in vivo*, since these processes occur during new blood vessel formation (Folkman, 1985).

Vascular ECs express the *c-met* receptor and are biologically responsive to SF (Rosen et al., 1990b,c, 1991b; Bussolino et al., 1992; Grant et al., 1993; Naidu et al., 1994). SF is chemotactic to ECs and stimulates random motility, as demonstrated in assays using microwell Boyden chambers (Rosen et al., 1990b, 1991b). SF induces migration of ECs from microcarrier beads to flat culture surfaces (Rosen et al., 1990b,c). In chemoinvasion assays, SF induces penetration of ECs through porous filters coated with a basement membrane matrix (Matrigel) (Rosen et al., 1991b). Maximal migration and invasion of various EC types (human umbilical vein ECs (HUVEC), calf pulmonary artery ECs (CPAE), bovine aortic ECs (BAEC), and bovine brain ECs (BBEC)) are typically observed at SF concentrations of 2-20 ng/ml. SF also induces large increases in expression of uPA activity by EC cultures (Rosen et al., 1991b; Grant et al., 1993). Most of the SF-induced uPA activity is cell-associated rather than secreted. Most of the cell-associated uPA is bound to uPA receptor on the cell surface, where it is well-positioned to mediate focal degradation of extracellular matrix proteins, a prerequisite for invasion (Saksela and Rifkin, 1988).

In addition to motility, SF stimulates DNA synthesis and proliferation of some EC cell lines (Rubin et al., 1991; Morimoto et al., 1991). Capillary tube formation is an independent property of ECs, not directly related to motility or proliferation (Grant et al., 1989). When ECs are plated onto a basement membrane surface (Matrigel), they stop dividing, extend long cytoplasmic processes, and begin to organize into capillary-like tubes. SF stimulates capillary-like tube formation in HUVEC and BBEC cultures by up to (5-10)-fold, as determined by computerized digital image analysis of stained cultures (Rosen et al., 1991b; Grant et al., 1993). Taken together, these findings indicate that SF can induce most or all of the phenotypic characteristics expected of ECs undergoing angiogenesis.

SF INDUCES PROLIFERATION OF SMOOTH MUSCLE CELLS AND PERICYTES

Psoriasis is a chronic inflammatory skin disease characterized by epidermal hyperplasia and neovascularization in dermal papillae and papillary dermis. Cells of the microvessel wall (pericytes, ECs) in psoriatic placques stain positively for c-met protein (Grant et al., 1993), suggesting that these cell types are potential target cells for SF. Smooth muscle cells in tumor microvasculature also express immunoreactive SF (Joseph et al., 1995). Recruitment of SMCs and pericytes (which are regarded as microvascular SMCs) is an essential component of angiogenesis. These cells stabilize newly formed vessels, contributing to termination of angiogenesis (Antonelli-Orlidge et al., 1989). Therefore, it is logical that SF itself might induce the influx and/or proliferation of SMCs at the appropriate time during the angiogenic response.

We found that cultured bovine retinal pericytes express c-met mRNA, consistent with the finding that pericytes express immunoreactive c-met protein in vivo. We further found that SF stimulates the proliferation of bovine aortic SMC and bovine retinal pericytes in vitro (Rosen and Goldberg, In Press). The maximum stimulation of proliferation of these cell types was about 2-fold, as compared with 1.6-fold for bovine capillary ECs. These maximal values were observed at 20-100 ng/ml of SF. These findings are consistent with the presumed role of SMCs in angiogenesis and the presence of SMCs in new microvessels induced by SF (see below).

SF INDUCES ANGIOGENESIS *IN VIVO*

We used two assays, the mouse Matrigel assay and the rat cornea assay, to demonstrate the ability of SF to induce new blood vessel formation in vivo (Grant et al., 1993; Naidu et al., 1994). In the first assay, SF was mixed with 0.5 ml of Matrigel in the liquid state at 4°C. The Matrigel was injected subcutaneously into XID nude beige mice or C57/BL mice. At body temperature, Matrigel forms a solid gel, retaining the SF and allowing prolonged exposure of the surrounding tissues to it. Animals were sacrificed after ten days, and the ingrowth of blood vessels into the Matrigel plugs was quantitated using computerized digital image analysis of histologic sections stained with Masson's trichrome. Angiogenesis assessed at Day 10 increased in a dose-dependent manner from 2-200 ng/ml of SF, up to four to five times control values. Responses were similar in nude mice and C57/BL mice. Inflammatory responses were not observed in nude mice at any SF dose and were found only at supramaximal SF doses ($\geq$ 2000 ng/ml) in C57/BL mice. Histologic sections prepared at early times (Days 2-3), revealed many SMC/pericyte-like cells present in the Matrigel. Furthermore, at higher doses of SF, histologic sections prepared on Day 10 revealed SMCs in some of the newly formed vessels in the Matrigel (Grant et al., 1993). Thus, SF-induced angiogenesis may be mediated through effects on both ECs and SMCs.

In the second assay, SF was dissolved in Hydron polymer; and dried Hydron pellets were placed in surgically created pockets about 1.5 mm from the limbus of the avascular rat cornea. Animals were perfused with colloidal carbon and sacrificed after seven days. The growth of new vessels from the limbus toward the pellet was assessed. In these assays, SF induced dose-dependent corneal neovascularization in a fashion similar to that observed in the mouse Matrigel assay. Native mouse SF and recombinant human SF induced equal angiogeneic responses. Maximal responses were observed at 100-200 ng of SF and were similar in intensity to that induced by 150 ng of basic FGF (Grant et al., 1993). Antibodies against SF blocked SF-induced angiogenesis but did not affect FGF-induced angiogenesis. Inflammatory responses, assessed by F4/80 immunostaining to detect monocyte/macrophage infiltration, were observed only at supramaximal doses of SF. These findings suggest that SF is as potent an inducer of angiogenesis as basic FGF.

SF AS A POTENTIAL TUMOR ANGIOGENESIS FACTOR

Various experimental studies suggest that angiogenesis is a critical requirement for local growth and metastasis of solid tumors (Folkman, 1992). Recent clinical studies suggest that tumor angiogenesis, indicated by increased numbers of microvessels in the tumor stroma, is a strong independent indicator of poor prognosis in patients with invasive breast cancer (Weidner et al., 1991b, 1992; Bosari et al., 1992; Toi et al., 1993). While physiologic angiogenesis in normal adult tissues (as occurs in wound healing, corpus luteum formation, placental implantation) is tightly regulated spatially and temporally, tumor angiogenesis is characterized by persistent, abnormal neovascularization. A modest number of growth factors and cytokines are capable of inducing angiogenesis in various *in vivo* and *in vitro* assay systems [eg., angiogenin, FGFs, EGF/TGFα, IL-8, PDECGF (platelet-derived endothelial cell growth factor), SF, TNFα, TGFß, VEGF]. These angiogenic factors may be produced by tumor cells, host stromal cells (eg., fibroblasts, SMCs), or infiltrating leukocytes (eg., lymphocytes, macrophages, mast cells) (Polverini, 1989; Leek et al., 1994). The mechanisms leading to angiogenesis in human cancers, the specific factor(s) involved, and the cell types that produce them are not well established.

Both SF and *c-met* appear to be up-regulated and down-regulated in well co-ordinated patterns during normal developmental and repair processes (Sonnenberg et al., 1993; Matsumoto and Nakamura, 1993; Joannidis et al., 1994). On the other hand, SF is chronically overexpressed in tumors, including breast and bladder carcinomas (Rosen et al., 1994b; Yamashita et al., 1994; Joseph et al., 1995). Overproduction of SF in tumors may be due to the accumulation of specific SF-inducing proteins (see below). A high titer of SF in extracts of primary invasive breast carcinomas was found to be a powerful *independent* predictor of relapse and death (Yamashita et al., 1994). In patients with transitional cell bladder cancers, higher titers of SF were found in high grade, muscle invasive cancers than in low grade non-invasive or superficially invasive cancers (Joseph et al., *In Press*). Patients in the former category usually fare poorly in comparison with patients in the latter category.

Since both SF content and tumor angiogenesis are strong independent prognostic indicators for breast carcinoma, we may speculate that SF functions as a breast cancer angiogen. If, indeed, SF functions as a tumor angiogen, then future studies should reveal a strong correlation between tumor SF content and the quantitative extent of tumor angiogenesis. Such a correlation may not be exact, since, as described above, various other factors may contribute to tumor angiogenesis. Moreover, several naturally occurring protein factors, including thrombospondin (TSP1) and platelet factor-4, function as inhibitors of angiogenesis (see below). It is also likely that SF may interact additively or synergistically with other angiogenesis-inducing factors, such as VEGF. Since many of these factors are found in tumors, the angiogenic phenotype of the tumor may be determined by a balance of pro-angiogenic and anti-angiogenic factors.

TUMORAL SF PRODUCTION

In vitro, the major SF-producing cell types are stromal cells, including fibroblasts, vascular smooth muscle cells, glial cells, macrophages, endothelial cells, and T lymphocytes

(Stoker et al., 1987; Rosen et al., 1989, 1994a; Noji et al., 1990; Shiota et al., 1992: Yanagita et al., 1992; Naidu et al., 1994). Many of these cell types are present within tumor stroma and so may contribute to the accumulation of SF in solid tumors. Epithelial and carcinoma cells generally produce little or no SF, although a small number of SF-producing epithelial cell lines have been reorted (Adams et al., 1991; Tsao et al., 1993). These cell lines produce relatively modest titers of SF compared with stromal cells.

We observed positive immunostaining of carcinoma cells, bladder wall smooth muscle, fibroblasts, and vascular wall cells (SMCs and ECs) for SF in transitional cell carcinomas of bladder (Joseph *et al.*, 1995). However, as is true for most carcinoma cell types, transitional carcinoma cells do not produce any SF *in vitro* (Joseph et al., 1995). Since SF is a soluble cytokine, the *in vivo* staining of carcinoma cells may result from uptake of the factor rather than direct synthesis. Alternatively, the tumor cells may lose the ability to synthesize SF during adaptation to cell culture.

Tumor-associated macrophages are thought to contribute to tumor angiogenesis by secretion of angiogenic cytokines such as TNFα (Polverini, 1989). Macrophages and macrophage-like cells express SF mRNA and produce immunoreactive and biologically active SF (Noji et al., 1990; Wolf et al., 1991; Yanagita et al., 1992; Inaba et al., 1993; Rosen et al., 1994a). Our findings suggest that monocytes normally produce little SF but acquire the ability to produce SF when they undergo conversion to macrophages (Rosen and Goldberg, *In Press*). Since the level of SF production may be related to the degree of macrophage activation, we expect that tumor macrophages, which are highly activated, might be expected to exhibit even higher levels of SF production than normal tissue-derived macrophages.

SCATTER FACTOR-INDUCING FACTORS

SF production by some normal human fibroblasts (eg., lung and gingival fibroblasts) is stimulated by the pro-inflammatory cytokines IL-1α, IL-1ß, and TNFα (Tamura et al., 1993) and inhibited by the anti-inflammatory cytokine TGFß (Godha et al., 1992). These cytokines may be found in tumors, and thus may regulate tumoral SF accumulation. However, recent studies suggest that other SF-regulating proteins are present in tumors. Mouse and human mammary carcinoma cells, which do not produce SF, produce soluble protein factors distinct from IL-1 and TNFα that stimulate SF mRNA and protein expression by fibroblasts and other stromal cell types (Seslar et al., 1993, Rosen et al., 1994a,b). At least two distinct SF-inducing factors (SF-IFs) are produced, a high molecular weight (>30 kDa) heat-labile SF-IF and a low molecular weight (<30 kDa) heat-stable SF-IF. A 10-30 kDa heat stable SF-IF activity called "injurin" appears in the serum of rats within several hours of liver injury (Matsumoto et al., 1992). Injurin may be homologous to the <30 kDa tumor-produced SF-IF, but its purification and further characterization were not described.

We purified and characterized a new 12 kDa SF-IF protein from a high producer clone of *ras*-oncogene transformed NIH2/3T3 mouse cells (Rosen et al., 1994a). This *ras*-3T3 SF-IF protein has properties similar to the <30 kDa factor secreted by breast carcinoma cells and to injurin. It acts at physiologic concentrations (20-400 picomolar) to stimulate SF mRNA and protein expression by up to 4-6 fold. While breast carcinoma cells produce both high and low molecular weight SF-IFs (Seslar et al., 1993; Rosen et al., 1994a,b), transitional cell bladder carcinomas produce mostly >30 kDa SF-IF (Joseph et al., 1995). A similar >30 kDa SF-IF is present in extracts from bladder cancers and in urine from bladder cancer patients. On the other hand, little or no SF-inducing activity is present in urine from control patients. These findings suggest that SF is overproduced in tumors via a tumor:stroma interaction in which tumor cells secrete factors (SF-IFs) that induce stromal cell SF production.

LINKAGE OF ANGIOGENESIS AND TUMOR SUPPRESSORS

Thrombospondin-1 (TSP1) is a large multidomain adhesive glycoprotein of the extracellular matrix that mediates cell:matrix interactions (Lawler, 1993). A portion of the TSP1 molecule with angio-inhibitory activity (GP140) is encoded by a putative tumor suppressor gene that is inactivated when BHK hamster cells are chemically transformed into a tumorigenic phenotype (Rastinjad et al., 1989; Good et al., 1990). GP140 and native TSP1 inhibit EC proliferation, EC migration, and angiogenesis induced by basic FGF, TNFα, and macrophage conditioned medium (Good et al., 1990; Taraboletti et al., 1990). Preliminary findings from our laboratory indicate that TSP1 also blocks SF-induced migration of capillary ECs and SF-induced neovascularization in the rat cornea (Polverini, P.J., and Rosen, E.M., unpublished results). SF binds to TSP1, but it is not clear how this binding interaction modulates SF and TSP1 bioactivity.

The mechanism of TSP1's anti-angiogenic activity has not been fully defined. However, most of this activity is concentrated in several 15-20 amino acid regions in the procollagen homology domain and type I repeats of the central stalk of the molecule (Tolsma et al., 1993). Thus, a critical mutation corresponding to one of these sites might render TSP1 unable to block angiogenesis. TSP1 accumulates in human tumors, such as breast cancers (Wong et al., 1992; Clezardin et al., 1993). However, its function in these tumors and its potential interaction with SF is unknown; nor is it known if TSP1 mutants are present in some of these tumors.

The p53 anti-oncogene encodes a nuclear phosphoprotein that generally regulates cell growth by transactivating or repressing transcription (Tominaga et al., 1992; Mack et al., 1993). Loss of p53 function in Li-Fraumeni cancer syndrome fibroblasts results in decreased expression of TSP1 and acquisition of an angiogenic phenotype; whereas transfection of a wild-type p53 expression construct into fibroblasts lacking a functional p53 allele results in up-regulation of TSP1 and loss of the angiogenic phenotype (Dameron et al., 1994). These findings suggest that tumor suppressor gene products can directly or indirectly regulate the angiogenic phenotype.

P53 inhibits transcription of the *bcl-2* proto-oncogene, whose protein product suppresses apoptosis (programmed cell death). P53 also induces transcription of *bax*, a gene whose protein product binds to *bcl-2* protein and inhibits its biologic activity (Miyashita et al., 1994). Both *bcl-2* overexpression and SF can overcome apoptosis associated with detachment of epithelial cells from their substratum (Frisch and Francis, 1994). Detachment of epithelial and vascular endothelial cells from the underlying basement membrane are early steps in tumor cell invasion and in angiogenesis, respectively. We have proposed several mechanisms by which loss of tumor suppressor gene function may promote SF-mediated invasion and angiogenesis in tumors (Rosen and Goldberg, 1995). For example, production of SF-IF(s) by tumor cells may be activated by a tumor suppressor gene mutation(s) (eg., p53), which leads to loss of normal transcriptional repression. Alternatively, or in addition, such mutations might lead to up-regulation of *c-met* expression, rendering the carcinoma cells better able to respond to SF.

CONCLUSIONS

Scatter factor (hepatocyte growth factor) is a stromal cell-derived cytokine that stimulates motility, proliferation, and morphogenesis of epithelia. These responses are transduced through the *c-met* proto-oncogene product, a transmembrane tyrosine kinase that functions as the SF receptor. SF is a potent angiogenic molecule, and its angiogenic activity is mediated primarily through direct actions on endothelial cells. These include stimulation of cell motility, proliferation, protease production, invasion, and organization into capillary-like tubes. SF is chronically overexpressed in tumors, suggesting that it may function as a tumor angiogenesis factor. SF production in tumors may be due, in part, to an abnormal tumor:stroma interaction, in which the tumor cells secrete factors (SF-IFs) that stimulate SF production by tumor-associated stromal cells. Recent studies suggest a linkage between tumor suppressors (anti-oncogenes) and inhibition of angiogensis. We hypothesize that tumor suppressor gene mutations may contribute to activation of an SF-IF → SF → c-met pathway, leading to an invasive and angiogenic tumor phenotype. Modulation of this pathway may, ultimately, provide clinically useful methods of enhancing or inhibiting angiogenesis.

ACKNOWLEDGEMENTS

Supported in part by the USPHS (CA64869). Dr. Rosen is an Established Investigator of the American Heart Association (AHA 90-195).

REFERENCES

Adams JC, Furlong RA, Watt FM. Production of scatter factor by ndk, a strain of epithelial cells, and inhibition of scatter factor activity by suramin. *J Cell Sci* 98: 385-394, 1991.

Antonelli-Orlidge A, Smith SR, D'Amore PA. Influence of pericytes on capillary endothelial cell growth. *Am Rev Resp Dis* 140: 1129-1131, 1989.

Bhargava M, Joseph A, Knesel J, Halaban R, Li Y, Pang S, Goldberg I, Setter E, Donovan MA, Zarnegar R, Michalopoulos GA, Nakamura T, Faletto D, Rosen EM. Scatter factor and hepatocyte growth factor: Activities, properties, and mechanism. *Cell Growth & Differen* 3:11-20, 1992.

Bosari S, Lee AK, DeLellis RA, Wiley BD, Heatley GJ, Silverman ML. Microvessel quantitation and prognosis in invasive breast carcinoma. *Hum Pathol* 23: 755-761, 1992.

Bottaro DP, Rubin JS, Faletto DL, Chan AM-L, Kmiecik TE, Vande Woude GF, Aaronson SA. Identification of the hepatocyte growth factor receptor as the *c-met* proto-oncogene product. *Science* 251: 802-804, 1991.

Bussolino F, DiRenzo MF, Ziche M, Bocchieto E, Olivero M, Naldini L, Gaudino G, Tamagnone L, Coffer A, Comoglio PM. Hepatocyte growth factor is a potent angiogenic factor which stimulates endothelial cell motility and growth. *J Cell Biol* 119: 629-641, 1992.

Clezardin P, Frappart L, Clerget M, Pechoux P, Delmas PD. Expression of thrombospondin and its receptors (CD36 and CD51) in normal, hyperplastic, and neoplastic human breast. *Cancer Res* 53: 1421-1430, 1993.

Dameron KM, Volpert OV, Tainsky MA, Bouck N. Control of angiogenesis in fibroblasts by p53 regulation of thrombospondin-1. *Science* 265: 1582-1590, 1994.

Folkman J. Tumor angiogenesis. *Adv Cancer Res* 43: 175-203, 1985.

Folkman J. The role of angiogenesis in tumor growth. *Semin Cancer Biol* 3: 65-71, 1992.

Frisch SM, Francis H. Disruption of epithelial cell-matrix interactions induces apoptosis. *J Cell Biol* 124: 619-626, 1994.

Gherardi E, Gray J, Stoker M, Perryman M, Furlong R. Purification of scatter factor, a fibroblast-derived basic protein which modulates epithelial interactions and movement. *PNAS USA* 86: 5844-5848, 1989.

Gohda E, Matsunaga T, Kataoka H, Yamamoto I. TGF-ß is a potent inhibitor of hepatocyte growth factor secretion by human fibroblasts. *Cell Biol Int Reports* 16: 917-926, 1992.

Gonzatti-Haces M, Seth A, Park M, Copeland T, Oroszlan S, Vande Woude GF. Characterization of the TPR-MET oncogene p65 and the MET protooncogene p140 protein tyrosine kinases. *PNAS USA* 85: 21-25, 1988.

Good DJ, Polverini PJ, Rastinejad F, Le Beau MM, Lemons RS, Frazier WA, Bouck NP. A tumor supressor-dependent inhibitor of angiogenesis is immunologically and functionally indistinguishable from a fragment of thrombospondin. *PNAS USA* 87: 6624-6628, 1990.

Grant DS, Tashiro KI, Segui-Real B, Yamada Y, Martin GR, Kleinman HK. Two different laminin domains mediate the differentiation of human endothelial cells into capillary-like structures in vitro. *Cell* 58:933-943, 1989.

Grant DS, Kleinman HK, Goldberg ID, Bhargava M, Nickoloff BJ, Polverini P, Rosen EM. Scatter factor induces blood vessel formation *in vivo*. *PNAS USA* 90: 1937-1941, 1993.

Halaban R, Rubin J, Funasaka Y, Cobb M, Boulton T, Faletto D, Rosen E, Chan A, Yoko K, White W, Cook C, Moellmann G. Met and hepatocyte growth factor/scatter factor signal transduction in normal melanocytes and melanoma cells. *Oncogene* 7: 2195-2206, 1992.

Huff JL, Jelinek MA, Borgman CA, Lansing TJ. The protooncogene c-sea encodes a transmembrane protein-tyrosine kinase related to the Met/HGF/SF receptor. *PNAS USA* 90: 6140-6144, 1994.

Inaba, M., Koyama, H., Hino, M., Okuno, S., Terada, M., Nishizawa, Y., Nishino, T., Morii, H. Regulation of release of hepatocyte growth factor from human promyelocytic leukemia cells, HL-60, by 1,25-dihydroxyvitamin D3, 12-O-tetradecanoylphorbol 13-acetate, and dibutyryl cyclic adenosine monophosphate. *Blood* 82: 53-59, 1993.

Joannidis M, Spokes K, Nakamura T, Faletto D, Cantley LG. Regional expression of hepatocyte growth factor/c-met in experimental renal hypertrophy and hyperplasia. *Am J Physiol* 267: F231-F236, 1994.

Joseph A, Weiss GH, Jin L, Fuchs A, Chowdhury S, O'Shaughnessy P, Goldberg ID, Rosen EM. Expression of scatter factor in human bladder carcinoma. *J Natl Cancer Inst* 87: 372-377, 1995.

Kan M, Zhang GH, Zarnegar R, Michalapoulos G, Myoken Y, McKeehan WL, Stevens JL. Hepatocyte growth factor-hepatopoietin A stimulates the growth of rat proximal tubule epithelial cells (rpte), rat non-parenchymal liver cells, human melanoma cells, mouse keratinocytes, and stimulates anchorage-independent growth of SV40-transformed rpte. *Biochem Biophys Res Commun* 174: 331-337, 1991.

Lawler J. Thrombospondin. *In:* "Guidebook to the extracellular matrix and adhesion proteins", Kreis T, Vale R, eds., Oxford University Press, Inc., New York, 1993, pp. 95-96.

Leek RD, Harris AL, Lewis CE. Cytokine networks in solid human tumors: regulation of angiogenesis. *J Leuk Biol* 56: 423-435, 1994.

Li Y, Bhargava MM, Joseph A, Jin L, Rosen EM, Goldberg ID. The effect of scatter factor and hepatocyte growth factor on motility and morphology of non-tumorigenic and tumor cells. *In Vitro Cell Dev Biol* 30A: 105-110, 1994.

Lokker NA, Mark MR, Luis EA, Bennett GL, Robbins KA, Baker JB, Godowski PJ. Structure-function analysis of hepatocyte growth factor: Identification of variants that lack mitogenic activity yet retain high affinity receptor binding. *EMBO J* 11: 2403-2410, 1992.

Mack DH, Vartikar J, Pipas JM, Laimins LA. Specific repression of TATA-mediated but not initiator-mediated transcription. *Nature* 363: 281-283, 1993.

Mars WM, Zarnegar R, Michalopoulos GK. Activation of hepatocyte growth factor by the plasminogen activators uPA and tPA. *Am J Pathol* 143: 949-958, 1993.

Matsumoto K, Nakamura T. Roles of HGF as a pleiotropic factor in organ regeneration. *In:* "Hepatocyte Growth Factor-Scatter Factor and the *c-Met* Receptor", Goldberg ID, Rosen EM, eds., Birkhauser-Verlag, Basel, 1993, pp. 225-250.

Matsumoto K, Tajima H, Hamanoue M, Kohno S, Kinoshita T, Nakamura T. Identification and characterization of "injurin", an inducer of expression of the gene for hepatocyte growth factor. *PNAS USA* 89: 3800-3804, 1992b.

Miyashita T, Krajewski S, Krajewska M, Wang HG, Lin HK, Liebermann DA, Hoffman B, Reed JC. Tumor suppressor p53 is a regulator of bcl-2 and bax gene expression in vitro and in vivo. *Oncogene* 9: 1799-1805, 1994.

Miyazawa K, Tsubouchi H, Naka D, Takahashi K, Okigaki M, Arakaki N, Nakayama S, Hirono S, Sakiyama O, Gohda E, Daikuhara Y, Kitamura N. Molecular cloning and sequence analysis of cDNA for human hepatocyte growth factor. *Biochem Biophys Res Commun* 163: 967-973, 1989.

Miyazawa K, Shimomura T, Kitamura A, Kondo J, Morimoto Y, Kitamura N. Molecular cloning and sequence analysis of the cDNA for a human serine protease responsible for activation of hepatocyte growth factor. Structural similarity of protease precursor to blood coagulation factor XII. *J Biol Chem* 268: 10024-10028, 1993.

Moghul A, Lin L, Beedle A, Kanbour-Shakir A, DeFrances MC, Liu Y, Zarnegar R. Modulation of c-MET proto-oncogene (HGF receptor) mRNA abundance by cytokines and hormones: evidence for rapid decay of the 8 kb c-MET transcript. *Oncogene* 9: 2045-2052, 1994.

Montesano R, Matsumoto K, Nakamura T, Orci L. Identification of a fibroblast-derived epithelial morphogen as hepatocyte growth factor. *Cell* 67: 901-908, 1991.

Morimoto, A., Okamura, K., Hamanaka, R., Sato, Y., Shima, N., Higashio, K., Kuwano, M. Hepatocyte growth factor modulates migration and proliferation of microvascular endothelial cells in culture. *Biochem Biophys Res Commun* 179: 1042-1049, 1991.

Naidu YM, Rosen EM, Zitnik R, Goldberg I, Park M, Naujokas M, Polverini PJ, Nickoloff BJ. Role of scatter factor in the pathogenesis of AIDS-related Kaposi's sarcoma. *PNAS USA* 91: 5281-5285, 1994.

Naldini L, Tamagnone L, Vigna E, Sachs M, Hartmann L, Birchmeier W, Daikuhara Y, Tsubouchi H, Blasi F, Comoglio PM. Extracellular proteolytic cleavage by urokinase is required for activation of hepatocyte growth factor/scatter factor. *EMBO J* 11: 4825-4833, 1992.

Nakamura T, Nishizawa T, Hagiya M, Seki T, Shimonishi M, Sugimura A, Shimizu S. Molecular cloning and expression of human hepatocyte growth factor. *Nature* 342: 440-443, 1989.

Noji S, Tashiro K, Koyama E, Nohno T, Ohyama K, Taniguchi S, Nakamura T. Expression of hepatocyte growth factor gene in endothelial and Kupffer's cells of damaged rat livers as revealed by *in situ* hybridization. *Biochem Biophys Res Commun* 173: 42-47, 1990.

Pepper MS, Matsumoto K, Nakamura T, Orci L, Montesano R. Hepatocyte growth factor increases urokinase-type plasminogen activator (u-PA) and u-PA receptor expression in Madin-Darby canine kidney epithelial cells. *J Biol Chem* 267: 20493-20496, 1992.

Polverini PJ. Macrophage-induced angiogenesis: A review. *Cytokines* Vol 1: 54-73, S. Karger, Basel, 1989.

Ponzetto C, Bardelli A, Zhen Z, Maina F, Zonca P, Giordano S, Graziani A, Panayoyou G, Comoglio PM. A multifunctional docking site mediates signalling and transformation by the HGF/SF receptor family. *Cell* 77: 261-271, 1994.

Ramakrishnan S, Xu FJ, Brandt SJ, Niedel JE, Bast RC Jr, Brown EL. Constitutive production of macrophage colony-stimulating factor by human ovarian and breast cancer cell lines. *J Clin Invest* 83: 921-926, 1989.

Rastinejad F, Polverini PJ, Bouck NP. Regulation of the activity of a new inhibitor of angiogenesis by a cancer suppressor gene. *Cell* 56: 345-355, 1989.

Ronsin C, Muscatelli F, Mattei M-G, Breathnach R. A novel putative receptor tyrosine kinase of the met family. *Oncogene* 8: 1195-1202, 1993.

Rosen EM, Goldberg ID. Scatter factor and angiogenesis. *Adv Cancer Res* 67, In Press.

Rosen EM, Goldberg ID, Kacinski BM, Buckholz T, Vinter DW. Smooth muscle releases an epithelial cell scatter factor which binds to heparin. *In Vitro Cell Dev Biol* 25: 163-173, 1989.

Rosen EM, Meromsky L, Setter E, Vinter DW, Goldberg ID. Smooth muscle-derived factor stimulates mobility of human tumor cells. *Invasion Metast* 10: 49-64, 1990a.

Rosen, EM, Meromsky L, Setter E, Vinter DW, Goldberg ID. Purification and migration-stimulating activities of scatter factor. *Proc Soc Exp Biol Med* 195: 34-43, 1990b.

Rosen EM, Meromsky L, Setter E, Vinter DW, Goldberg ID. Quantitation of cytokine-stimulated migration of endothelium and epithelium by a new assay using microcarrier beads. *Exp Cell Res* 186: 22-31, 1990c.

Rosen EM, Goldberg ID, Liu D, Setter E, Donovan MA, Bhargava M, Reiss M, Kacinski BM. Tumor necrosis factor stimulates epithelial tumor cell motility. *Cancer Res* 57: 5315-5321, 1991a.

Rosen EM, Grant D, Kleinman H, Jaken S, Donovan MA, Setter E, Luckett PM, Carley W. Scatter factor stimulates migration of vascular endothelium and capillary-like tube formation. *In*: "Cell Motility Factors", Goldberg ID, Rosen EM, eds., Birkhauser-Verlag, Basel, 1991b, pp 76-88.

Rosen EM, Jaken S, Carley W, Setter E, Bhargava M, Goldberg ID. Regulation of motility in bovine brain endothelial cells. *J Cell Physiol* 146: 325-335, 1991c.

Rosen EM, Joseph A, Jin L, Rockwell S, Elias JA, Knesel J, Wines J, McClellan J, Kluger MJ, Goldberg ID, Zitnik R. Regulation of scatter factor production via a soluble inducing factor. *J Cell Biol* 127: 225-234, 1994a.

Rosen EM, Knesel J, Goldberg ID, Bhargava M, Joseph A, Zitnik R, Wines J, Kelley M, Rockwell S. Scatter factor modulates the metastatic phenotype of the EMT6 mouse mammary tumor. *Int J Cancer* 57: 706-714, 1994b.

Rubin JS, Chan AM-L, Bottaro DP, Burgess WH, Taylor WG, Cech AC, Hirschfield DW, Wong J, Miki T, Finch PW, Aaronson SA. A broad spectrum human lung fibroblast-derived mitogen is a variant of hepatocyte growth factor. *PNAS USA* 88: 415-419, 1991.

Saksela O, Rifkin DB. Cell-associated plasminogen activation: regulation and physiologic functions. *Annu Rev Cell Biol* 4: 93-126, 1988.

Seslar SP, Nakamura T, Byers SW. Regulation of fibroblast hepatocyte growth factor/scatter factor expression by human breast carcinoma cell lines and peptide growth factors. *Cancer Res* 53: 1233-1238, 1993.

Shiota G, Rhoads DB, Wang TC, Nakamura T, Schmidt EV. Hepatocyte growth factor inhibits growth of hepatocellular carcinoma cells. *PNAS USA* 89: 373-377, 1992.

Sonnenberg E, Meyer D, Weidner KM, Birchmeier C. Scatter factor/hepatocyte growth factor and its receptor, the c-met tyrosine kinase, can mediate a signal exchange between between mesenchyme and epithelia during mouse development. *J Cell Biol* 123: 223-235, 1993.

Stoker M, Gherardi E, Perryman M, Gray J. Scatter factor is a fibroblast-derived modulator of epithelial cell mobility. *Nature* 327: 238-242, 1987.

Tamura M, Arakaki N, Tsoubouchi H, Takada H, Daikuhara Y. Enhancement of human hepatocyte growth factor production by interleukin-1 alpha and -1 beta and tumor necrosis factor-alpha by fibroblasts in culture. *J Biol Chem* 268: 8140-8145, 1993.

Taraboletti G, Roberts D, Liotta LA, Giavazzi R. Platelet thrombospondin modulates endothelial cell adhesion, motility, and growth: a potential angiogenesis regulatory factor. *J Cell Biol* 111: 765-772, 1990.

Toi M, Kashitani J, Tominaga T. Tumor angiogenesis is an independent prognostic indicator in primary breast carcinoma. *Int J Cancer* 55: 371-374, 1993.

Tolsma SS, Volpert OV, Good DJ, Frazier WA, Polverini PJ, Bouck N. Peptides derived from two separate domains of the matrix protein thrombospondin-1 have antiangiogenic activity. *J Cell Biol* 122: 497-511, 1993.

Tominaga O, Hamelin R, Remvikos Y, Salmon RJ, Thomas G. P53 from basic research to clinical applications. *Crit Rev Oncogenesis* 3: 257-282, 1992.

Tsao MS, Zhu H, Giaid A, Viallet J, Nakamura T, Park M. Hepatocyte growth factor/scatter factor is an autocrine factor for human bronchial epithelial and lung carcinoma cells. *Cell Growth Differen* 4: 571-579, 1993.

Tsarfaty I, Resau JH, Rulong S, Keydar I, Faletto D, Vande Woude GF. The *met* proto-oncogene receptor and lumen formation. *Science* 257: 1258-1261, 1992.

Wang MH, Ronsin C, Gesnel M-C, Coupey L, Skeel A, Leonard EJ, Breathnach R. Identification of the ron gene product as the receptor for the human macrophage stimulating protein. *Science* 266: 117-119, 1994.

Weidner KM, Behrens J, Vandekerckhove J, Birchmeier W. Scatter factor: Molecular characteristics and effect on invasiveness of epithelial cells. *J Cell Biol* 111: 2097-2108, 1990.

Weidner KM, Arakaki N, Vandekereckhove J, Weingart S, Hartmann G, Rieder H, Fonatsch C, Tsubouchi H, Hishida T, Daikuhara Y, Birchmeier W. Evidence for the identity of human scatter factor and human hepatocyte growth factor. *PNAS USA* 88: 7001-7005, 1991a.

Weidner N, Semple JP, Welch WR, Folkman J. Tumor angiogenesis and metastasis - correlation in invasive breast carcinoma. *New Engl J Med* 324: 1-8, 1991b.

Weidner N, Folkman J, Pozza F, Bevilaqua P, Allred EN, Moore DH, Meli S, Gasparini G. Tumor angiogenesis: a new significant and independent prognostic indicator in early stage breast carcinoma. *JNCI* 84: 1875-1887, 1992.

Weidner KM, Sachs M, Birchmeier W. The met receptor tyrosine kinase transduces motility, proliferation, and morphogenic signals of scatter factor/hepatocyte growth factor in epithelial cells. *J Cell Biol* 121: 145-154, 1993.

Wolf, H.K., Zarnegar, R., Michalopoulos, G.K. Localization of hepatocyte growth factor in human and rat tissues: an immunohistochemical study. *Hepatology* 14: 488-494, 1991.

Wong SY. Purdie AT, Han P. Thrombospondin and other possible related matrix proteins in malignant and benign disease. An immunohistochemical study. *Am J Pathol* 140: 1473-1482, 1992.

Yamashita J, Ogawa M, Yamashita S, Nomura K, Kuramoto M, Saishoji T, Sadahito S. Immunoreactive hepatocyte growth factor is a strong and independent predictor of recurrence and survival in human breast cancer. *Cancer Res* 54: 1630-1633, 1994.

Yanagita K, Nagaike M, Ishibashi H, Niho Y, Matsumoto K, Nakamura T. Lung may have endocrine function producing hepatocyte growth factor in response to injury of distal organs. *Biochem Biophys Res Commun* 182: 802-809, 1992.

Yoshimura T, Yuhki N, Wang MH, Skeel A, Leonard EJ. Cloning, sequencing, and expression of human macrophage stimulating protein (MSP, MST1) confirms MSP as a member of the family of kringle proteins and locates the MSP gene on chromosome 3. *J Biol Chem* 268: 15461-15468, 1993.

THE ROLE OF THROMBIN IN ANGIOGENESIS

Michael E. Maragoudakis

University of Patras Medical School, Dept. of Pharmacology

261 10 Patras, Greece

INTRODUCTION

A relation between blood coagulation and cancer growth and metastasis was first observed by Trousseau in 1872 (Trousseau, 1872). Many investigators have subsequently verified the frequency of spontaneous coagulation in cancerous patients and provided clinical and pathological data supporting the notion of a systemic activation of the blood coagulation cascade in patients with cancer. For example, measurements of the circulating fibrinopeptides have shown that patients with cancer have inappropriately high intravascular coagulation and firbinolysis (Rickles & Edwards, 1983). In addition, it has been shown that tumor cells interact with platelets, leucocytes and endothelial cells as well as the thrombin and plasmin generating systems, all of which influence clot formation (Silverstein & Nachman, 1992). Furthermore, some tumors express the transmembrane protein tissue factor, which when exposed to circulating factor VII activates factor X leading to generation of thrombin and the formation of fibrin, which is throught to be important in tumor growth. Other tumors release mucin and cystein proteases, which activate factor X (Sloan et al., 1986).

These findings provide considerable insight for the molecular mechanisms involved in Trousseau's spontaneous coagulation induced by cancer. However, they can not explain the mechanism by which the generation of thrombin may influence tumor progression and metastasis.

Indeed more recent evidence show that thrombin-treated tumor cells showed a drastic increase in their ability for pulmonary metastasis (up to 156-fold) and thrombin *in vivo* enhances dramatically the number of colonies formed by injection of B16 melanoma cells in mice. In fact, following successful experiments with anti-coagulants in animal tumor models, several clinical trials are in progress using warfarin in human cancers (Tapparelli et al., 1993).

The mechanism of anti-coagulants in the suppression of cancer growth and metastasis is not clear. Thrombin has a multitude of effects both enzymatic and cellular actions on a variety of cell types (platelets, neurones, tumor, heart, smooth muscle, leucocytes and endothelial cells) (Tapparelli et al., 1993). None of all these known effects of thrombin can provide the basis for an explanation for the role of thrombin in tumor growth and metastasis.

Molecular, Cellular, and Clinical Aspects of Angiogenesis
Edited by Michael E. Maragoudakis, Plenum Press, New York, 1996

A plausible explanation may be our recent finding that thrombin is a potent promoter of angiogenesis. This effect of thrombin is independent of clotting activity (Tsopanoglou, et al., 1993). Angiogenesis has been shown by many investigators to play a pivotal role in tumor growth and metastasis (Folkman, 1985 and 1995). Our finding, therefore, of the angiogenic effect of thrombin, suggests a new role of thrombin in the metastatic spread of cancer, which is based on its ability to promote tumor angiogenesis. In addition, thrombin may play a role in wound healing and inflammation as well as other conditions, where angiogenesis is thought to be involved (Maragoudakis, 1995). In this paper will be summarized the results on the angiogenic action of thrombin in three model systems of angiogenesis and these findings will be discussed in relation to possible therapeutic applications.

EXPERIMENTAL PROCEDURES

The chick chorioallantoic membrane (CAM) system initially described by Folkman (1985) and modified as previously reported (Maragoudakis et al., 1988) was used. The CAM is a highly vascularized transparent tissue, which allows for morphological evaluation of vascular density, and it is widely used as a model system for studying angiogenesis (Auerbach et al., 1991). We have developed techniques where by measuring the rate of basement membrane collagen biosynthesis we can evaluate quantitatively angiogenesis (Maragoudakis et al., 1988). The rationale for this is based on the fact that new blood vessels, at least for the early stages of their formation, are bare endothelial cell tubes supported by the basement membrane, which is formed by these proliferating endothelial cells. It is logical, therefore, to expect a correlation between the rate of basement membrane synthesis and the extent of angiogenesis. Both the endothelial cell proliferation and the biosynthesis of new basement membrane are extremely slow process in the absence of angiogenesis. Basement membrane collagen type IV is the major component of basement membrane, which can be measured as collagenase digestible protein with low background activity in the absence of angiogenesis. Thus measuring newly-synthesized collagenase-digestible proteins provides for a sensitive and quantitative index of angiogenesis. This method has been validated by computer-assisted image analysis and also by the morphological method of Harris-Hooker by counting blood vessels (Maragoudakis et al., 1995).

Another system used in these studies was the *in vitro* Matrigel system of angiogenesis as described by Grant et al., 1989 and Haralabopoulos et al., 1994. In this system human umbilical cord endothelial cells (HUVECs) plated on Matrigel form tube-like structures within 8 hours of incubation. This allows for studies on the expression of angiogenic phenotype of endothelial cells *in vitro*.

For the *in vivo* results the method described by Passaniti et al. (1992) and Haralabopoulos et al. (1994) was used. This involves injection of Matrigel sc in mice and observing the formation of new vessels within the transparent Matrigel plug. The advantage of this method is that Matrigel, which is liquid at $0°$-$5°C$, can be easily injected and forms a gel at $35°C$. This does not require surgery, minipumps, sponges etc. Because Matrigel is produced by EHS tumors in C57BL6N mice, it does not elicit immune response, since the same mice are used for injection. In the absence of angiogenic agents the Matrigel plug appears clear at the end of the experiment (about 2 weeks).

RESULTS AND DISCUSSION

Alpha-thrombin is a glycosylated trypsin-like serine proteinase generated from the circulating plasma prothrombin. Upon antocatalytic and factor Xa cleavage the functional two chain molecule of a-thrombin is generated consisting of 36 and 259 amino acid residue for A & B chain respectively for human thrombin. A & B chains are covalently interconnected by a disulfide bridge, the B chain having the catalytic domain, which has homology to other pancreatic and trypsin like proteinases. In addition to catalytic domain α-thrombin posesses an exosite, quite distant from the catalytic residues, which is important for cleavage of fibrinogen to fibrin. This "fibrin-(ogen)-recognizing exocite" or "anion binding exocite" is involved in binding to other proteins such as hirudin, fibrin peptides, negatively charged cell surfaces etc. (Bode et al., 1992). An important site for thrombin specificity is the "apolar binding site" located close to catalytic center. These exocites help in aligning protein substrates and inhibitors of thrombin for more efficient interactions. The anion-binding exocite is essential for clotting activity.

Alpha-thrombin upon further proteolytic or autolytic cleavage is converted to γ-thrombin, which has lost all clotting activity (Fenton, 1988). Gama-thrombin lacks the fibrinogen-recognizing exocite. Gama-thrombin, like α-thrombin is also a potent cell activator. Platelet activation, effects on shape and permeability of endothelial cells, Ca^{2+} release, prostacyclin biosynthesis, release of plasminogen activator are only a few of the effects exerted by both α- and γ-thrombin. The important difference between α- and γ-thrombin, however, is that γ-thrombin lacks the anion binding exosite, therefore, it can not form fibrin or participate in the blood clotting cascade (Fenton, 1988).

We have shown previously that both α- and γ-thrombin causes a dose-dependent promotion of angiogenesis in the CAM system. As with other actions of γ-thrombin a 10-15 fold higher concentration of γ-thrombin than α- thrombin is required to obtain the maximum angiogenic response (about 80% over that in controls) (Tsopanoglou et al., 1993). The maximum angiogenic response was obtained with 1 IU (8.4 pmoles) of α- thrombin or 130 pmoles of γ-thrombin. With higher doses of either α- or γ-thrombin there is a drop in angiogenic effect, as with many of the actions of thrombin, which is probably the result of desensitization of the receptor. We have used the various agents summarized in Table 1 in our studies on the mechanism of promotion of angiogenesis by thrombin.

It was shown that hirudin, which binds both the catalytic and the anion binding exocites, thus inactivating thrombin, abolishes the angiogenic action of thrombin, when hirudin was used in combination with thrombin. Hirudin by itself was without effect on CAM angiogenesis (Tsopanoglou et al., 1993).

Similarly, Ro-464260, which is a novel synthetic inhibitor of thrombin (Gast et al., 1995) abolishes the angiogenic activity of thrombin in the CAM, while by itself Ro464260 has no effect on angiogenesis.

Heparin, which binds to thrombin and accelerates its destruction by antithrombin III also cancels out the angiogenic action of thrombin when it was used in combination with thrombin (Tsopanoglou et al., 1993).

P-PACK thrombin is chemically modified thrombin in which the catalytic site is inactivated. This analog lacks catalytic activity but retains the anion binding exocite. This analog of thrombin had no significant effect on angiogenesis in the CAM, indicating that binding of the anion binding exosite alone is not sufficient to promote angiogenesis. However, when P-PACK-thrombin was combined with either α- or γ-thrombin in the same experiment, it prevented the stimulatory effect of thrombin (Tsopanolgou et al., 1993). P-PACK thrombin presumable interferes with binding of α-thrombin via its catalytic site. Binding to the anion binding exocite may orient the thrombin molecule to carry out its enzymatic function or binding to as yet unidentified sites that initiate the events related to promotion of angiogenesis.

Table 1. Agents used for elucidation of the mechanism by which α-thrombin promotes angiogenesis

Agent	Characteristics
α-thrombin	Has the catalytic and the exosites required for fibrinogen binding and clotting activity.
γ-thrombin	Has the catalytic site but lacks the anion binding exosite. Can not form fibrin and lacks clotting activity.
P-PACK-thrombin	Is a chemically modified thrombin in which the catalytic site is inactivated.
Hirudin	Binds both the catalytic and anion binding exocites of thrombin.
Heparin	Binds to thrombin and accelerates its destruction by antithrombin III.
TRAP	A synthetic peptide representing the new N-terminus of thrombin receptor, after cleavage by thrombin. This peptide mimics many of the actions of thrombin (Tapparelli et al., 1993).
Ro-46-4260	A novel synthetic inhibitor of thrombin (Gast et al., 1995).

The aforementioned results in the CAM were obtained using the biochemical assay of angiogenesis based on the rate of collagenous protein biosynthesis (Maragoudakis et al., 1988). They were confirmed, however, by morphological evaluation of angiogenesis using computer-assisted image analysis (Tsopanoglou et al., 1993 and Maragoudakis et al., 1995).

Thrombin was also shown to promote the angiogenic phenotype of human umbilical cord endothelial cells (HUVECs) in the Matrigel system *in vitro* (Haralabopoulos et al., 1994). In this system HUVECs plated on Matrigel form tube-like structures within 8-16 hours. In the presence of 0.1-0.3 IU of α-thrombin the area covered by tube-like structures is increased 80% over that in controls. This promotion of tube formation is evident only when the fetal calf serum in the culture medium is reduced from 10% to 2%. These experiments further support the view that the promotion of the angiogenic phenotype by thrombin is a phenomenon independent of its blood clotting activity (Tsopanoglou et al., 1993). Furthermore, the appearance of angiogenic phenotype does not require the presence of other cell types such as pericytes, smooth muscle cells or inflammatory cells.

The angiogenesis promoting effect of thrombin is also evident in the Matrigel system *in vivo* as described by Passaniti et al. (1992) and outlined in the experimental procedures. When 1.0 ml Matrigel is injected sc in C57BL6N mice, it solidifies into a transparent plug, which remains clear for more than two weeks. However, when the Matrigel is mixed before the injection with angiogenic substances there is an infiltration of cells and new complete blood vessels are formed. When α-thrombin

(0.1-0.3 IU/ml Matrigel) was incorporated into the Matrigel we have obtained a dramatic increase in blood vessel formation within the Matrigel plug, which was dose dependent.

We conclude from these experiments that thrombin has a specific and potent effect on angiogenesis which is independent of fibrin formation and involves the catalytic site of thrombin.

Recent studies on the cloned thrombin receptor reveal that it is a member of the seven transmembrane receptor family. The extracellular amino terminal extention of the receptor posesses a cleavage site by thrombin. Upon proteolytic cleavage by thrombin a proteolytic fragment of the receptors is released and a new N-terminus is unmasked. This N-terminus peptide acts a tethered receptor agonist by binding to an as yet unidentified domain of the cloned receptor to effect cell activation. This specific proteolysis of the receptor by thrombin provides all the necessary information for receptor activation. The cleavage site has close similarities to the cleavage sites of protein C, factor VIII and fibrinogen Aα chain by thrombin (Tapparelli et al., 1993).

A synthetic peptide TRAP$_{1-14}$ representing the N-terminus of the activated thrombin receptor is a full agonist for activation of the cloned thrombin receptor. It has been shown that TRAP can mimic the effects of thrombin in the promotion of cellular events such as secretion, mitogenesis, second messenger, generation etc. However, not all actions of thrombin can be seen by substituting TRAP for thrombin eg. tyrosine kinase, endothelial cell activation etc. (Tapparelli et al., 1993).

We have shown that TRAP, which bypass the requirement for proteolytic cleavage of the receptor by thrombin and mimics many of the actions of thrombin, also promotes angiogenesis in the CAM (Maragoudakis et al., 1995). However, the maximum angiogenesis promoting effect obtained by TRAP is less than that of thrombin (80% stimulation by thrombin as compared to 50% stimulation of angiogenesis by TRAP at 7 pmoles). Like thrombin at higher doses TRAP also shows a decrease in the angiogenesis promoting effect.

We conclude from these studies that promotion of angiogenesis by thrombin is a receptor mediated event which is distinct from the clotting mechanism. The question then arises by which signal transduction mechanism is the effect of receptor activation is mediated.

The demonstration that thrombin stimulates the proliferation of microvascular endothelial cells (Carney, 1992) could be the first explanation for the angiogenic effect of thrombin. However, our *in vitro* experiments of HUVECs in Matrigel show that thrombin can induce the angiogenic phenotype in the absence of cell proliferation.

The mechanism by which thrombin stimulates cell proliferation of endothelial cells remains to be elucidated. It has been shown that thrombin causes the production of PDGF-like molecules and this activity can be inhibited by agents that increase c-AMP levels. This stimulation by thrombin of the production of PDGF-like molecules requires the proteolytic activity of thrombin (Carney, 1992). Our finding, that TRAP, which is a synthetic peptide, representing the high affinity domain of thrombin receptor, is angiogenic in the CAM, leads to an interesting question. Can TRAP, which does not have proteolytic activity stimulate the formation of PDGF- like molecules by receptor binding only and what is the transduction mechanism involved?

Both α- and γ-thrombin has been shown to activate phospholipase C (PLC), leading to inositol thriphosphate (IP$_3$) generation and PKC activation (Fig. 1). The activation of phosphoinositol-specific membrane PLC by thrombin causes hydrolysis of phosphatidyl 4,5-bis-phosphase and the generation of the short-lived second messenger IP_3 and diacylglycerol (DAG). IP_3 regulates the Ca^{2+} by mobilizing the release of Ca^{2+} from internal stores and indirectly by stimulating Ca^{2+} entry (Fig. 1) (Garcia et al., 1992). Thrombin stimulation of inositol phosphate turn over in endothelial cells depends upon the presence of an active catalytic site. Both α- and γ-thrombin activate PLC resulting in IP_3 generation and activation of PKC, whereas DIP- and P-PACK-thrombin are inactive. Also DIP- and P-PACK-thrombin do not increase (Ca^{2+}) in HUVEC. These events can be seen as early as 5 seconds after the addition of thrombin. Similar responses were obtained with other cell types, while non-responsive cells to thrombin showed little or no stimulation by thrombin of IP_3 synthesis. These studies indicate direct involvement of phosphoinositol turnover in the cellular actions of thrombin in which the catalytic site is essential, since γ-thrombin, which retains the esterase activity stimulates PLC and phosphoinositol turnover to the same extent as α-thrombin, but P-PACK thrombin is inactive (Brock & Capasso, 1988, Jaffe et al., 1987, Carney, 1992).

In addition, thrombin causes a potent stimulation of arachidonic acid release and PGI_2 synthesis. PGI_2 is the primary metabolite of arachidonic acid produced by cultured endothelial cells and it has profound effects in vascular permeability, smooth muscle cell proliferation and hemostasis. Both α- and γ- thrombin (but not DIP or P-PACK thrombin) increases arachidonic acid production as a result of PLC and phospholipase A_2 (PLA_2) activation. It is evident from these studies that as with other functions of thrombin the stimulation of PGI_2 synthesis requires the proteolytic activity of thrombin. The rate limiting step in the release of arachidonic acid and the synthesis of PGI_2 is the hydrolysis of phosphatidylcholine to lysophosphatidylcholine by PLA_2. This enzyme is very sensitive to flunctuations of cytosolic Ca^{2+} (Carney, 1992).

Many investigators have shown that DAG participates in the regulation of PGI_2 synthesis and endothelial cell permeability. This is accomplished by virtue of the ability of DAG to activate and translocate PKC from the cytosolic to the membrane compartment. The stimulation of DAG production by thrombin is rapid and occurs within 15 seconds. In addition PKC activation appears to regulate PGI_2 synthesis by a negative feed back inhibition of PLC and PLA_2. As shown schematically in Fig. 1 thrombin-induced PGI_2 synthesis involves two Ca^{2+}-dependent phospholipases (PLC and PLA_2). Mobilization of Ca^{2+} by thrombin results from activation of PLC and the transient generation of IP_3. This events precede the production of arachidonic acid by PLA_2 and the synthesis of PGI_2 (Garcia et al., 1992).

The involvement of G proteins in the transduction mechanisms of thrombin is suggested from many studies. For example pertusis toxin inhibits thrombin-induced activation of PLC and cell proliferation but has no similar effects on other mitogenic factors such as EGF, which functions through tyrosine kinase activation (Carney, 1992).

Fig. 1 is a hypothetical model for the cellular events that may be involved in the activation of endothelial cells by thrombin leading to stimulation of angiogenesis. It is evident from the aforementioned transduction mechanisms that are involved, following the activation of thrombin receptor, that activation of PKC by thrombin plays a key role. Activation of PKC is one of the early events of thrombin action and as shown in Fig. 1 leads to cellular responses, which are related to angiogenesis.

We have shown that activation of PKC by phorbol esters (PMA) or DAG causes a marked stimulation of angiogenesis. On the contrary, inhibitors of PKC such as Ro318220 causes an inhibition of angiogenesis both the basal and that stimulated by PMA or DAG (Tsopanolgou et al., 1993). Furthermore, it was shown that activation of c-AMP-dependent protein kinase (PKA) causes suppression of angiogenesis

(Tsopanoglou et al., 1994). We conclude from these studies that activation of PKC and PKA have opposing effects in the modulation of angiogenesis. This cross-talk phenomenon between the two major signalling pathways seems to be functional in many other physiological processes (Tsopanoglou et al., 1994).

A plausible explanation, therefore, for the mechanism by which thrombin promotes angiogenesis could be the activation of PKC. This is in agreement with our findings that inhibitors of PKC, such as Ro 318220, completely abolish the angiogenesis- promoting effect of thrombin (Maragoudakis et al., 1994).

In addition to activation of PKC, other mechanisms may be involved in the promotion of angiogenesis by thrombin. Factors that are essential for vascular proliferation such as PAF, plasminogen activator, metalloproteinases etc. may be generated as a result of thrombin action.

In conclusion we have demonstrated that thrombin promotes angiogenesis by activation of the cloned receptor. The catalytic site of thrombin is essential for this action and the transduction mechanism is probably via activation of PKC. This action of thrombin is independent of clotting activity and can be mimicked by TRAP, a peptide representing the N-terminus of the activated thrombin receptor. Promotion of the angiogenic phenotype by thrombin can be demonstrated *in vitro* by plating endothelial cells in Matrigel.

These data suggest a new role of thrombin not only in the metastatic spread of cancer but also in inflammation and wound healing. This new role is based on the ability of thrombin to promote angiogenesis. The dependence of tumor growth and

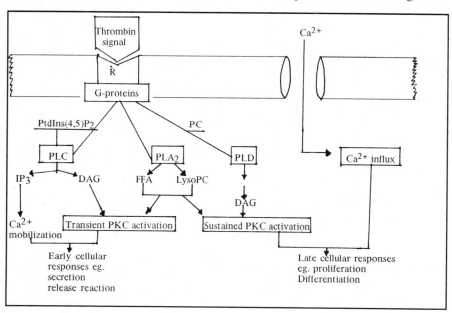

Fig. 1 Model for signal induced degradation of membrane phospholipids that may be involved in the endothelial cell activation by thrombin for the expression of angiogenic phenotype.
The abbreviations used are: RtdIns (4,5) P2 (phosphatidylinositol 4,5-biphosphate); PC (phosphatidylcholine); IP3 (inositol 1,4,5 triphosphate); DAG, (diacylglycerol); FFA (arachidonic acid); LysoPC (lysophosphotidylcholine); PLC (phospholipase C); PLA2 (phospholipase A2); PLD (phospholipase D); PKC (protein kinase C)

metastasis on new blood vessel formation is well documented (Folkman, 1985). Angiogenesis is also part of the inflammatory process (Folkman 1995). Proinflammatory effect of thrombin involves the activation of both endothelial cells and monocytes/macrophages. This leads to increased vascular permeability expression of adhesion molecules and leucocyte extravasation. These effects of thrombin can be blocked by hirudin. It is likely that local thrombin generation by macrophage-like cells, which is evident in synovial tissue of rheumatoid arthritic patients, is responsible for inflammatory angiogenesis (Tapparelli et al., 1993). Therefore, anti-thrombotic therapy may prove beneficial not only to cancer therapy but also to inflammatory conditions. In that respect specific inhibitors of the angiogenic action of thrombin, which may not have effects on blood coagulation, may be important new anti-tumor and anti-inflammatory agents.

On the other hand, angiogenesis is essential in wound healing. Thrombin promotes angiogenesis and proliferation of fibroblasts and smooth muscle cells. These events are blocked by hirudin. The potential, therefore, exists for therapeutic application of non-thrombogenic analogs of thrombin or thrombin-mimetic peptides in situations where promotion of wound healing is desirable. This may be a novel therapeutic approach in non-healing wounds or ulcers in diabetics or older patients.

ACKNOWLEDGEMENT

This original work described in this paper was supported in part by a Collaborative Research Grant #CRG940677 from NATO.
I thank Anna Marmara for typing the manuscript.

REFERENCES

Auerbach, R., Auerbach, W., and Polokowski, I., 1991, Assays for angiogenesis: review, Pharmacol. Ther. 51:1-11.

Bode, W., Huber, R., Rydel, T. and Tulinsky, A., 1992, X-ray crystal structures of human a-thrombin and of the human thrombin-hirudin compley, In: Thrombin Structure and Function, L.J. Berliner ed., Plenum Press, NY, pp. 3-61.

Brock, T.A. and Capasso, E.A., 1988, Thrombin and histamine activate phospholipase C in human endothelial cells via a phorbol ester-sensitive pathway, J. Cell Physiol. 136:54-62.

Carney, D.H, 1992, Postclotting cellular effects of thrombin mediated by interaction with high-affinity thrombin receptors, In: Thrombin Structure and Function, Berliner J.L. ed., Plenum Press, pp. 351-396.

Fenton, J.W. II, 1988, Regulation of thrombin generation and functions, Semin. Thromb. Haemostas. 14:234-240.

Folkman J., 1985, Tumor angiogenesis, Adv. Cancer Res. 43:172-203.

Folkman, J., 1985a, Toward an understanding of angiogenesis, Search and Discovery, Prospect. Biol. Med. 29:10-35.

Folkman, J., 1995, Angiogenesis in cancer, vascular, rheumatoid and other disease, Nature Med. 1:27-31.

Garcia, G.N., Aschner, J.L., and Malik, A.B., 1992, Regulation of thrombin-induced endothelial cell barier dysfunction and prostaglandin synthesis, In: Thrombin Structure and Function, Berliner, J.L. ed., Plenum Press, pp. 397-430.

Gast, A., Tschopp, T.B., Schmid, G., Hilpert, K., Ackermann, J., 1995, Inhibition of clot-bound and free (fluid-phase thrombin) by a novel synthetic thrombin inhibitor (Ro-46-6240), recombinant hirudin and heparin in human plasma, Blood Coagulation & Fibrinolysis 5:879-887.

Grant, D.S., Tashiro, K.I., Segui-Real, B., Yamada, Y., Martin, G.R., and Kleinman, H.K., 1989, Two different laminin domains mediate the differentiation of human endothelial cells into capillary-like structures *in vivo, Cell* 58:933-943.

Haralabopoulos G.C., Grant, D.S., Kleinman, H.K., Lelkes, P.I., Papaioannou, S.P., and Maragoudakis, M.E., 1994, Inhibitors of basement membrane collagen biosynthesis prevent endothelial cell alignment in Matrigel *in vitro* and angiogenesis *in vivo, Lab. Invest.* 71(4):575-581.

Jaffe, E.A., Grulich, J. Weksler, B.B., Hampel, G., and Watanabe, K., 1987, Correlation between thrombin-induced prostacyclin production and inositol triphosphate and cytosolic free calcium levels in cultured human endothelial cells, *J. Biol. Chem.* 137:8557-8565.

Maragoudakis, M.E., Sarmonika, M., Panoutsacopoulou, M., 1988, Rate of basement membrane biosynthesis as an index of angiogenesis, *Tissue and Cell* 20:531-539.

Maragoudakis, M.E., Tsopanoglou, N.E., Haralabopoulos, G.C., Sakkoula, E., Pipili-Synetos, E. and Missirlis, E., 1994, Regulation of angiogenesis: The role of protein kinase, nitric oxide, thrombin and basement membrane synthesis, In: *Angiogenesis: Molecular Biology, Clinical Aspects,* edited by. M. E. Maragoudakis, P. Gullino, P. Lelkes, Plenum Press, pp. 125-134.

Maragoudakis, M.E., Tsopanoglou, N.E., Sakkoula, E., and Pipili-Synetos, E.,1995, On the mechanism of promotion of angiogenesis by thrombin, *FASEB J.* 9:A587

Maragoudakis, M.E., Haralabopoulos, G.C., Tsopanoglou, N.E., and Pipili-Synetos, E., 1995, Validation of collagenous protein synthesis as an index for angiogenesis with the use of morphological methods, *Microvascular Res.,* 50:215-222.

Maragoudakis, M.E., 1995, Angiogenesis: An overview of regulation and potential application, In: *Vascular Endothelium, Responses to Injury,* edited by J. Catravas, U. Ryan, Plenum Press (in press).

Passaniti, A., Taylor, R.M., Pili, R., Guo, Y., Lowrey, P.V., Haney, J.A.m Daniel, R.R., Grawd, D.S., and Martin, G.R., 1992, A simple quantitative method for assessing angiogenesis and antiangiogenic agents using reconstituted basement membrane heparin, fibroblast growth factor, *Lab. Invest.* 67:519-528.

Ricles, F.R., Edwards, R.L., 1983, Activation of blood coagulation in cancer: Trousseau's syndrome revisited, *Blood* 64:14-31.

Silverstein, R.L. and Nachman, R.L., 1992, Cancer and clotting - Trousseau's warning, *N. Engl. J. Med.* 327:1163-1164.

Sloan, B.F., Rozhin, J., Johnson, K., Taylor, H., Crissman, J. D., and Honn, K. V., 1986, Cathepsin B: Association with plasma membrane in metastatic tumors, *Proc. Natl. Acad. Sci.* (USA) 83:2483-2487.

Tapparelli, C., Metternich, R., Ehrhardt, C., and Cook, N.S., 1993, Synthetic low molecular weight thrombin inhibitors: molecular design and pharmacological prolife, *TIPS* 14:366-376.

Tapparelli, C., Metternich, R., and Cook, N.S., 1993, Structure and function of thrombin receptors, *TIPS* 14:426-428.

Trousseau, A., 1872, Phlegmasia alba dolens in Trousseau A. Lectures in clinical medicine, delivered in Hotel-Dieu, Paris, London, *New Sydenham Society,* 281-295.

Tsopanoglou, N.E., Pipili-Synetos, E., and Maragoudakis, M.E., 1993, Thrombin promotes angiogenesis by a mechanism independent of fibrin formation, *Am. J. Physiol.* 264 (Cell Physiol. 33):C1302-1307.

Tsopanoglou, N.E., Pipili-Synetos E. and Maragoudakis, M.E., 1993, Protein kinase C involvement in the regulation of angiogenesis. *J. Vasc. Res.* 30:202-208.

Tsopanoglou, N.E., Haralabopoulos, G.C., and Maragoudakis, M.E., 1994, Opposing effects on modulation of angiogenesis by protein kinase C and cyclic AMP-mediated pathways. *J. Vasc. Res.* 31:195-204.

THE ROLE OF THROMBOSPONDIN IN ANGIOGENESIS

Luisa A. DiPietro[1] and Peter J. Polverini[2]

[1]Burn and Shock Trauma Institute
Department of Surgery
Loyola University Medical Center
Maywood, IL 60153

[2]Laboratory of Molecular Pathology
School of Dentistry
University of Michigan
Ann Arbor, MI   48109

INTRODUCTION

Thrombospondin 1 (TSP1) is a large multi-functional extracellular matrix molecule that is a member of a family of glycoproteins. TSP1 is a prominent component of the alpha granule of platelets, and is produced by numerous cell types, including endothelial cells, fibroblasts, macrophages, monocytes, keratinocytes, and some tumor cells (Frazier, 1987; Lawler, 1986). TSP1 has been described to have many physiologic functions, including involvement in cell migration, recognition, and binding. The multi-functional nature of TSP1 is illustrated by the complex molecular structure (Figure 1). TSP1 exists as a trimer of three identical subunits, each with a molecular weight of 180 kd. Each subunit is dumb-bell shaped, containing globular NH2- and COOH terminal domains joined by a central stalk (Galvan et al., 1985; Lawler et al., 1985). The NH2- domain has heparin binding activity, while the COOH domain contains a cell attachment region. The stalk can be further divided into four regions: 1) a region containing the cysteines involved in interchain disulfide bonding, 2) a region homologous to procollagen type I, 3) a region with three properdin (Type 1) repeats, and 4) a region with three EGF-like (Type 2) repeats.

Molecular, Cellular, and Clinical Aspects of Angiogenesis
Edited by Michael E. Maragoudakis, Plenum Press, New York, 1996

105

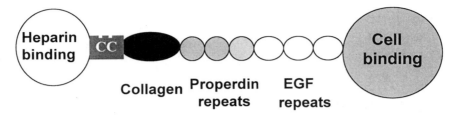

Figure 1. Schematic of the molecular structure of the 180kd subunit of thrombospondin 1. Each mature 540 kd TSP1 molecule is composed of three covalently linked subunits. Each subunit contains an NH terminal globular domain that is heparin binding, as well as a COOH terminal globular domain that is cell binding. The stalk region contains a cysteine containing area, a region homologous to Type I procollagen, three properdin-like repeats, and three EGF-like repeats.

TSP1 AND ENDOTHELIAL CELLS

The production and function of TSP1 has been extensively studied in endothelial cells. TSP1 production by endothelial cells in culture was first demonstrated by Bornstein and colleagues in 1981 (McPherson et al., 1981). Early studies of the interaction between TSP1 and endothelial cells demonstrated that TSP1 was anti-adhesive for these cells, as TSP1 was shown to reduce focal adhesions in culture (Lahav, 1988; Murphy-Ullrich and Höök, 1989). The rate of production of TSP1 by endothelial cells varies with the *in vitro* phenotype. Proliferating endothelial cells actively synthesize and secrete TSP1, while endothelial cells that have undergone *in vitro* angiogenesis and have assumed a capillary-like formation exhibit a significant decrease in TSP1 production (Canfield et al., 1990; Iruela-Arispe et al., 1991b). These observations suggest that TSP1 might be functionally important to the process of angiogenesis. TSP1, produced by proliferating cells, would allow for the loss of focal adhesions during cell proliferation. As new vessels form, TSP1 production would decrease to allow for the formation of stable cell to cell contacts.

THROMBOSPONDIN 1 AS AN INHIBITOR OF ANGIOGENESIS

In 1989, Bouck and colleagues, working in a tumor cell system, described the linkage of the production of an inhibitor of angiogenesis to the expression of a tumor suppressor gene (Rastinajed et al., 1989). This inhibitor was subsequently determined to be a truncated form of TSP1 (Good et al., 1990). In these studies, purified TSP1 inhibited endothelial cell chemotaxis, and also potently inhibited *in vivo* angiogenesis in the rat corneal bioassay. Because angiogenesis is a requirement of solid tumor growth, this finding suggested that the down-regulation of TSP1 production in cells might play an important role in tumorigenesis.

This singular finding sparked an increased interest in the anti-angiogenic activity of TSP1 both *in vitro* and *in vivo*. TSP1 inhibits endothelial cell proliferation and chemotaxis, and inhibits angiogenesis *in vitro* (Bagavandoss and Wilks, 1990; Iruela-Arispe et al., 1991a; Taraboletti et al., 1992). Conversely, when TSP1 production by endothelial cells is inhibited by antisense RNA, cells become increasingly responsive to both chemotactic and angiogenic stimuli (DiPietro et al., 1994). To date, more than ten different studies have substantiated the anti-angiogenic properties of purified TSP1.

THROMBOSPONDIN 1 PRODUCTION AND FUNCTION *IN VIVO*

Although the anti-angiogenic properties of TSP1 have been clearly demonstrated, paradoxical and conflicting data regarding the role of this protein *in vivo* have been reported. Firstly, the role of TSP1 in solid tumor growth, a situation where active angiogenesis is required, is not yet clear. A down-regulation of TSP1 synthesis, in concert with the development of a tumorigenic and angiogenic phenotype, has been documented in multiple different systems (Good et al., 1990; Dameron et al., 1994; Zabrenetzky et al., 1994). In one such system, the transfection of a vector that produced high levels of TSP1 inhibited tumor progression (Weinstat-Saslow et al., 1994). These studies indicate that TSP1 may play a negative regulatory role in the development of the angiogenic phenotype that is required for tumor growth. In contrast, other investigations have described the active production of TSP1 by tumor cells both *in vitro* and *in vivo*. TSP1 production has been described in breast cancer, melanoma, squamous cell carcinoma, and osteosarcoma cells (Clezardin et al., 1989;. Varani et al., 1987; Apelgren and Bumol, 1989; Pratt et al., 1989). In at least two systems, the production of TSP1 by the tumor line appears to positively influence the metastatic capability of the cell line, suggesting a positive functional role for this protein in tumor progression (Castle et al., 1991; Tuszynski et al., 1987). The conclusion from the data must be that the role of TSP1 in tumorigenesis is varied.

Investigations of TSP1 production in areas of inflammation also appear to be inconsistent. Studies by Nickoloff et al. (1994) have shown that whereas normal, non-angiogenic, keratinocytes produce TSP1, angiogenic keratinocytes from psoriatic skin show a seven-fold decrease in production of the protein. These results suggest that TSP1 production would be diminished in inflammatory lesions. However, the production of TSP1 within inflammatory lesions with active angiogenesis, such as rheumatoid arthritis and atheroslcerosis has been documented (Botney et al., 1992; Koch et al., 1993). Finally, we and others have shown that activated macrophages, a cell type that is itself angiogenic, produce TSP1 (Jaffe et al., 1985; Ostergaard and Flodgaard, 1992; DiPietro and Polverini, 1993; Varani et al., 1991).

As stated above, many *in vitro* investigations have substantiated the anti-angiogenic effect of TSP1. Yet a few recent investigations have demonstrated an angiogenesis promoting effect for TSP1 in *in vitro* and *in vivo* assays (BenEzra et al, 1993; Nicosia and Tuszynski., 1994). These studies do not necessarily contradict previous findings, but instead provide a greater understanding of the function of TSP1. Studies by Nicosia and Tuszynski suggest that angiogenic or anti-angiogenic properties of TSP1 depend upon the gene (Rastinajed et al., 1989). This inhibitor was subsequently determined to be a truncated form of TSP1 (Good et al., 1990). In these studies, purified TSP1 inhibited endothelial cell chemotaxis, and also potently inhibited *in vivo* angiogenesis in the rat corneal bioassay. Because angiogenesis is a requirement of solid tumor growth, this finding suggested that the down-regulation of TSP1 production in cells might play an important role in tumorigenesis.

This singular finding sparked an increased interest in the anti-angiogenic activity of TSP1 both *in vitro* and *in vivo*. TSP1 inhibits endothelial cell proliferation, and inhibits angiogenesis *in vitro* (Bagavandoss and Wilks, 1990; Iruela-Arispe et al., 1991a). Conversely, when TSP1 production by endothelial cells is inhibited by antisense RNA, cells become increasingly responsive to both chemotactic and angiogenic stimuli (DiPietro et al., 1994). To date, more than ten different studies have substantiated the anti-angiogenic properties of purified TSP1.

THROMBOSPONDIN 1 PRODUCTION AND FUNCTION *IN VIVO*

Although the anti-angiogenic properties of TSP1 have been clearly demonstrated, paradoxical and conflicting data regarding the role of this protein *in vivo* have been reported. Firstly, the role of TSP1 in solid tumor growth, a situation where active angiogenesis is required, is not yet clear. A down-regulation of TSP1 synthesis, in concert with the development of a tumorigenic and angiogenic phenotype, has been documented in multiple different systems (Good et al., 1990; Dameron et al., 1994; Zabrenetzky et al., 1994). In one such system, the transfection of a vector that produced high levels of TSP1 inhibited tumor progression (Weinstat-Saslow et al., 1994). These studies indicate that TSP1 may play a negative regulatory role in the development of the angiogenic phenotype that is required for tumor growth. In contrast, other investigations have described the active production of TSP1 by tumor cells both *in vitro* and *in vivo*. TSP1 production has been described in breast cancer, melanoma, squamous cell carcinoma, and osteosarcoma cells (Clezardin et al., 1989;. Varani et al., 1987; Apelgren and Bumol, 1989; Pratt et al., 1989). In at least two systems, the production of TSP1 by the tumor line appears to positively influence the metastatic capability of the cell line, suggesting a positive functional role for this protein in tumor progression (Castle et al., 1991; Tuszynski et al., 1987). The conclusion from the data must be that the role of TSP1 in tumorigenesis is varied.

Investigations of TSP1 production in areas of inflammation also appear to be inconsistent. Studies by Nickoloff et al. (1994) have shown that whereas normal, non-angiogenic, keratinocytes produce TSP1, angiogenic keratinocytes from psoriatic skin show a seven-fold decrease in production of the protein. These results suggest that TSP1 production would be diminished in inflammatory lesions. However, the production of TSP1 within inflammatory lesions with active angiogenesis, such as rheumatoid arthritis and atheroslcerosis has been documented (Botney et al., 1992; Koch et al., 1993). Finally, we and others have shown that activated macrophages, a cell type that is itself angiogenic, produce TSP1 (Jaffe et al., 1985; Ostergaard and Flodgaard, 1992; DiPietro and Polverini, 1993; Varani et al., 1991).

As stated above, many *in vitro* investigations have substantiated the anti-angiogenic effect of TSP1. Yet a few recent investigations have demonstrated an angiogenesis promoting effect for TSP1 in *in vitro* and *in vivo* assays (BenEzra et al, 1993; Nicosia and Tuszynski., 1994). These studies do not necessarily contradict previous findings, but instead provide a greater understanding of the function of TSP1. Studies by Nicosia and Tuszynski suggest that angiogenic or anti-angiogenic properties of TSP1 depend upon the molecular form of the molecule. Bound TSP1 promotes *in vitro* angiogenesis, probably by direct interactions with smooth muscle cells. This interaction appears to induce the smooth muscle cells to secrete soluble angiogenic factors. Additional studies by Tolsma et al. (1994) with TSP1 peptides suggest that the anti-angiogenic effect is concentration sensitive. Taken together, these results provide a hypothesis for the varied *in vivo* observations. The influence of TSP1 on angiogenesis may depend upon the predominant molecular form and concentration within the local environment.

TSP1 AND WOUND REPAIR

The numerous functions ascribed to TSP1, as well as the production of TSP1 by activated macrophages, led us to study the *in vivo* production of TSP1 in normal dermal wound repair (DiPietro et al., 1995). Normal wound repair represents a well-defined example of inflammation that includes macrophage infiltration, physiologic angiogenesis, and fibroplasia. In this investigation, the local production of TSP1 was assessed in full thickness excisional dermal wounds, a wound type that is characterized by significant inflammation and granulation tissue formation prior to resolution. Northern analysis revealed a moderate increase in TSP1 mRNA level in the early phase of wound repair (Figure 2A). *In situ* hybridization studies indicated that inflammatory cells containing high levels of TSP1 mRNA were present within the early wound bed, but not within normal skin (Figure 2B). Histologic examination suggested that these cells were macrophages. Our findings are complementary to those of Reed et al. (1993), who described TSP1 production in incisional wounds.

The early wound environment has been documented to be remarkably angiogenic, and this observation therefore provide evidence for alternate roles for this protein. TSP1 within

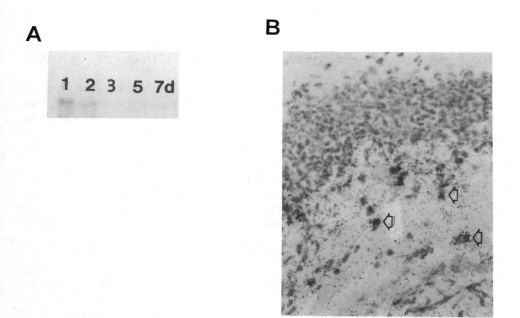

Figure 2. (A)Northern analysis and (B) *in situ* hybridization (day 1 wound) of dermal wound tissue for TSP1 mRNA. In (A), time after wounding is indicated at the top. In (B), arrows indicated labeled macrophages.

the early wound may be quickly bound and sequestered. Alternatively, macrophage derived TSP1 within wounds may have an alternate role, one unrelated to its anti-angiogenic properties. One possibility is that TSP1 is important to macrophage infiltration into the wound. TSP 1 has been shown to promote monocyte chemotaxis (Mansfield and Suchard, 1994) and TSP1 facilitates macrophage migration through endothelial cell layers *in vitro* (Huber et al.,1992). Another possible role for TSP1 in wound repair is suggested by the recent studies of Schultz-Cherry and Murphy-Ullrich, 1993. These investigators have shown that TSP1 can activate TGF-β from its latent form. TGF-β production has been described in dermal wounds (Kane et al., 1991). Because activated TGF-β can promote both fibrous tissue repair and capillary regrowth, the production of TSP1 in the wound may serve to augment the repair response via TGF- β activation (Roberts et al., 1986; Mustoe et al., 1987).

The functional role of local TSP1 production within wounds has been further studied in TSP1 knockout mice (Polverini et al., 1995). Wound repair in these mice was both significantly delayed and prolonged, manifesting as increased granulation tissue formation, prolonged neovascularization, and sustained monocyte/ macrophage infiltration. However, with time, the wounds of the knockout mice fail to resolve, suggesting that TSP1 production in the late stages of repair might aid in down-regulating the angiogenic response. The introduction of pure TSP1 into wounds in knock-out mice is able to partially reverse these effects, resulting in a more rapid onset and earlier resolution of wound healing (Polverini et al., unpublished observations).

Thus, the function of TSP1 in the wound repair model would be hypothesized to be two-fold. In the early phase of repair, TSP1 synthesis by macrophages would facilitate the repair process, probably by local interaction with macrophages or other cells. As repair progresses, soluble forms of TSP1 would function as an anti-angiogenic agents and cause a cessation of neovascularization.

A MODEL FOR THROMBOSPONDIN 1 FUNCTION IN ANGIOGENESIS

The multiple and complex roles of TSP1 in angiogenesis of TSP1 may seem difficult to reconcile. However, by considering the molecular nature of TSP1, as well as information concerning the function of this molecule, a model for influence of TSP1 on the angiogenic response can be proposed (Figure 3). The model incorporates two important pieces of information regarding this molecule. First, soluble TSP1, as well as specific soluble fragments of TSP1, is anti-angiogenic (Good et al., 1990; Tolsma et al., 1994). Secondly, bound TSP1 promotes the angiogenic response (Nicosia and Tuszynski , 1994).

The model proposes that the dominant function of TSP1 is directed by the local environment into which it is secreted (Figure 3). Secreted trimeric TSP1 is a large molecule of 540 kd, and as such, is unlikely to be freely diffusible. In contrast, the angiogenic cytokines that are produced by macrophages or tumor cells are generally small and freely diffusible. An overall inhibitory influence of TSP1 on the angiogenic environment would only occur if TSP1 remained soluble, or was locally cleaved into small diffusible fragments. In contrast, if TSP1 was quickly bound to cells or the extracellular matrix, angiogenesis might be facilitated. A local proteolytic environment that favored the degradation of secreted TSP1 into active, diffusible peptides might be anti-angiogenic. If instead, most of the secreted TSP1 would remained intact, it would promote the angiogenic process via cellular interactions. The model proposed here allows for a dual influence of TSP1, and would explain many of the apparently conflicting observations regarding this protein.

The ability of TSP1 to bind growth factors lends yet another dimension to the role of this molecule in the angiogenic response and may also be included in the model. As mentioned above, TSP1 can bind and activate latent TGF-β; this activated TGF-β is potently angiogenic *in vivo* (Roberts et al., 1987). Therefore TSP1, if produced in concert with TGF-β, could promote the development of a positive angiogenic milieu. In addition, TSP1 may bind other growth factors. The role that this binding capacity plays in modulating an angiogenic response is currently unclear.

FUTURE DIRECTIONS

The *in vitro* interactions of TSP1 with endothelial cells have been extensively examined, and have provided important clues regarding the function of this molecule.

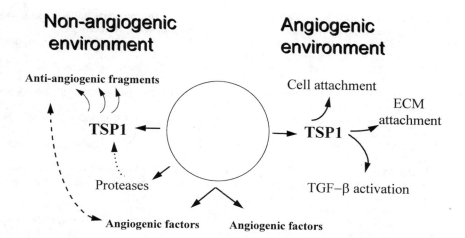

Figure 3. A model for the action of TSP1 in angiogenic and non-angiogenic environments. For details, see the text.

Further studies must clearly be directed at an *in vivo* understanding of TSP1 function. The healing wound may provide a model in which to examine TSP1. One extremely informative and yet unanswered question is that of the specific location of this protein in *in vivo* settings. Within the early wound, does the molecule bind to endothelial cells, macrophages, smooth muscle cells, keratinocytes or the ECM? Is TSP1 intact in the wound, or is it degraded into fragments? If degraded, are the fragments bioactive? These questions are difficult to address, and will require creative experimental approaches. Nevertheless, answers to these questions will greatly increase our knowledge of the angiogenic process.

ACKNOWLEDGEMENTS

This work was supported by the Dr. Ralph and Marion C. Falk Foundation (LAD) and NIH grants GM50875 (LAD) and HL39926 (PJP).

REFERENCES

Apelgren, L.D., and Bumol, T.F., 1989, Biosynthesis and secretion of thrombospondin in human melanoma cells, *Cell Biol. Intl. Rep.* 13: 189-195.

Bagavandoss, P., and Wilks, J.W., 1990, Specific inhibition of endothelial cell proliferation by thrombospondin, *Biochem.Biophys. Res. Comm.* 170: 867-872.

BenEzra, D., Griffin, B.W., Maftzir, G., and Aharonov, O., 1993, Thrombospondin and *in vivo* angiogenesis induced by basic fibroblast growth factor or lipopolysaccharide, *Invest. Ophthamol. Vis. Sci.* 34: 3601-3608.

Botney, M.D., Kaiser, L.R., Cooper, J.D., Mecham, R.P., Parghi, D., Roby, J., and Parks, W.C., 1992, Extracellular matrix protein gene expression in atherosclerotic hypertensive pumonary arteries, A*m. J. Pathol.* 140: 357-364.

Canfield, A.E., Boot-Hadford, P., and Schor, A.M., 1990, Thrombospondin gene expression by endothelial cells in culture is modulated by cell proliferation, cell shape, and the substratum, *Biochem. J.* 268: 225-230.

Castle, V., Varani, J., Fligiel, S., Prochownik, E., and Dixit, V., 1991, Anti-sense mediated reduction in thrombospondin reverses the malignant phenotype of a human squamous carcinoma, *J. Clin. Invest.* 87: 1883-1888.

Clezardin, P., Jouishomme, H., Chavassieux, P., and Marie, P.J., 1989, Thrombospondin is synthesized and secreted by human osteoblasts and osteosarcoma cells, *Eur. J. Biochem.* 181: 721-726.

Dameron, K.M., Volpert, O.V., Tainsky, M.A., and Bouck, N., 1994, Control of angiogenesis in fibroblasts by p53 regulation of thrombospondin-1, *Science* 265:1582-1584.

DiPietro, L.A. and Polverini, P.J., 1993, Angiogenic macrophages produce the angiogenic inhibitor thrombospondin 1, *Am. J. Pathol.* 143: 678-684.

DiPietro, L.A., Nebgen, D.R., and Polverini, P.J., 1994, Down-regulation of endothelial cell thrombospondin 1 enhances *in vitro* angiogenesis, *J. Vasc. Res.* 31: 178-185.

DiPietro, L.A., Nissen, N.N., Gamelli, R.L., Koch, A.E., and Polverini, P.J., 1995, Inhibition of wound repair by topical antisense thrombospondin 1 oligomers, FASEB J. 9: A696.

Frazier, W. A., 1987, Thrombospondin: a modular adhesive glycoprotein of platelets and nucleated cells, *J. Cell Biol.* 105: 625-32.

Galvan, N.J., Dixit, V.M., O'Rourke, K.M., Santoro, S.A., Grant, G.A., and Frazier, W.A., 1985, Mapping of epitopes for monoclonal antibodies against human platelet thrombospondin with electron microscopy and high sensitivity amino acid sequencing, *J. Cell Biol.* 101: 1434-1441.

Good, D.J., Polverini, P.J., Rastinejad, F., LeBeau, M.M., Lemons, R.S., Frazier, W.A. and Bouck, N.P., 1990, A tumor suppressor-dependent inhibitor of angiogenesis is immunologically and functionally indistinguishable from a fragment of thrombospondin, *Proc. Nat. Acad. Sci. USA.* 87: 6624-6628.

Huber, A.R., Ellis, S., Johnson, K.J., Dixit, V.M., and Varani, J., 1992, Monocyte diapedesis through an *in vitro* vessel wall construct: inhibition with monoclonal antibodies to thrombospondin, *J. Leukoc. Biol.* 52: 524-528.

Iruela-Arispe, M.L., Bornstein, P., and Sage, H., 1991a, Thrombospondin exerts an antiangiogenic effect on cord formation by endothelial cells *in vitro, Proc. Natl. Acad. Sci. USA* 88: 5026-5030.

Iruela-Arispe, M.L., Hasselar, P., and Sage, H. 1991b, Lab. Invest. 64: 174-186.

Jaffe, E.A., Ruggiero, J.T., and Falcone, D.J., 1985, Monocytes and macrophages synthesize and secrete thrombospondin, *Blood* 65: 79-84.

Kane, C.J.M., Hebda, P.A., Mansbridge, J.N., and Hanawalt, P.C., 1991, Direct evidence for spatial and temporal regulation of transforming growth factor $\beta 1$ expression during cutaneous wound healing, *J. Cell. Physiol.* 148: 157-173.

Koch, A.E., Friedman, J., Burrows, J.C., Haines, G.K., and Bouck, N.P., 1993, Localization of the angiogenesis inhibitor thrombospondin in human synovial tissues, *Pathobiology* 61: 1-6.

Lahav, J., 1988, Thrombospondin inhibits adhesion of endothelial cells, *Exp.Cell Res.*177: 199-204.

Lawler, J., Derick, L.H., Connolly, J.E., Chen, J.H., and Chao, F.C., 1985, The structure of human platelet thrombospondin, J. Biol. Chem. 260: 3672-3772.

Lawler, J., 1986, The structural and functional properties of thrombospondin, *Blood* 67: 1197-1209.

Mansfield, P.J., and Suchard, S.J., 1994, Thrombospondin promotes chemotaxis and haptotaxis of human peripheral monocytes, *J. Immunol.* 153: 4219-4229.

McPherson, J.H., Sage, H., and Bornstein, P., 1981, Isolation and characterization of a glycoprotein secreted by aortic endothelial cells in culture: apparent identity with platelet thrombospondin, J. Biol. Chem. 256: 11330-11336.

Murphy-Ullrich, J.E., and Höök, M., 1989, Thrombospondin modulates focal adhesions in endothelial cells, *J. Cell Biol.* 109: 1309-1319.

Mustoe, T.A., Pierce, G.F., Thomason, A., Gramates, P., Sporn, M.B., and Deuel, T.F., 1987, Accelerated healing of incisional wounds by transforming growth factor-β, *Science* 236: 1333-1336.

Nickoloff, B.J., Mitra, R.S., Varani, J., Dixit, V.M., and Polverini, P.J., 1994, Aberrant production of interleukin-8 and thrombospondin-1 by psoriatic keratinocytes mediates angiogenesis, *Am. J. Pathol.* 144: 820-828.

Nicosia, R.F., and Tuszynski, G.P., 1994, Matrix-bound thrombospondin promotes angiogenesis *in vitro, J. Cell Biol.* 124: 183-193.

Ostergaard E., and Flodgaard, H., 1992, A neutrophil-derived proteolytic inactive elastase homologue (hHBP) mediates reversible contraction of fibroblasts and endothelial cell monolayers and stimulates monocyte survival and thrombospondin secretion, *J. Leukoc. Biol.* 51: 316-323.

Polverini, P.J., DiPietro, L.A., Dixit, V.M., Hynes, R.O., and Lawler, J., 1995, Thrombospondin 1 knockout mice show delayed organization and prolonged neovascularization of skin wounds, *FASEB J.* 9: A272.

Pratt, D.A., Miller, W.R., and Dawes, J., 1989, Thrombospondin in malignant and non-malignant breast tissue, *Eur. J. Cancer Clin. Oncol.* 25: 343-350.

Rastinejad, F., Polverini, P.J., and Bouck, N.P., 1989, Regulation of the activity of a new inhibitor of angiogenesis by a cancer suppressor gene, *Cell* 56: 345-355.

Reed, M.J., Puolakkainen, P., Lane, T.F., Dickerson, D., Bornstein, P., and Sage, E.H., 1993, Differential expression of SPARC and thrombospondin 1 in wound repair: immunolocalization and *in situ* hybridization, *J. Histochem. Cytochem.* 41: 1467-1477.

Roberts, A.B., Sporn, M.B., Assoian, R.K., Smith, J.M., Roche, N.S., Wakefield, L.M., Heine, U.I., Liotta, L.A., Falanga, V., Kehrl, J.H., et al., 1986, Transforming growth factor type beta: rapid induction of fibrosis and angiogenesis *in vivo* and stimulation of collagen formation *in vitro*, Proc. Natl. Acad. Sci. USA. 83: 4167-4171.

Schultz-Cherry, S. and Murphy-Ullrich, J.E., 1993, Thrombospondin causes activation of latent transforming growth factor-β secreted by endothelial cells by a novel mechanism, *J. Cell. Biol.* 122: 923-932.

Taraboletti, G., Belotti, D., and Giavazz, R., 1992, Thrombospondin modulates basic fibroblast growth factor activities on endothelial cells, EXS 61: 210-213.

Tolsma, S.S., Volpert, O.V., Good, D.J., Frazier, W.A., Polverini, P.J., and Bouck, N., 1993, Peptides derived from two separate domains of the matrix protein thrombospondin 1 have anti-angiogenic activity, *J. Cell Biol.* 122: 497-511.

Tuszynski, G.P., Gasic, T.B., Rothman, V.L., Knudsen, K.A., and Gasic, G.J., 1987, Thrombospondin, a potentiator of tumor cell metastasis, *Cancer Res.* 47: 4130-4133.

Varani, J., Carey, T.E., Fligiel, S.E.G., McKeever, P.E., and Dixit, V.M., 1987, Tumor type-specific differences in cell-substrate adhesion among human tumor cell lines, *Int. J. Cancer* 39: 397-403.

Varani J., Stoolman, L., Wang, T., Schuger, L., Flippen, C., Dame, M., Johnson, K.J., Todd, R.F., Ryan, U.S., and Ward, P.A., 1991, Thrombospondin production and thrombospondin-mediated adhesion in U937 cells, *Exp. Cell Res.* 195: 177-182.

Weinstat-Saslow, D.L., Zabrenetzky, V.S., VanHoutte, K., Frazier, W.A., Roberts, D.D., and Steeg, P.S., 1994, Transfection of thrombospondin 1 complementary DNA into human breast carcinoma cell line reduces primary tumor growth, metastatic potential, and angiogenesis, *Cancer Res.* 54: 6504-6511.

Zabrenetzsky, V., Harris, C.C., Steeg, P., and Roberts, D.D., 1994, Expression of the extracellular matrix molecule thrombospondin inversely correlates with malignant progression in melanoma, lung, and breast carcinoma cell lines, Int. J. Cancer 59: 191-195.

SIGNAL TRANSDUCTION PATHWAYS AND THE REGULATION OF ANGIOGENESIS

Michael E. Maragoudakis

University of Patras Medical School, Dept. of Pharmacology

261 10 Patras, Greece

INTRODUCTION

All through the 1970s the long search for an elusive angiogenic factor has caused skepticism for the concept proposed by Folkman (1972) that tumors secrete promoters of angiogenesis. In 1983 Folkman and his colleagues have succeeded in isolating a heparin binding factor and a year later Guilleman's laboratory described the amino acid sequence of b-FGF (Folkman, 1985). Today we have numerous factors that have been reported to either promote or suppress angiogenesis. Many of them have been sequenced and their genes cloned. A partial list of angiogenic substances are shown in Table 1. The relative contribution of these factors in the control of physiological and pathological angiogenesis remains to be elucidated. All the angiogenic factors listed in Table 1 do not satisfy to the same extent the critiria for

Table 1. Endogenous substances with effects on vascular proliferation.

Promoters of angiogenesis	α-FGF, b-FGF, TGFa, PDECGF, ESAF, TNFa, TAF, VEGF, Prostaglandins (PGE_1, PGE2), Leucotrienes (LTC_4, TD_4), Angiotropin, Angiogenin, Adenosine-Histamine, Lactic acid, Gangliosides, Thrombin, PMA, Plasminogen activator, Hyaluronan fragments, Cu^{2+}.
Suppressors of angiogenesis	TGF_β, a-interferon, TNF_γ, $TIMP_1$, $TIMP_2$, Thrombospondin, Protamine, Hydrocortisone-plus heparin, Cartillage extract, Angiostatin, 1,25-dihydroxy-Vitamin D, Nitric oxide (NO).

Molecular, Cellular, and Clinical Aspects of Angiogenesis
Edited by Michael E. Maragoudakis, Plenum Press, New York, 1996

115

an *in vivo* angiogenic factor such as VEGF. For VEGF it has been shown that it specifically stimulates mitogenic activity and invasion into extracellular matrix and tube formation of endothelial cells only. In addition VEGF promotes angiogenesis in vivo and there is a temporal and spatial formation of VEGF in all types of angiogenesis (embryonic, ovulation and cancer). m-RNA for the receptor of VEGF and binding to the receptor during angiogenesis is demonstrated only in the proliferating endothelial cells. Furthemore, inhibition of the production of VEGF or the expression of the receptor leads to inhibition of angiogenesis and tumor growth.

The multiplicity of the factors involved in the promotion or suppression of angiogenesis in model systems has led to the concept that a sensitive balance must exist among them so that under physiological conditions angiogenesis is maintained in a quiescent state. However, the ability for initiation and activation of the complex angiogenic cascade is ever present in all tissues. It is likely that the overproduction or activation of any one of the many endogenous promoters can tip the balance and activate angiogenesis. Similarly, the disappearance or the inactivation of endogenous suppressors of angiogenesis can set off the balance and promote angiogenesis (eg. angiostatin). On the contrary overproduction or activation of angiosuppressors will stop angiogenesis. In tumors, for example, many factors can contribute to activation of angiogenesis: the angiogenic substances released by tumor cells, the proteolytic enzymes that degrade extracellular matrix and release sequestered b-FGF, laminin, peptides, etc., the invasion of macrophages and mast cells which release angiogenic substances, the accumulation of thrombin and fibrin as a result of bleeding etc. (Folkman and Shing, 1992, Tsopanoglou et al., 1993).

Activation of angiogenesis can take place in every tissue at a moments notice in situations such as wound healing and inflammation. In that respect angiogenesis is analogous to blood coagulation. Both blood coagulation and angiogenesis require a multiplicity of factors which are involved in the events of a complex cascade. However, in angiogenesis the interrelation and interaction of the various factors has not been elucidated as yet. The many endogenous suppressors and promoters of angiogenesis act through receptors each connected to a particular transduction system. Like blood coagulation, angiogenesis is tightly regulated and involves the interaction of endothelial cells with other cell types such as pericytes, smooth muscle cells, mast cells, inflammatory cells and extracellular matrix proteins. The control of these events both temporally and specially results in the sensitive regulation of angiogenesis under physiological conditions.

It is reasonable to assume that the complexity of such a system requires that overall controls of the multiple transduction mechanism must exist so that a sensitive and effective regulation of neovascularization or disappearance of blood capillaries is maintained under physiological or pathological conditions.

Signal transduction systems in eucaryotic cells

In eucaryotic cells there are five principal signalling systems through which most extracellular agonists exert their effects on cells by activating or inhibiting transmembrane signalling system that control the production of second messengers (Cohen, 1992). These second messengers modulate the activities of protein kinases and protein phosphatases, which catalyse phorphorylation (or dephosphorylation) of serine or threonine and occasionally tyrosine residues of cellular proteins. These events cause conformational changes in proteins leading to physiological responses that are characteristic of a particular agonist. The diversity of the actions of agonists can be explained by the isoforms of the kinases and phosphatases and by the presence or absence of the target proteins on which they act, as well as their intracellular location.

Table 2 summarizes the second messengers involved in the activation of protein kinases, which phosphorylate many cellular proteins with broad or restricted specificity. Agonists such as insulin and growth factors bind to receptors that are protein tyrosine kinases and become autophosphorylated upon activation. This is followed by activation of many protein Ser/Thr. kinases and phosphorylation of intracellular proteins. However, the molecular mechanisms leading from the activation of receptor protein tyrosine kinase to Protein Ser/Thr. kinase activation have not been elucidated. It is possible that a generation of a novel second messenger "X" is involved (Cohen, 1992).

Table 2. Principal signalling systems that operate in eucaryotic cells.

Signal by extracellular agonists	Membrane receptor (R)	Second messenger	Activation of	
1	R	c-AMP $\longrightarrow$	PKA	
2	R	c-GMP $\longrightarrow$	PKG	Multifunctional protein kinases
3	R	Ca^{2+} $\longrightarrow$	Calmodulin	Dedicated protein kinases
4	R	Diacylglycerol $\longrightarrow$ PKC		
5	Receptor protein tyrosine kinase	"X" $\longrightarrow$	Protein Ser/ Thr. kinases	

The abbreviations used are: PKA, c-AMP-dependent protein kinase; PKG, c-GMP-dependent protein kinase; PKC, protein kinase C; "X", hypothetical second messenger. Modified from Cohen (1992).

Since the discovery of nitric oxide (NO) seven years ago, its involvement in several distinct signalling pathways have been well established (Knowles and Moncada, 1992). NO is the smallest molecule and the first gas, known to act as a biological mediator in mammals. It is now known that NO is released from a variety of cell types and is involved in an evergrowing number of biological functions such as vasodilatation (Amezcua et al., 1989), cytotoxicity (Hibbs et al., 1988), inhibition of protein synthesis (Nakaki et al., 1990) neurotransmission (Snyder and Bredt, 1992), inhibition of platelet aggregation (Radomski et al., 1990) etc. In mammals, NO is formed enzymatically from a terminal quanidino-nitrogen of the amino acid L-arginine (Palmer et al., 1988, Schmidt et al., 1988) by a gene family of NO synthases (NOS) (Forsterman et al., 1991). L-citrulline is formed as a co-product of this reaction. NOS (I and II isoforms), are constitutivelly expressed in blood vessels and brain and are Ca^{2+}-Calmodulin dependent. In macrophages, NOS (II isoform) is expressed after immunological activation and is Ca^{2+}-independent/Calmodulin-dependent (Palmer and Moncada, 1988; Mayer et al., 1989). The main effect of low concentrations of NO is stimulation of the heme protein soluble guanylate cyclase to produce cGMP (Schmidt et al., 1993). High concentrations of NO which are generated by the immunologically induced NOS-II do not involve guanylate cyclase activation (Schmidt et al., 1992). Under those circumstances NO can interact with almost every cellular protein and in many cases these reactions with NO lead to loss of function of the entire cell or of particular enzymes. Activated macrophages, which possess a NOS-II yielding high amounts of NO, may utilize the above reactions as non-specific immune defence mechanism against bacterial protozoan and possibly viral infections.

The aforementioned signalling systems do not operate independently of one another. Instead they interact with each other at many levels so that different extracellular signals may have additive, synergistic or antagonistic effects depending on the situation. For example, it has been reported that NO and NO-generating agents induce a reversible inactivation of PKC activity and phorbol ester binding (Gopalakrishna et al., 1993). The intergration of the extracellular signals takes place at the level of protein kinases and phosphatases and their protein substrates.

Regulation of angiogenesis by the signal transduction pathways of PKA and PKC.

We began to study the roles of the two major signalling pathways c-AMP-dependent protein kinase (PKA) and PKC with the understanding that as mentioned above all the signalling systems do not operate independently but interact with each other in every conceivable way.

PKA and PKC mediate the effects of growth factors, hormones, neurotransmitters and oncogenes on cell proliferation and cellular responses. We have demonstrated that they are also involved in the signal transduction mechanisms that regulate angiogenesis (Tsopanoglou et al., 1993 and 1994).

The physiological importance of PKC in the regulation of cellular functions is well documented in many systems. It is now clear that at least seven subtypes of PKC can be distinguished, one of which is expressed only in the central nervous system. These subtypes of PKC are derived from multiple genes and from splicing of a single m-RNA transcript. PKC subtypes are located in particular cell types and intracellular locations and show subtle differences in their mode of activation, their sensitivity to Ca^{2+} and their catalytic activity (Iwamoto et al., 1992).

Activation of PKC is an integral part of the signal-induced degradation of various membrane phosholipids, which is catalysed by phospholipase A, C and D. Phospholipase C via a receptor-activated G-protein generates diacylglycerol (DG) and activates PKC. At the same time the generation of IP_3 causes the release of Ca^{2+} from intracellular Ca^{2+} stores and activates Ca^{2+} receptive proteins. Thus the cellular responses elicited by PKC activation are separate from and synergistic to those related to activation by the increase in intracellular Ca^{2+}. The endogenously generated DG and phorbol esters, which share common structural features with DG, activate PKC. The hydrolysis of inositol phospholipids was initially thought to be the only mechanism leading to activation of PKC. More recent studies suggest that DG from several other sources can be generated and activate PKC. For example, DG may be generated from phosphatidylcholine at a relatively later stages of cellular responses to growth factors.

We have demonstrated that PKC plays a key role in the signal transduction mechanisms that regulate angiogenesis (Tsopanoglou et al., 1993 and 1994). In the CAM system of angiogenesis we have shown that activators of PKC such as 4-β-phorbol-12-myristate-13-acetate (4-β-PMA) or 1,2-dioctanoyl-glycerol (DiC$_8$) cause a dose-dependent promotion of angiogenesis. The effect is specific since the inactive analogs of 4-α-PMA and diolein, which do not activate PKC, are not promoters of angiogenesis. In addition, the specific inhibitor of PKC, Ro-318220, causes a dose-dependent suppression of basal angiogenesis and also reverses the angiogenesis-promoting effect of 4-β-PMA and DiC$_8$ to levels below control. This suggests that both basal and stimulated angiogenesis are controlled by PKC activity. Similar conclusions were reached from experiments with the *in vitro* system of angiogenesis using endothelial cells on Matrigel (Tsopanoglou et al., 1994). In the Matrigel system endothelial cells from human umbilical cord (HUVECS) form tube-like structures. The PKC inhibitor Ro-318220 prevents tube formation. These results suggest that PKC activation may be one of the intracellular signals for switching capillary endothelial cells from the proliferative into a differentiated state expressed by their ability to form tube-like structures.

The tumor promoting effects of phorbol esters are well known (Castagna et al., 1982) and their primary target is PKC. This phenomenon may be related to their angiogenesis-promoting effect (Tsopanoglou et al., 1993). In addition PKC has been recently shown to be inactivated by NO (Gopalakrishna et al., 1993). Since NO is a suppressor of angiogenesis (Pipili-Synetos et al., 1993 and 1994), the fact that it causes also inactivation of PKC further supports the importance of the latter in the signalling events of the angiogenic cascade.

Contrary to PKC activation, when activators of PKA were used in the CAM or the Matrigel system of angiogenesis resulted in suppression of angiogenesis. Thus Forskolin or Sp-diastereoisomer of adenosine-3´,5´-cyclic monophosphothionate (Sp-c-AMPS) caused inhibition on basal angiogenesis. The inactive analog Rp-c-AMPS, which is a competitive inhibitor of c-AMP, was without effect. When angiogenesis was stimulated by 4-β-PMA both Forskolin and Sp-c-AMPS could reverse this stimulatory effect. We conclude from these studies that activators of PKC and PKA have opposing effects on angiogenesis (Tsopanoglou et al., 1994). In the regulation of angiogenesis, therefore, we are dealing with a cross-talk phenomenon between the two major signalling pathways. This appears to be a general phenomenon in many biological systems and Houslay (1991) has reviewed the wealth of evidence available in the literature, which suggest a complex and cell type specific modulation of c-AMP signalling pathways by PKC. Both increases and decreases in functioning have been observed.

The role of NO-signal transduction system in the regulation of angiogenesis.

Recently we have presented evidence that NO is an endogenous anti-angiogenic mediator (Pipili-Synetos et al., 1994 and 1995). The involvement of NO in the regulation of angiogenesis has been demonstrated in the CAM system of angiogenesis (Maragoudakis et al., 1988) and the Matrigel tube formation assay of angiogenesis *in vitro* (Harabopoulos et al., 1994). It was shown that sodium nitroprusside (SNP), which releases spontaneously NO, caused a dose-dependent inhibition of angiogenesis in the CAM and stimulated the release of c-GMP. In the Matrigel system both c-GMP and SNP inhibited tube formation. The competitive inhibitors of NO synthase, L-NMMA and L-NAME stimulated angiogenesis in the CAM in a dose dependent fashion. On the other hand the inactive analogs D-NMMA and D-NAME were without effect on angiogenesis. L-arginine, the precursor of the endogenously formed NO, had a modest antiangiogenic effect by itself, but when it was used in combination with either L-NMMA or L-NAME it abolished their angiogenic effect. Both NO-releasing agents, such as SNP and isosorbide mononitrate (ISMN) or NO-preserving agents such as superoxide dismutase (SOD) prevented the stimulation of angiogenesis by a-thrombin and phorbol esters (PMA), (Tsopanoglou et al., 1994, Pipili-Synetos et al., 1994 and 1995).

All these experiments suggest that NO plays the role of an effective antiangiogenic mediator. The mechanism for the antiangiogenic effect of NO remains to be elucidated. It does not seem to involve direct effect on endothelial cells, since NO-releasing agents, such as ISMN, had not effect on the proliferation rate of endothelial cells from various sources (Pipili-Synetos et al., 1995). A likely target for

NO might be the blood platelets (Radomski and Moncada, 1991).

This newly described biological action of NO on angiogenesis suggests a possible involvement of NO in tumor growth and metastasis. Indeed it was shown recently that the NO-releasing vasodilator ISMN exhibited a marked suppression of tumor size and pulmonary metastatic foci of Lewis lung carcinoma in mice (Pipili-Synetos et al., 1995). The antitumor effect of ISMN is most likely related to its antiangiogenic action, since there is no direct cytotoxic effects of ISMN on endothelial cells from different sources or on Lewis lung carcinoma cells in culture (Pipili-Synetos, 1995). It remains to be seen whether these observations on the antitumor effects of ISMN can be obtained using other NO-releasing agents and other tumor models. If this holds true, then the possibility exists for new therapeutic indication for old established drugs, such as nitrovasodilators, as anticancer agents.

In addition to the role of NO in angiogenesis there are many reports in the literature associating NO and cancer: Mantley et al (1994) has shown that release of NO correlates with killing of tumor cells by activated murine macrophages. Cui et al. (1994) has suggested that NO mediates the macrophase-induced apoptosis in tumor cells. The release of NO is inversely correlated with the ability of K1735 murine melanoma cells to metastasize to lungs (Dong et al., 1994). The receptor-mediated release of NO is inhibited by the tumor promoting phorbol esters (Severn et al., 1992). This suggests a negative regulation of NO release by PKC activation. Conversely, Gopalakrishna et al. (1993) have shown that NO and NO-releasing agents cause a reversible inactivation of PKC and phorbol ester binding to PKC. Since activation of PKC promotes angiogenesis (Tsopanoglou et al., 1993 and 1994), the inactivation of PKC by NO may be the underlying molecular mechanism for the antiangiogenic effect of NO.

All the aforementioned experimental evidence point to a possible role of NO as a negative regulator of tumor growth and metastasis. However, there are reports pointing to a tumor-promoting effect of NO. Maeda et al. (1994) showed that NO enhances vascular permeability in solid tumors, which might facilitate tumor metastasis. Others have suggested that NO along with other oxygen radicals is involved in carcinogenesis as result of chronic inflammation (Oshima and Barsch, 1994).

DISCUSSION

We conclude from these studies that activation of PKC and PKA modulate angiogenesis with opposing effects. The role of NO as negative regulator of angiogenesis and the interrelation of NO-release with PKC activity is probably important in tumor growth and metastasis. The relative importance of these transduction mechanisms and their short and long term effects upon cellular functioning may shed light into the understanding of the complex phenomenon of angiogenesis in health and disease.

These type of studies on the regulation of angiogenesis may also have practical implications in the development of both inhibitors and promoters of angiogenesis for potential clinical applications (Folkman, 1995). Given the multiplicity of angiogenic factors present in any given tissue, it is unlikely that inhibition of one of them eg. b-FGF could be effective and a practical approach to suppress angiogenesis. The tissue most likely will overproduce another factor to compensate for the missing one. A more realistic strategy could be to intervene with critical components of the cellular signalling system that activates angiogenesis. The PKC and NO signal transduction systems may be such a target. PKC has pleiotropic effects and as such represents a complicated target for drug development. However, the fact that many isoenzymes of

PKC exist, that mediate different cellular processes, suggests the possibility of identifying the specific isoenzyme of PKC that activates angiogenesis. Selective inhibitors to that enzyme are likely to be useful antiangiogenic and anti-tumor agents. In view of the fact that phorbol esters are tumor promoters and activators of PKC, many investigators have embarked a search for novel PKC inhibitors for anticancer therapy (Powiss, 1991). It is of interest that tamoxifen, the non-steroidal antiestrogen, and doxorubicin, both widely used anti-cancer drugs are inhibitors of PKC activity (O'Brian et al., 1988; Hannun et al., 1989).

Therefore, based on these observations and given the molecular heterogeneity of PKC family of enzymes and their functional divergence, PKC may be an attractive target for developing angiosuppressors with antitumor properties.

ACKNOWLEDGEMENTS

This work was supported in part by a NATO Collaborative Research Grant #CRG 940677 and the Greek Ministry of Science and Technology.
I thank Anna Marmara for typing the manuscript.

REFERENCES

Amezcua, J.L., Palmer, R.M.J., De Souza, B.M. and Moncada, S., 1989, Nitric oxide synthesized from L-arginine regulates vascular tone in the coronary circulation of the rabbit, *Br. J. Pharmacol.*, 97:1119-1124.

Castagna, M., Takai, Y., Kaibuchi, K., Sano, K., Kikkawa, U., and Nishizuka, Y., 1982, Direct activation of calcium activated, phospholipid dependent protein kinase by tumor promoting phorbol esters, *J. Biol. Chem.*, 257:7847-7851.

Cohen, P., 1992, Signal integration at the level of protein kinases, protein phosphateses and their substrates, *TIBS*, 17:408-413.

Cui, S.J., Reichner, J.S., Mateo, R.B., and Albina, J.E., 1994, Activated murine macrophages induce apoptosis in tumor cells through nitric oxide-dependent or -independent mechanisms, *Cancer Res.*, 54:2462-2467.

Dong, Z.Y., Starolesky, A.H., Qi, X.X., Xie, K.P., and Fidler, I.J., 1994, Inverse correlation between expression of nitric oxide synthase activity and production of metastasis in K-1735 murine melanoma cells. *Cancer Res.*, 54:789-793.

Folkman J., 1972, Anti-angiogenesis: New concept for therapy of solid tumors, *Ann. Surg.*, 409-416.

Folkman, J., 1985, Towards an understanding of angiogenesis, Search and Discovery, *Prospect. Biol. Med.* 29:10-35.

Folkman, J. and Shing, Y., 1992, Angiogenesis: Minireview, *J. Biol. Chem.*, 267:10931-10932.

Folkman, J., 1995, Angiogenesis in cancer, vascular, rheumatoid and other disease, *Nature Medicine* 1:27-31.

Forstermann, U., Schmidt, H.H.H.W., Pollock, J.S., Sheng, H., Mitchell, J.A., Warner, T.D., Nakane, M. & Murad, F. ,1991, Commentary. Isoforms of nitric oxide synthase. Characterization and purification from different cell types, *Biochem. Pharmacol.*, 42(10):1849-1857.

Iwamoto, Y., Koide, H., Ogita, K., and Nishizuka, Y., 1992, The protein kinase C family for the regulation of cellular functions, *Bio Med. Reviews*, 1:1-6.

Gopalakrishna, R., Hai Chen, Z., and Gundimeda, U., 1993, Nitric oxide and nitric oxide generating agents induce a reversible activation of protein kinase C activity and phorbol ester binding, *J. Biol. Chem.*, 268:27180-27185.

Hannun, Y.A., Foflesong, R.J., and Bell, R.M., 1989, The adriamycin-iron (III) complex is a potent inhibitor of protein kinase C. *J. Biol. Chem.*, 264:9960-9966.

Haralabopoulos, G.C., Grant, D.S., Kleinman, H.K., Lelkes, P.I., Papaioannou, S.P. and Maragoudakis, M.E., 1994, Inhibitors of basement collagen biosynthesis prevent endothelial cell alignment in Matrigel *in vitro* and angiogenesis *in vivo*. *Lab. Invest.* 71(4):575-582.

Hibbs, J.B., Taintor, R.R, Vavrin, Z., and Rachlin, E.M., 1988, Nitric oxide: A cytotoxic activated macrophage effector molecule, *Biochem. Biophys. Res. Commun.*, 157:87-94.

Houslay, M.D., 1991, "Crosstalk": A pivotal role of protein kinase C in modulating relationships between signal transduction pathways, *Eur. J. Biochem.*, 195:9-27.

Knowles, R.G., Moncada, S., 1992, Nitric oxide as a signal in blood vessels, *TIBS* 17:399-402.

Maeda, H., Noguchi, Y., Sato, K., and Akaike, T., 1994, Enhanced vascular permeability in solid tumor is mediated by nitric oxide and inhibited by both new nitric oxide scavenger and nitric oxide synthase inhibitor, *Japan. J. Cancer Res.*, 85:331-334.

Mantley, C. L., Perera, P.Y., Salkowski, C.A., and Vogel, S.N., 1994, Taxol provides a second signal for murine macrophages tumoricidal activity, *J. Immunol.*, 152: 825-831

Maragoudakis, M.E., Sarmonika, M., Panoutsakopoulou, M, 1988, Inhibition of basement membrane biosynthesis prevent angiogenesis, *J. Pharm. & Exper. Ther.*, 244:729-733.

Maragoudakis, M.E., Sarmonika, M., and Panoutsakopoulou, M., 1988, Rate of basement membrne biosynthesis as an index to angiogenesis, *Tissue & Cell*, 20 (4):531-539.

Mayer, B., Schmidt, K., Humbert, P., and Bohme, E., 1989, Biosynthesis of endothelium derived relaxing factor: a cytosolic enzyme in porcine aortic endothelial cells Ca^{++} dependently converts L-arginine into an activator of soluble guanylate cyclase, *Biochem. Biophys. Res. Commun.* 164:678-685.

Nakaki, T., Makayama, M., and Kto, R., 1990, Inhibition by nitric oxide and nitric oxide-producing vasodilators of DNA synthesis in vascular smooth muscle cells, *Eur. J. Pharmacol.*, 189:347-353.

O'Brian, C.A.,Housey, G.M., and Weinstein, I.B., 1988, Specific and direct binding of protein kinase C to an immobilized tamoxifen analog, *Cancer Res.* 48:3626-3629.

Oshima, H., and Barsch, H., 1994, Chronic inflammatory processes as cancer risk factors. Possible role of nitric oxide in carcinogenesis, *Mutation Res.*, 305:253-264.

Palmer, R.M.J., Ashton, D.S., and Moncada, S., 1988, Vascular endothelial cells synthesize nitric oxide from L-arginine, *Nature* (Lond.), 333:664-666.

Pipili-Synetos, E., Sakkoula, E., Haralabopoulos, G., Andriopoulou, P., Peristeris P., and Maragoudakis, M.E, 1994, Evidence that nitric oxide is an endogenous antiangiogenic mediator, *Br. J. Pharmacol.*, 111:194-202.

Pipili-Synetos, E., Papageorgiou, G., Sakkoula, E., Sotiropoulou, G., Fotsis, T., Karakiulakis, G., and Maragoudakis, M.E., 1995, Inhibition of angiogenesis, tumor growth and metastasis by the NO-releasing vasodilators, isosorbide mononitrate and dinitrate, *Br. J. Pharmacol.*, 116:1829-1834.

Powiss, G., 1991, Signalling targets for anticancer drug development. *Trends in Pharm. Sci.*, 12:188-194.

Radomski, M.W., Palmer, R.M.J., and Moncada, S., 1990, An L- arginine/nitric oxide pathway present in human platelets regulates aggregation, *Proc. Natl. Acad. Sci. U.S.A.*, 87:5193-5197.

Schmidt, H.H.H.W.,1992, NO., CO and OH, Endogenous soluble guanylate cyclase-activating factors, *FEBS Letters*, 307 (1):102-107.

Schmidt, H.H.H.W., Lohman, S., and Walter, U., 1993, Minireview, The nitric oxide and c GMP signal transduction system: regulation and mechanism of action, *Biochim. Biophys. Acta* 1178:153-175.

Severn, A., Wakelam, M.J.O., Liew, F.Y., 1992, The role of protein kinase C in the induction of nitric oxide synthesis by murine macrophages, *Biochim. Biophys. Res. Com.* 188:997-1002.

Snyder, S.H., and Bredt, D.S., 1992, Biological roles of nitric oxide, *Scientific American*, May, 28-35.

Tsopanoglou, N.E., Pipili-Synetos, E., and Maragoudakis, M.E., 1993, Thrombin promotes angiogenesis by a mechanism independent of fibrin formation, *Am. J. Physiol.* 264 (Cell Physiol. 33):C1302-1307.

Tsopanoglou, N.E., Pipili-Synetos E. and Maragoudakis, M.E., 1993, Protein kinase C involvement in the regulation of angiogenesis. *J. Vasc. Res.*, 30:202-208.

Tsopanoglou, N.E., Haralabopoulos, G.C., and Maragoudakis, M.E., 1993, Opposing effects on modulation of angiogenesis by protein kinase C and cyclic AMP-mediated pathways. *J. Vasc. Res.* 31:195-204.

ANGIOGENESIS IN THE FEMALE REPRODUCTIVE ORGANS

Lawrence P. Reynolds,[1] Dale A. Redmer,[1]
Anna T. Grazul-Bilska,[2] S. Derek Killilea,[3]
and R. M. Moor[4]

[1]Department of Animal & Range Sciences,
[2]Cell Biology Center, and [3]Department of
Biochemistry, North Dakota State University,
Fargo, ND 58105, U.S.A., and
[4]Department of Development and Signalling,
Babraham Institute, Babraham,
Cambridge CB2 4AT, U.K.

INTRODUCTION

Angiogenesis refers to the formation of new blood vessels, or neovascularization, and is an essential component of tissue growth and development (Folkman and Klagsbrun, 1987; Hudlicka, 1984; Klagsbrun and D'Amore, 1991; Shepro and D'Amore, 1984). The angiogenic process begins with capillary proliferation and culminates in the formation of a new microcirculatory bed, composed of arterioles, capillaries and venules (Folkman and Klagsbrun, 1987; Hudlicka, 1984; Klagsbrun and D'Amore, 1991; Shepro and D'Amore, 1984). The initial component of angiogenesis, capillary proliferation, has been shown to consist of at least three processes: 1) fragmentation of the basal lamina of the existing vessel, 2) migration of endothelial cells (the primary cell type comprising capillaries) from the existing vessel toward the angiogenic stimulus, and 3) proliferation of endothelial cells (Folkman and Klagsbrun, 1987; Klagsbrun and D'Amore, 1991; Shepro and D'Amore, 1984). Neovascularization is completed by formation of capillary lumina and differentiation of the newly formed capillaries into arterioles and venules (Hudlicka, 1984; Klagsbrun and D'Amore, 1991).

In most adult tissues, capillary growth occurs only rarely, and the vascular endothelium represents an extremely stable population of cells with a low mitotic rate (Denekamp, 1984; Hudlicka, 1984; Klagsbrun and D'Amore, 1991). Angiogenesis does occur in adults, however, during tissue repair, such as in the healing of wounds or fractures (Hudlicka, 1984; Klagsbrun and D'Amore, 1991). Based upon these observations, angiogenesis in normal adult tissues has been likened to processes such as blood clotting, which must remain in a constant state of readiness yet must be held in check for long periods of time (Folkman and Klagsbrun, 1987).

Molecular, Cellular, and Clinical Aspects of Angiogenesis
Edited by Michael E. Maragoudakis, Plenum Press, New York, 1996

Angiogenesis is therefore thought to be regulated by angiogenic as well as anti-angiogenic factors (Folkman and Klagsbrun, 1987; Hudlicka, 1984).

Rampant or persistent capillary growth is associated with numerous pathological conditions, including tumor growth, retinopathies, hemangiomas, fibroses, and rheumatoid arthritis (Beranek, 1988; Folkman and Klagsbrun, 1987; Hudlicka, 1984; Klagsbrun and D'Amore, 1991). Folkman and co-workers (reviewed by Folkman and Klagsbrun, 1987) demonstrated that development of new capillaries is requisite for sustained growth of tumors. In addition, vascular endothelial cells of growing tumors exhibit an extremely high mitotic rate compared with endothelial cells of most normal tissues (Denekamp, 1984). Conversely, insufficient capillary growth occurs in several disease states, including delayed wound healing, nonhealing fractures, and chronic varicose ulcers (Burgos et al., 1989; Folkman and Klagsbrun, 1987). Because regulation of the angiogenic process could become the primary method of treatment for these disorders (Beranek, 1988; Burgos et al., 1989; Denekamp, 1984; Folkman and Klagsbrun, 1987), and because capillary growth occurs infrequently in normal adult tissues, most of the work on angiogenesis and its regulation has focused on pathological or developmental processes (Folkman and Klagsbrun, 1987; Hudlicka, 1984; Klagsbrun and D'Amore, 1991).

The female reproductive organs (i.e., ovary, uterus, and placenta) contain some of the few adult tissues that exhibit periodic growth and regression. In addition, growth and regression of these tissues is extremely rapid (Jablonka-Shariff et al., 1993, 1994; Nicosia et al., 1995; Reynolds et al., 1994; Zheng et al., 1994; Ricke et al., 1995a). In fact, the labelling index, which is an index of the rate of cellular proliferation, of early-cycle corpora lutea (CL) is comparable to that of regenerating liver or rapidly growing tumors (Grisham, 1962; Denekamp, 1984; Jablonka-Shariff et al., 1993; Zheng et al., 1994; Christenson and Stouffer, 1995; Nicosia et al., 1995; Ricke et al., 1995a). Uterine and placental tissues also exhibit impressive growth during gestation (Meschia, 1983; Ferrel, 1989; Reynolds et al., 1990). It is not surprising, therefore, that female reproductive tissues are some of the few adult tissues in which angiogenesis occurs as a normal process (Barcroft, 1947; Bassett, 1943; Findlay, 1986; Hudlicka, 1984; Klagsbrun and D'Amore, 1991; Ramsey, 1989; Reynolds et al., 1992).

Rapid growth and regression of female reproductive tissues is accompanied by equally rapid changes in rates of blood flow (Ferrell, 1989; Ramsey, 1989; Reynolds, 1986; Reynolds and Redmer, 1995). When fully functional, ovarian, uterine and placental tissues receive some of the greatest rates of blood flow, on a weight-specific basis, of any tissues in the body (Reynolds, 1986; Reynolds and Redmer, 1995). In addition, vascular endothelial cells of these tissues exhibit mitotic rates equal to or greater than those of tumor endothelial cells (Denekamp, 1984; Gaede et al., 1985; Jablonka-Shariff e t al., 1993; Christenson and Stouffer, 1995). Because the tissues of the female reproductive system are so dynamic, they provide a unique model for studying regulation of angiogenesis during growth, differentiation and regression of normal adult tissues.

PATTERNS OF GROWTH AND VASCULAR DEVELOPMENT

Ovarian Tissues

The extensive development of ovarian vascular beds was noted by early investigators, who recognized that capillary growth may be of primary importance in the growth and selection of ovulatory follicles as well as in the subsequent development and function of corpora lutea (Andersen, 1926; Bassett, 1943; Clark, 1900). These studies showed that the capillary network of preovulatory follicles was more extensive than that of other follicles (Andersen, 1926; Clark,

1900). Based upon this observation, it was proposed that the initiation and maintenance of follicular growth depends upon development of the follicular microvasculature.

More recent evidence has led to renewed interest in angiogenesis as a regulator of follicular growth and atresia (Greenwald and Terranova, 1988; Moor and Seamark, 1986). For instance, Zeleznik *et al.* (1981) found that all preovulatory follicles of monkeys had similar concentrations of gonadotropin binding sites; however, only the follicle destined to ovulate (the "dominant" follicle) became heavily labelled after intravenous injection of radiolabelled gonadotropin. This selective in vivo uptake of gonadotropin was associated with increased vascularity of the dominant follicle (Zeleznik *et al.*, 1981), and was observed even before the size of the dominant follicle differed from that of other follicles (DiZeraga and Hodgen, 1980).

Morphometric measurements have shown not only increased vascularity of dominant follicles, but also decreased vascularity of follicles undergoing atresia (Moor and Seamark, 1986; Zeleznik *et al.*, 1981). Greenwald (1989) recently showed that reduced DNA synthesis of follicular endothelial cells, in association with reduced follicular vascularity, was one of the earliest signs of atresia. Moor and co-workers (reviewed in Moor and Seamark, 1986) found that atretic follicles will regenerate when placed into culture, and suggested that reduced vascularity of atretic follicles in vivo may limit their access to nutrients, substrates and tropic hormones, thereby maintaining these follicles in an atretic state.

Vascular development of the ovarian follicle becomes even more impressive after ovulation, in association with development of the CL (Andersen, 1926; Bassett, 1943; Clark, 1900). The capillary network of the mature CL is so extensive, in fact, that the majority of parenchymal (steroidogenic) cells are adjacent to one or more capillaries (Dharmarajan *et al.*, 1985; Zheng *et al.*, 1993). The mature CL also receives most of the ovarian blood supply, and ovarian blood flow is highly correlated with progesterone secretion (Magness *et al.*, 1983; Niswender and Nett, 1988; Reynolds, 1986). Conversely, inadequate luteal function has been associated with decreased luteal vascularization (Jones *et al.*, 1970), and several investigators have suggested that reduced ovarian blood flow plays a role in luteal regression (Niswender and Nett, 1988; Reynolds, 1986).

Uterus

The mucosal, innermost layer of the uterus is termed the endometrium. In primates, growth of the endometrial vasculature begins during the proliferative (follicular) phase and continues throughout the secretory (luteal) phase of the menstrual cycle (Meschia, 1983, Ramsey, 1989). Associated with endometrial capillary proliferation is increased DNA synthesis of vascular endothelial cells (Ferenczy *et al.*, 1979). During the follicular phase, growth of the spiral arteries, which supply the subepithelial capillary plexus of the primate endometrium, is thought to occur in response to follicular estrogens (Meschia, 1983; Ramsey, 1989).

Endometrial growth and vascular development in response to systemic concentrations of ovarian steroids was demonstrated in the classic experiments of Markee (Markee, 1932; Markee, 1940). In these studies, endometrial explants of rabbits or monkeys were transplanted to the anterior chamber of the eye. Not only did these explants quickly recruit a vascular supply, but they also underwent cyclic periods of growth and regression that were associated with changes in systemic concentrations of steroids. More recently, Abel (1985) observed rapid vascularization and growth of human endometrial explants that were transplanted to the hamster cheek pouch. Other than descriptive histological observations (Abel, 1985; Eckstein and Zuckerman, 1952; Markee, 1932; Markee, 1940), quantitative data on the effects of ovarian steroids on uterine vascularity are sparse. Nevertheless, ovarian steroids may play a role in regulating vascular growth and development of uterine tissues, a possibility which we will address later in this review.

The rate of blood flow to uterine tissues varies regularly throughout the nonpregnant cycle, being greatest at or just before ovulation and least during the luteal phase of the cycle (Meschia, 1983; Reynolds, 1986). These regular patterns of uterine blood flow are temporally associated with the ratio of estrogen to progesterone in systemic blood (Meschia, 1983; Reynolds, 1986). In addition, administration of estrogen to ovariectomized animals stimulates an increase in uterine blood flow in those mammalian species studied to-date (Meschia, 1983; Reynolds, 1986). Progesterone, while having little direct effect, seems to modulate the responsiveness of the uterine vascular bed to vasodilatory agents (Reynolds, 1986).

We recently have evaluated growth and vascular development of the uterus in cyclic (nonpregnant) ewes and also in ovariectomized, steroid-treated ewes (Reynolds *et al.*, 1992; *Reynolds et al., 1993b*). In cyclic ewes, fresh and dry weights of uterine horns were ≈ 40% greater at estrus (day 0 = estrus; length of the estrous cycle = 16 days) than at day 8 after estrus (*Reynolds et al., 1993b*). Fresh and dry weights of uterine horns remained constant from days 8-15 after estrus. Total microvascular volume followed a pattern similar to that of tissue growth, being greatest at estrus and least on days 8-15 after estrus, such that microvascular volume density (i.e., microvascular volume as a percentage of tissue volume) of endometrial tissues did not differ among days of the estrous cycle, and averaged 8.5 ± 0.2% (*Reynolds et al., 1993b*). These data indicate that, in the endometrium, growth of vascular tissues is highly coordinated with growth of nonvascular tissues, since the density of microvasculature remained constant throughout the estrous cycle.

In ovariectomized ewes, exogenous estradiol and/or progesterone caused dramatic increases in uterine fresh and dry weights (Johnson *et al.*, 1993; Reynolds *et al.*, 1995). For example, within 24-48 h, uterine fresh and dry weights increased by 2.5-3-fold in ovariectomized ewes receiving estradiol implants. Additionally, steroid treatment restored endometrial histology and protein secretion to levels similar to those of intact ewes (*Reynolds et al., 1993b*). As we had observed for cyclic ewes, microvascular volume followed a similar pattern to that of uterine weight, and therefore microvascular volume density remained constant (*Reynolds et al., 1993b*). Thus, endometrial vascular and nonvascular growth again were highly coordinated, since the density of the microvasculature did not change during the steroid-induced uterine growth response.

Placenta

In eutherian mammals, the definitive placenta consists of maternal (endometrial) and fetal (chorioallantoic) tissues and is the site of physiological exchange between the maternal and fetal systems (Ramsey, 1982; Reynolds and Redmer, 1995). Placental vascular growth begins early in pregnancy and continues throughout gestation (Barcroft, 1947; Ferrel, 1989; Meschia, 1983; Reynolds *et al.*, 1990; Reynolds and Redmer, 1995). Associated with placental vascular growth is a continual and dramatic increase in rates of uterine and umbilical blood flows throughout gestation (Meschia, 1983; Reynolds *et al.*, 1986; Reynolds and Redmer, 1995). Increased blood flow to placental tissues satisfies the steadily increasing metabolic demands of fetal growth (Meschia, 1983; Ferrell, 1989; Reynolds and Redmer, 1995). The importance of placental vascular development in supporting fetal growth and development has long been recognized (Barcroft, 1947; Ferrell, 1989; Meschia, 1983; Teasdale, 1976; Reynolds and Redmer, 1995). Inadequate placental vascular development may be a major contributor to embryonic wastage and reduced birth weights, which are major socioeconomic problems associated with pregnancy (Ferrell, 1989; Meegdes, 1988).

During early pregnancy in ewes, interdigitation of fetal membranes (presumptive cotyledon) with maternal caruncles, which is analogous to implantation, begins by days 20-25 after mating but is not well established until day 30 (Assheton, 1906; Boshier, 1969; Wimsatt,

1950). Until day 30, therefore, the metabolic demands of conceptus development, including growth of the embryo and fetal membranes as well as embryonic organogenesis, are highly dependent upon the uterine lumenal environment (Green and Winters, 1945; Koos and Olson, 1989; Nephew et al., 1989). It is not surprising, therefore, that the majority of embryonic loss, representing ≈ 30% of potential offspring, occurs during early pregnancy (Edey, 1969). In association with the metabolic demands of the developing conceptus, a marked increase in uterine blood flow is observed during early pregnancy in ewes (Greiss and Anderson, 1970; Reynolds et al., 1984).

We recently have observed not only growth of the uterus, but also a large (> 50%) increase in the density of the endometrial microvasculature, by day 24 after mating in ewes (Reynolds and Redmer, 1992; Reynolds et al., 1993b). When the increases in uterine weight and microvascular density are both accounted for, total endometrial microvascular volume increased by approximately two-fold during early pregnancy. By day 14 after mating, the ovine conceptus begins to produce estrogens, and uterine lumenal content of estradiol is elevated (Nephew et al., 1989; Reynolds et al., 1984). Thus increased uterine weight, vascularity, and blood flow during early pregnancy may be stimulated by estrogens secreted by the conceptus. Additional factors, however, probably are involved in these uterine responses, since conceptus and endometrial tissues produce angiogenic and other growth factors during early pregnancy (Ko et al., 1991; Millaway et al., 1989), and since, as discussed above, the density of endometrial microvasculature remains relatively constant in the nonpregnant uterus.

ANGIOGENIC FACTORS

Ovarian Tissues

Initially, it was reported that follicular tissue from rabbits treated with pregnant mare serum gonadotropin (PMSG, which has been used to stimulate follicular growth in numerous species; Greenwald and Terranova, 1988) had no influence on angiogenesis in the corneal pocket assay (Gospodarowicz and Thakral, 1978). Ovarian tissue from PMSG-treated rats, however, caused an angiogenic response in the CAM assay (Koos and LeMaire, 1983). Makris et al. (1984) reported that extracts of the thecal layer, but not the granulosa cell layer, of porcine follicles stimulated migration and proliferation of endothelial cells in vitro. This observation was potentially significant, since only the thecal layer of the ovarian follicle is vascularized, whereas the granulosa layer is avascular (Greenwald and Terranova, 1988; Moor and Seamark, 1986). Later reports indicated that media conditioned by granulosa cells stimulated migration (Redmer et al., 1985a) and proliferation (Koos, 1986) of endothelial cells.

Thus, early reports on angiogenic activity of ovarian follicles were inconclusive, and the relationships among stage of follicular development, gonadotropin treatment, and follicular angiogenic activity were not known. To evaluate these relationships, we stimulated preovulatory follicular growth in ewes by inducing luteal regression or by treatment with follicle-stimulating hormone, and thus were able to obtain follicles at known stages of development (Taraska et al., 1989). In this study, media conditioned by thecal tissues stimulated proliferation of endothelial cells regardless of the stage of follicular development. Granulosa cells, however, produced an endothelial mitogen only when they were obtained from non-atretic follicles just before ovulation. We recently have reported similar data for bovine preovulatory follicles (Redmer et al., 1991). Based upon these observations, we concluded that production of angiogenic factors by granulosa cells may help maintain the vasculature, and thereby the health of the preovulatory follicle. Thecal production of angiogenic factors, however, seems to be independent of the stage of preovulatory development or follicular status.

The endothelial mitogen produced by preovulatory bovine follicles is retained on heparin-affinity columns and elutes with high-salt (3.0 M NaCl) buffer (Fricke, 1991). This observation is significant because previously identified heparin-binding growth factors include the fibroblast growth factors (FGF) and the vascular endothelial growth factors (VEGF), both of which are potent angiogenic factors (Burgess and Maciag, 1989; Folkman and Klagsbrun, 1987; Ferrara, 1993; Klagsbrun and Soker, 1993). Messenger RNA for FGF-2 has been found in extracts of bovine but not rat granulosa cells (Koos and Olson, 1989; Neufeld *et al.*, 1987). Additionally, VEGF mRNA has been found in rat granulosa cell extracts (Koos, 1995). Determining the role of these growth factors in follicular angiogenesis will, however, require further investigation.

The CL is a transient endocrine gland that is formed from the cells of the ovarian follicle after ovulation (Niswender and Nett, 1988). In mammals, the principle function of the CL is to secrete the pro-gestational hormone, progesterone, during the nonpregnant cycle as well as during pregnancy (Niswender and Nett, 1988). In fact, the concentration of progesterone in systemic blood is used as an index of luteal function (Niswender and Nett, 1988; Hansel *et al.*, 1973; Goodman, 1988). In the nonpregnant animal, progesterone inhibits pituitary gonadotropin secretion and thereby regulates the length of the estrous/menstrual cycle (Niswender and Nett, 1988). During pregnancy, progesterone relaxes the uterine smooth muscle, stimulates uterine growth and secretory activity, influences maternal metabolism and mammary development, and is a precursor for other gestational steroids, all of which serve to support the developing embryo/fetus as well as the resulting offspring (Niswender and Nett, 1988; Heap and Flint, 1984; Catchpole, 1991). Thus, inadequate luteal function during pregnancy leads to death of the embryo/fetus in most mammalian species (Niswender and Nett, 1988; Heap and Flint, 1984; Catchpole, 1991; Niswender *et al.*, 1985).

As mentioned in the Introduction, the CL is one of the few adult tissues that exhibits regular periods of growth and development (Hudlicka, 1984; Reynolds *et al.*, 1992). In addition, growth and development of the CL are extremely rapid. For example, immediately after ovulation the CL of ewes weighs ≈30-40 mg. By day 12 of the estrous cycle (i.e., 12 days after ovulation), the CL reaches a maximum weight of ≈750 mg, which represents a 20-fold increase in tissue mass over a 12-day period (Jablonka-Shariff *et al.*, 1993). Similar rapid growth has been reported for the bovine and porcine CL (Zheng *et al.*, 1994; Ricke *et al.*, 1995a). Based upon these studies, the doubling time for luteal tissue mass during this rapid growth phase is ≈60-70 h. Such rapid growth is equaled only by the fastest growing tumors (Baserga, 1985). In contrast with tumor growth, however, growth of the CL is a self-limiting and highly ordered process (12). By midcycle, growth of the CL has ceased and luteal blood flow and progesterone secretion are maximal (Jablonka-Shariff *et al.*, 1993; Hossain *et al.*, 1979).

Luteal tissue and luteal tissue extracts from several species have been shown to stimulate angiogenesis (in vivo assays) as well as endothelial cell migration and proliferation (Gospodarowicz and Thakral, 1978; Jakob *et al.*, 1977; Koos and LeMaire, 1983; Makris *et al.*, 1984; Redmer *et al.*, 1988a; Grazul-Bilska *et al.*, 1993). In addition, CL from cynomologous monkeys were shown to produce a factor that stimulates migration of endothelial cells (Redmer *et al.*, 1985b). Bovine and ovine CL obtained from three stages of the estrous cycle (early cycle [developing CL]; mid-cycle [mature CL]; late cycle [regressing CL]) produce angiogenic activity, as confirmed by in vivo (chicken chorioallantoic membrane) and in vitro (endothelial cell migration and proliferation) assays (Grazul-Bilska *et al.*, 1992a; Redmer *et al.*, 1988a). In addition, bovine and ovine CL produce endothelial mitogen(s) throughout pregnancy (Grazul-Bilska *et al.*, 1992b, 1995; Redmer *et al.*, 1988b), confirming an earlier report of potent angiogenic activity of CL from pregnant cows (Koos and LeMaire, 1983). These observations indicate that CL produce angiogenic factor(s) at all stages of luteal development. We have hypothesized that angiogenic factors may, therefore, be involved not only in vascular growth during luteal development but also in maintaining luteal vascular beds in mature CL (Grazul-Bilska *et al.*, 1992b, 1995). Angiogenic factors, such as FGF, also could have other roles in differentiation and function of the CL (Burgess and Maciag, 1989). In this regard, an interesting possibility is that FGF may "protect" luteal cells from undergoing cell death, since a recent report has indicated that FGF inhibits apoptosis in oligodendroglial cells (Yasuda *et al.*, 1995). In addition, we have recently found that FGF receptor-1 (FGFR-1) is present in luteal parenchymal cells at high levels during early-mid cycle but is dramatically reduced during luteal regression (Figure 1). We have therefore hypothesized that FGF affects not only luteal cell proliferation but also luteal cell death, and thereby regulate luteal cell turnover and function.

The majority of the angiogenic activity in media conditioned by bovine and ovine CL binds to heparin-affinity columns (Grazul-Bilska *et al.*, 1992a, 1992b, 1993, 1995). These angiogenic activities can be partially immunoneutralized with antibody against FGF-2 but not with FGF-1 antibody (Grazul-Bilska *et al.*, 1992a, 1992b, 1993, 1995; Zheng *et al.*, 1993; Ricke *et al.*, 1995b; Doraiswamy *et al.*, 1995). FGF-2 also has been detected in bovine and ovine tissues and conditioned media by immunohistochemical and immunoblot procedures, respectively (Grazul-Bilska *et al.*, 1992a, 1992b, 1993, 1995; Zheng *et al.*, 1993; Ricke *et al.*, 1995b; Doraiswamy *et al.*, 1995). In addition, expression of mRNA for FGF-2 in bovine CL follows a pattern similar to that of angiogenic activity (Redmer *et al.*, 1988a; Stirling *et al.*, 1991). It is likely, therefore, that FGF-2 is a major angiogenic factor in CL.

We have found that bovine CL from the estrous cycle and ovine CL from early pregnancy each produce at least four heparin-binding endothelial mitogens with similar affinities for heparin (Grazul-Bilska *et al.*, 1993b, 1995). We also have good evidence that endothelial chemoattractants are present in luteal-conditioned media from cows, sheep and pigs (Redmer *et al.*, 1988a; Redmer *et al.*, 1987; Doraiswamy *et al.*, 1995; Grazul-Bilska et al., 1992b, 1995; Ricke, Redmer and Reynolds, unpublished). In addition, Rone and Goodman (Rone and Goodman, 1990) have reported that a factor present in rat luteal-conditioned media is chemotactic but not mitogenic for endothelial cells and is not a FGF. It is therefore apparent that multiple angiogenic factors, some of which are distinct from known FGF, are produced by CL and may play an important role in luteal angiogenesis. Thus, in CL as in other tissues, the coordinated growth of vascular and nonvascular components probably is regulated by multiple factors (Hudlicka, 1984; Adair et al., 1990).

Recently, we have shown that antibody against VEGF will immunoneutralize the endothelial chemoattractant activity produced by early cycle but not mid or late cycle ovine CL (Doraiswamy *et al.*, 1995). In addition, a portion (approximately 40%) of the endothelial mitogenic activity produced by porcine CL was neutralized by VEGF antibody (Ricke *et al.*, 1995b). We also have shown by ribonuclease protection assay that luteal levels of VEGF mRNA are 3.5-fold greater early in the estrous cycle (day 2-4) than during the mid or late cycle (days 8-15; Redmer, Reynolds and Moor, unpublished observations). These observations suggest a role for VEGF in vascularization of the developing CL. Yan *et al.* (1993) also have demonstrated the presence of VEGF mRNA in luteinized human granulosa cells, Garrido et al. (1993) reported that LH can induce VEGF transcription in bovine granulosa cells, and others have found VEGF in corpus luteum (Phillips et al., 1990). We have therefore hypothesized that VEGF, in addition to FGF-2, are probably important regulators of luteal vascular development.

Luteinizing hormone, which is luteotropic (Niswender and Nett, 1988), stimulates production of an endothelial chemoattractant by CL from nonpregnant cows (Redmer *et al.*, 1988a; Redmer *et al.*, 1987). This effect of luteinizing hormone is blocked by the luteolytic hormone prostaglandin F2α, even though prostaglandin F2α has no effect by itself. The effects of luteinizing hormone and prostaglandin $F_2\alpha$ on luteal production of angiogenic factors are similar to the effects of these hormones on expression of mRNA for FGF-2 (Stirling *et al.*,

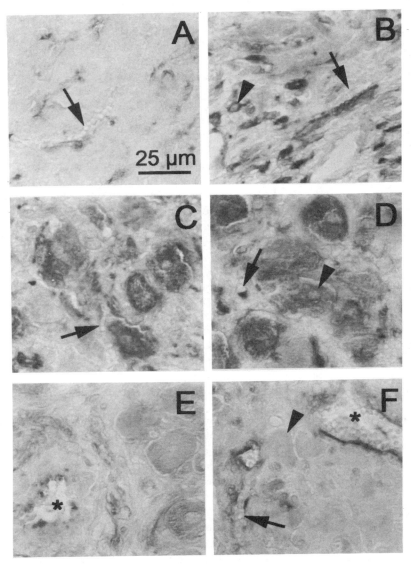

Figure 1. Immunohistochemical localization of FGFR-1 (flg) in ovine CL from the early (A, B), mid (C, D), and late (E, F) luteal stages of the estrous cycle. Note intense localization of FGFR in smaller (arrows) and larger (*) microvessels at all stages. Also note intense staining for FGFR in parenchymal cells (arrowheads) of some early and most mid luteal stage CL, but not in those of the late luteal stage (regressing) CL. All micrographs are at the same magnification; size bar is shown in (A) only.

1990, 1991). These findings emphasize the potential importance of hormones and growth factors in regulating luteal vascular growth and development.

Uterine and Placental Tissues

Media conditioned by uterine tissues from nonpregnant cows and ewes stimulate migration and proliferation of endothelial cells (Millaway *et al.*, 1989; Reynolds and Redmer, 1988a, 1988b; Zheng, 1995). Production of angiogenic factors by endometrial tissues of ovariectomized ewes can be modulated by in vivo treatment with progesterone and estradiol (Reynolds and Redmer, 1988b). The endothelial mitogen produced by ovine endometrium is >100 kDa based upon ultrafiltration, and is heat-labile (Millaway *et al.*, 1989; Zheng, 1995). We have found that the majority of this mitogenic activity binds to heparin (Reynolds *et al.*, 1992; *Reynolds et al., 1993b*; Zheng, 1995). This observation is significant *because*, as discussed above, FGF and VEGF are potent angiogenic factors which have been found in other tissues of the female reproductive system. In addition, estradiol has been shown to stimulate FGF-2 and VEGF production by human endometrial adenocarcinoma cells in vitro (Presta, 1988; Charnock-Jones *et al.*, 1993), and to stimulate VEGF expression in the rat uterus (Cullinan-Bove and Koos, 1993). Furthermore, FGF and VEGF proteins, mRNA and receptors have been found in the human and rat uterus (Charnock-Jones *et al.*, 1993; Cullinan-Bove and Koos, 1993; Ferriani *et al.*, 1993).

Ovine endometrial tissues produce endothelial mitogen(s) between days 12 and 40 after mating (Millaway *et al.*, 1989; Reynolds *et al.*, 1992), coincident with the dramatic uterine growth and vascular development that we have already described (Reynolds and Redmer, 1992). The majority of this endothelial mitogen binds to heparin-affinity columns and exhibits at least two peaks of activity upon elution with a salt gradient (Reynolds *et al.*, 1992; *Reynolds et al., 1993b*; Zheng, 1995). The major peak of mitogenic activity, which we have designated H3, elutes at about 1.9 M NaCl, which corresponds with the elution profile of FGF-2 (Burgess and Maciag, 1989). Several observations, however, indicate that H3 is distinct from FGF-2 (Reynolds *et al.*, 1992; *Reynolds et al., 1993b*; Zheng, 1995). Whereas FGF-2 is highly mitogenic for BALB/3T3 cells, H3 is not (*Reynolds et al., 1993b*; Zheng, 1995). When subjected to ultrafiltration or to polyacrylamide gel electrophoresis under denaturing conditions, H3 appears to be greater than 70 kDa, whereas FGF-2 is less that 50 kDa (Burgess and Maciag, 1989; *Reynolds et al., 1993b*; Zheng, 1995). Mitogenic activity of H3 was increased 2.5-fold by addition of 50 µg/ml heparin, whereas mitogenic activity of FGF-2 is unaffected by heparin (Burgess and Maciag, 1989; Zheng, 1995). In addition, FGF-2 was not detected in ovine endometrial-conditioned media by using immunoblot or immunoneutralization procedures, even though both procedures readily detected FGF-2 in luteal-conditioned media (Grazul-Bilska *et al.*, 1992a; Zheng, 1995). We have, therefore, suggested that H3 may represent a novel heparin-binding endothelial mitogen. Additionally, H3 may represent a large molecular weight form of FGF-2, since multiple forms of FGF-2 have been isolated from human placenta (Moscatelli *et al.*, 1988), and we have detected FGF-2 in ovine endometrial tissues by using immunohistochemistry (Reynolds and Redmer, unpublished observations). This suggestion seems reasonable because the presence of high molecular weight, immunoreactive FGF-2 in serum has been reported for several species (Baird *et al.*, 1986). In addition, although FGF-1 and FGF-2 are synthesized without a signal peptide and therefore do not appear to be secreted proteins, they have been found in the extracellular matrix in a variety of tissues (Grazul-Bilska *et al.*, 1992a; Vlodavsky *et al.*, 1990) and also, as cited above, in the circulation. Thus, it seems likely that H3 produced by the ovine endometrium is a secreted form of FGF.

More recently, we also have characterized a second heparin-binding endothelial mitogen, designated H2, which elutes from heparin-agarose at about 1.2 M NaCl. Although H2 is a relatively poor endothelial mitogen, it is a potent endothelial chemoattractant (Zheng *et al.*, 1995a). Although it binds to heparin with similar affinity to that of VEGF and also is similar in size to VEGF (13-16 kD by SDS-PAGE), immunoneutralizing antibody to VEGF has little effect on the endothelial chemoattractant activity of H2 (Zheng *et al.*, 1995a). Identification of H2 will therefore require further investigation.

Angiogenic activity of placental tissues from human, bovine and ovine sources has been evaluated by using in vivo (CAM) and in vitro (endothelial protease production, migration and proliferation) assays (Millaway *et al.*, 1989; Moscatelli *et al.*, 1988; Reynolds *et al.*, 1987). In cows and ewes, these angiogenic factors are produced primarily by maternal placental (endometrial) and not fetal placental tissues (Millaway *et al.*, 1989; Reynolds and Redmer, 1988a). It therefore seems that maternal placental tissues may direct placental vascularization. If this hypothesis is correct, factors which influence maternal placental production of angiogenic factors could have a significant effect on placental size, transport and/or blood flow, and thereby on fetal growth and development. Such factors include maternal genotype, multiple fetuses, inadequate maternal nutrition, and environmental heat stress (Ferrell, 1989; Reynolds and Redmer, 1995).

Throughout most of gestation, fetal placental tissues of ewes and cows also produce factor(s) which inhibit endothelial cell migration and proliferation (Millaway *et al.*, 1989; Reynolds and Redmer, 1988a). We have suggested that the target of these fetal placental anti-angiogenic factors is the maternal placental (uterine) vasculature, where they may function to limit vascular development (*Reynolds et al., 1993a*; Reynolds and Redmer, 1995). This proposal seems reasonable since, as mentioned above, angiogenesis in normal adult tissues, such as the uterus, must be held in check to prevent development of a pathological condition resulting from rampant capillary growth. In addition, the proposal that fetal anti-angiogenic factors may limit maternal placental vascular development is consistent with the observation that the fetal genome regulates placental size until late in gestation (Ferrell, 1991). The presence of anti-angiogenic factors in fetal placental tissues would not be expected to have an adverse effect on fetal placental development, since fetal placental vascular growth is a developmental process, sometimes termed vasculogenesis, which may occur independently of angiogenic factors (Ramsey, 1982; Hudlicka, 1984).

During a brief period late in gestation (about day 120 after mating), the ovine fetal placenta was found to produce endothelial mitogens (Millaway *et al.*, 1989), consistent with the previously reported increase in the number of fetal but not maternal placental endothelial cells during this same period (Teasdale, 1976; Ferrell, 1989; Reynolds and Redmer, 1995). It appears that this mitogenic activity is represented primarily by FGF-2 and VEGF (Zheng *et al.*, 1995b). This observation agrees with a report of VEGF expression by ovine placenta in late pregnancy (Ebaugh et al., 1994) as well as expression of VEGF protein and receptor by human placenta (Barleon et al., 1994; Charnock-Jones et al., 1994; Jackson et al., 1994).

CONCLUSIONS

Angiogenic factors have been identified in tissues of the female reproductive system. The angiogenic factors which have been characterized are primarily heparin-binding molecules. Some of these factors appear to be similar to previously identified FGF and VEGF, whereas the identity of others is presently unknown. The identity of anti-angiogenic factors, which are produced by placental tissues and probably also by ovarian tissues (Folkman and Klagsbrun, 1987; Millaway *et al.*, 1989; Reynolds and Redmer, 1988a; Reynolds *et al.*, 1987), remains to be determined. Further isolation and characterization of these factors will be important not only for understanding reproductive function, but also because these tissues are a rich source of angiogenic and growth factors (Tiollier *et al.*, 1990). Placental extracts already have been used to treat chronic varicose ulcers, and may be useful in treatment of other conditions such as delayed wound healing (Burgos *et al.*, 1989).

Production of angiogenic factors during growth and development of follicular, luteal, and placental tissues has been characterized. The precise relationships among angiogenic factor production and developmental changes in the vascularity of these tissues, however, need to be further defined. In addition, the effects of angiogenic or anti-angiogenic factors on ovarian or placental vascular development and function have not been evaluated. The recent development of techniques to rapidly and quantitatively evaluate vascular growth and development (e.g., Weidner *et al.*, 1991; Reynolds and Redmer, 1992; Jablonka-Shariff *et al.*, 1993; Zheng *et al.*, 1993) should allow these relationships to be determined. Once these relationships are known, regulation of reproductive processes with angiogenic or anti-angiogenic factors may be possible. The tissues of the female reproductive system should be especially amenable to this type of manipulation because, as discussed in the Introduction, their endothelial cells are unique among adult tissues in having a high mitotic rate. Reproductive disorders that potentially could be treated by regulating the angiogenic process include anovulatory cycles, subnormal luteal function, follicular and luteal cysts, choriocarcinomas, and inadequate placental development leading to embryonic loss or subnormal fetal growth (Findlay, 1986; Reynolds *et al.*, 1992).

Although we have found that production of angiogenic factors by the female reproductive tissues is responsive to tropic hormones, much remains to be learned about the physiological role of these factors in reproductive processes. This is an especially important area of research because of its implications for the control of fertility as well as for regulation of angiogenesis in other normal and pathological processes, which at present is poorly understood (Folkman and Klagsbrun, 1987; Klagsbrun and D'Amore, 1991). Nonetheless, because of their phenomenal rates of growth and vascular development, the tissues of the female reproductive organs are ideal for studying the regulation of angiogenesis in normal adult tissues.

ACKNOWLEDGEMENTS

Work from the authors' laboratories was supported in part by grants from the U.S. National Science Foundation (RII8610675 and ERH9108770), National Institutes of Health (HD22559-01-06), and Department of Agriculture (87-CRCR-1-2573, 93-37208-9278, and 93-37203-9271). We wish to acknowledge the expert technical assistance of J.D. Kirsch and K.C. Kraft.

REFERENCES

Abel, M.H., 1985, Prostanoids and menstruation. In: Mechanisms of menstrual bleeding, edited by Baird, D.T., Michie, E.A., pp 139-156. Raven Press, New York.

Adair, T.H., Gay, W.J., and Montani, J.-P., 1990, Growth regulation of the vascular system: evidence for a metabolic hypothesis. Am. J. Physiol. 259:R393-R404.

Andersen, D.H., 1926, Lymphatics and blood vessels of the ovary of the sow. Contrib. Embryol. 88:107-123.

Assheton, R., 1906, Phil. Trans. Roy. Soc., London, , Ser. B. 198:143-200.

Baird, A., Esch, F., Mormede, P., Ueno, N., Ling, N., et al., 1986, Molecular characterization of fibroblast growth factor: distribution and biological activities in various tissues. Recent Prog. Horm. Res. 42:143-205.

Barcroft, J., 1947, Researches on Pre-Natal Life. Charles C. Thomas, Springfield, IL.

Barleon, B., Hauser, S., Schollmann, C., Weindel, K., Marme, D., Yayon, A., and Weich, H.A., 1994, Differential expression of the two vegf receptors flt and kdr in placenta and vascular endothelial cells, J. Cell. Biochem. 54: 56-66.

Baserga R. *The Biology of Cell Reproduction*. Harvard University Press, Cambridge; 1985.

Bassett, D.L., 1943, The changes in the vascular pattern of the ovary of the albino rat during the estrous cycle. Am. J. Anat. 73:251-291.

Beranek, J.T., 1988, Antiangiogenesis comes out of its shell. Cancer J. 2:87-88.

Boshier, D.P., 1969, A histological and histochemical examination of implantation and early placentome formation in sheep. J. Reprod. Fert. 19:51-61.

Burgess, W.H., and Maciag, T., 1989, The heparin-binding, fibroblast, growth factor family of proteins. Annu. Rev. Biochem. 58:575-606.

Burgos, H., Herd, A., and Bennett, J.P., 1989, Placental angiogenic and growth factors in the treatment of chronic varicose ulcers: preliminary communication. J Roy. Soc. Med. 82:598-99.

Catchpole HR. Hormonal mechanisms in pregnancy and parturition. In: Cupps PT, ed. *Reproduction in Domestic Animals, 4th edn.* Academic Press, New York; 1991:361-383.

Charnock-Jones, D.S., Sharkey, A.M., Rajput-Williams, J., Burch, D., Schofield, J.P., Fountain, S.A., Boocock, C.A., and Smith, S.K., 1993, Identification and localization of alternately spliced mRNAs for vascular endothelial growth factor in human uterus and estrogen regulation in endometrial carcinoma cell lines, Biol. Reprod. 48: 1120-1128.

Charnock-Jones, D.S., Sharkey, A.M., Boocock, C.A., Ahmed, A., Plevin, R., Ferrara, N., and Smith, S.K., 1994, Vascular endothelial growth factor receptor localization and activation in human trophoblast and choriocarcinoma cells, Biol. Reprod., 51: 524-530.

Christenson, L.K., and Stouffer, R.L., 1995, Endothelial cell proliferation in the primate corpus luteum, FASEB J. 9: A543, Abstr. 3146, .

Clark, J.G., 1900, The origin, development and degeneration of the blood vessels of the human ovary. John Hopkins Hosp. Rep. 9:593-676.

Cullinan-Bove, K., and Koos, R.D., 1993, Vascular endothelial growth factor/vascular permeability factor expression in the rat uterus: Rapid stimulation by estrogen correlates with estrogen-induced increases in uterine capillary permeability and growth, Endocrinology 133: 829-837.

Denekamp, J., 1984, Vasculature as a target for tumour therapy. In: Progress in applied microcirculation. edited by Hammersen, F., Hudlicka, O,, Vol. 4, , pp 28-38. Karger, Basel.

Dharmarajan, A.M., Bruce, N.W., and Meyer, G.T., 1985, Qualitative ultra- structural characteristics relating to transport between luteal cell cytoplasm and blood in the corpus luteum of the pregnant rat. Am. J. Anat. 172:87-99.

DiZerega, G.S., and Hodgen, G.D., 1980, Fluorescence localization of luteinizing hormone/human chorionic gonadotropin uptake in the primate ovary. II. changing distribution during selection of the dominant follicle. J. Clin. Endocrinol. Metab. 51:903-907.

Doraiswamy, V., A.T. Grazul-Bilska, W.A. Ricke, D.A. Redmer and L.P. Reynolds. 1995. Immunoneutralization of angiogenic activity from ovine corpora lutea, CL, with antibodies against fibroblast growth factor (FGF, -2 and vascular endothelial growth factor, VEGF, . Biol. Reprod. 52(Suppl. 1):112.

Ebaugh, M.J., Singh, M., Brace, R.A., anc Cheung, C.Y., 1994, Vascular endothelial growth factor (VEGF) gene expression in ovine placenta and fetal membranes. Proc. 41st Annu. Mtg. Soc. Gynecol. Invest., Abstr. P198.

Eckstein, P., and Zuckerman, S., 1952, Changes in the accessory reproductive organs of the non-pregnant female. In: Marshall's physiology of reproduction, edited by Parker, A.S., 3rd ed., Vol. I, pp 543-654. Longman's Green, NY.

Edey, T.N., 1969, Prenatal mortality in sheep: a review. Anim. Breed. Abstr. 37:173-90.

Ferenczy, A., Bertrand, and Gelfand, M.M., 1979, Proliferation kinetics of human endometrium during the normal menstrual cycle. Am. J. Obstet. Gynecol. 133:859-867.

Ferrara, N., 1993, Vascular endothelial growth factor, *Trends Cardiovasc. Med.* 3: 244-250.

Ferrell, C.L., 1989, Placental regulation of fetal growth. In: Animal growth regulation, edited by Campion, D.R., Hausman, G.J., Martin, R.J., pp 1-19. Plenum, New York.

Ferrell, C.L., 1991, Maternal and fetal influences on uterine and conceptus development in the cow: I. growth of tissues in the gravid uterus. J. Anim. Sci. 69:1945-53.

Ferriani, R.A., Charnock-Jones, D.S., Prentice, A., Thoman, E.J., and Smith, S.K., 1993, Imminohistochemical localization of acidic and basic fibroblast growth facotrs in normal human endometrium and endometriosis and the detection of their mRNA by polymerase chain reaction, Human Reprod. 8: 11-16.

Findlay, J.K., 1986, Angiogenesis in reproductive tissues. J. Endocrinol. 111:357-366.

Folkman, J., and Klagsbrun, M., 1987, Angiogenic factors. Science 233:442-447.

Fricke, P.M., 1991, Regulators of follicular growth and development in the cow, M.S. Thesis, North Dakota State University, Fargo, ND, USA.

Gaede, S.D., Sholley, M.M., and Quattropani, S.L., 1985, Endothelial mitosis during the initial stages of corpus luteum neovascularization in the cycling adult rat. Am. J. Anat. 172:173-180.

Garrido, C., Simon, S., and Gospodarowicz, D., 1993, Transcriptional regulation of vascular endothelial growth factor gene expression in ovarian granulosa cells, Growth Factors 8:109-117.

Goodman RL. Neuroendocrine control of the ovine estrous cycle. In: Knobil E, Neil J, eds. *The Physiology of Reproduction.* Raven Press, New York 1988:1929-1969

Gospodarowicz, D., and Thakral, K.K., 1978, Production of a corpus luteum angiogenic factor responsible for proliferation of capillaries and neovascularization of the corpus luteum. Proc. Natl. Acad. Sci. USA 75:847-851.

Grazul-Bilska, A.T., Redmer, D.A., Killilea, S.D., Kraft, K.C., and Reynolds, L.P., 1992a, Production of mitogenic factor(s, by ovine corpora lutea throughout the estrous cycle. Endocrinology 130:3625-3632.

Grazul-Bilska, A.T., Reynolds, L.P., Slanger, W.D., and Redmer, D.A., 1992b, Production of heparin-binding angiogenic factor(s, by bovine corpora lutea during pregnancy. J. Anim. Sci. 70:254-262.

Grazul-Bilska, A.T., D.A. Redmer, S.D. Killilea, J. Zheng and L.P. Reynolds, 1993, Initial characterization of endothelial mitogens produced by bovine corpora lutea from the estrous cycle. Biochem. Cell Biol. 71:270-277.

Grazul-Bilska, A.T., Redmer, D.A., Killilea, S.D., and Reynolds, L.P., 1995, Characterization of mitogenic factors produced by ovine corpora lutea of early pregnancy, Growth Factors, In press, .

Green, W.W., and Winters, L.M., 1945, Prenatal development of sheep. MN Agri. Exp. Sta. Tech. Bull. 169:1-36.

Greenwald, G.S., 1989, Temporal and topographic changes in DNA synthesis after induced follicular atresia. Biol. Reprod. 41:175-181.

Greenwald, G.S., and Terranova, P.F., 1988, Follicular selection and its control, edited by Knobil, E., Neill, J., et al., pp 387-445. Raven Press, New York.

Greiss, F.C., and Anderson, S.G., 1970, Uterine blood flow during early ovine pregnancy. Am. J. Obstet. Gynecol. 106:30-38.

Grisham, J.W., 1962, A morphologic study of deoxyribonucleic acid synthesis and cell proliferation in regenerating rat liver; Autoradiography with thymidine-H^3. Cancer Res. 22:842-849.

Hansel W, Concannon PW, Lukaszewska JH. Corpora lutea of the large domestic animals. Biol Reprod. 1973; 8:222-245.

Heap RB, Flint APF. Pregnancy. In: Austin CR, Short RV, eds. Reproduction in Mammals, 2nd edn. Book 3: Hormonal Control of Reproduction. Cambridge University Press, Cambridge 1984:153-194.

Hossain MI, Lee CS, Clarke IJ, O'Shea JD. Ovarian and luteal blood flow, and peripheral plasma progesterone levels, in cyclic guinea pigs. J Reprod Fertil. 1979; 57:167-174.

Hudlicka, O., 1984, Development of microcirculation: capillary growth and adaptation. In: Handbook of Physiology, edited by Renkin, E.M., Michel, C.C.,, Sec. 2, Vol. IV, Part 1, , pp 165-216. Waverly Press, Baltimore.

Jablonka-Shariff, A., A.T. Grazul-Bilska, D.A. Redmer and L.P. Reynolds. 1993. Growth and cellular proliferation of ovine corpora lutea throughout the estrous cycle. Endocrinology 133:1871-1879.

Jablonka-Shariff, A., P.M. Fricke, A.T. Grazul-Bilska, L.P. Reynolds and D.A. Redmer. 1994. Size, number, cellular proliferation, and atresia of gonadotropin-induced follicles in ewes. Biol. Reprod. 51:531-540.

Jackson, M.R., Carney, E.W., Lye, S.J., Ritchie, J.W.K., 1994, Localization of two angiogenic growth factors (PDECGF and VEGF, in human placentae throughout gestation. Placenta 15: 341-353.

Jakob, W., Jentzsch, K.D., Mauersberger, B., and Oehme, P., 1977, Demonstration of angiogenesis-activity in the corpus luteum of cattle. Exp. Path. 13:231-236.

Johnson, M.L., Y. Ma and L.P. Reynolds. 1993. Influence of ovarian steroids on uterine growth in ewes. Biol. Reprod. 48, Suppl. 1, :162.

Jones, G.S., Maffezzoli, R.D., Strott, C.A., Ross, G.T., and Kaplan, G., 1970, Pathophysiology of reproductive failure after clomiphene-induced ovulation. Am. J. Obstet. Gynecol. 108:847-867.

Klagsbrun, M., and D'Amore, P.A., 1991, Regulators of angiogenesis. Annu. Rev. Physiol. 53:217-39.

Klagsbrun, M., and Soker, S., 1993, Vegf/vpf - the angiogenesis factor found, Curr. Biol. 3: 699-702.

Ko, Y., Young, L.C., Ott, T.L., Davis, M.A., Simmen, R.C.M., Bazer, F.W., and Simmen, F.A., 1991, Insulin-like growth factors in sheep uterine fluids: Concentrations and relationship to ovine trophoblast protein-1 production during early pregnancy. Biol. Reprod. 45:135-142.

Koos, R.D., 1986, Stimulation of endothelial cell proliferation by rat granulosa cell-conditioned media. Endocrinology 119:481-489.

Koos, R.D., 1995, Increased expression of vascular endothelial growth/permeability vactor in the rat ovary following an ovulatory gonadotropin stimulus: Potential roles in follicle rupture, Biol. Reprod. 52: 1426-1435.

Koos, R.D., and LeMaire, W.J., 1983, Evidence for an angiogenic factor from rat follicles. In: Factors regulating ovarian function, edited by Greenwald, G.S., Terranova, P.F., pp 191-195. Raven Press, New York.

Koos, R.D., and Olson, C.E., 1989, Expression of basic fibroblast growth factor in the rat ovary: detection of mRNA using reverse transcription-polymerase chain reaction amplification. Molec. Endocrinol. 3:2041-2048.

Magness, R.R., Christenson, R.K., and Ford, S.P., 1983, Ovarian blood flow throughout the estrous cycle and early pregnancy in sows. Biol. Reprod. 28:1090-1096.

Makris, A., Ryan, K.J., Yasymizu, T., Hill, C.L., and Zetter, B.R., 1984, The nonluteal porcine ovary as a source of angiogenic activity. Endocrinology 115:1672-1677.

Markee, J.E., 1932, An analysis of the rhythmic vascular changes in the uterus of the rabbit. Am. J. Physiol. 100:374-383.

Markee, J.E., 1940, Menstruation in intraocular endometrial transplants in the rhesus monkey. Contrib. Embryol. 28:221-308.

Meegdes, H.L.M., Ingenhoes, R., Peeters, L.L.H., and Exalto, N., 1988, Early pregnancy wastage: relationship between chorionic vascularization and embryonic development. Fertil. Steril. 49:216-220.

Meschia, G., 1983, Circulation to female reproductive organs. In: Handbook of physiology, edited by Shepherd, J.T., Abboud, F.M., Sec 2, Vol. III, Part 1, pp 241-269. Amer. Physiol. Soc., Bethesda.

Millaway, D.S., Redmer, D.A., Kirsch, J.D., Anthony, R.V., and Reynolds, L.P., 1989, Angiogenic activity of maternal and fetal placental tissues of ewes throughout gestation, J. Reprod. Fertil. 86: 689-696.

Moor, R.M., and Seamark, R.F., 1986, Cell signaling, permeability, and microvasculatory changes during antral follicle development in mammals. J. Dairy Sci. 69:927-943.

Moscatelli, D., Joseph-Silverstein, J., Presta, M., and Rifkin, D.B., 1988, Multiple forms of an angiogenesis factor: Basic fibroblast growth factor. Biochimie 70:83-87.

Nephew, K.P., K.E. McClure, and W.F. Pope, 1989, Embryonic migration relative to maternal recognition of pregnancy in sheep. J. Anim. Sci. 67:999-1005.

Neufeld, G., Ferrara, N., Schweigerer, L., Mitchell, R., and Gospodarowicz, D., 1987, Bovine granulosa cells produce basic fibroblast growth factor. Endocrinology 121:597-603.

137

Nicosia, S.V., Diaz, J., Nicosia, R.F., Saunders, B.O., and Muro-Cacho, C., 1995, Cell proliferation and apoptosis during development and aging of the rabbit corpus luteum, Ann. Clin. Lab. Sci. 25: 143-157.

Niswender, G.D., and Nett, T.M., 1988, The corpus luteum and its control. In: The physiology of reproduction, edited by Knobil, E., Neill, J., et al., pp 489-525. Raven Press, New York.

Niswender, G.D., Schwall, R.H., Fitz, T.A., Farin, C.E., and Sawyer, H.R., 1985, Regulation of luteal function in domestic ruminants: New concepts. Rec. Prog. Horm. Res. 41:101-151.

Phillips, H.S., Hains, J., Leung, D.W., and Ferrara, N., 1990, Vascular endothelial growth factor is expresses in rat corpus luteum, *Endocrinology* 127:965-967.

Presta, M., 1988, Sex hormones modulate the synthesis of basic fibroblast growth factor in human endometrial adenocarcinoma cells: implications for the neovascularization of normal and neoplastic endometrium. J. Cell. Physiol. 137:593-597.

Ramsey, E.M., 1982, The placenta, human and animal. Praeger, New York.

Ramsey, E.M., 1989, Vascular anatomy. In: Biology of the uterus, edited by Wynn, R.M., Jollie, W.P., pp 57-68. Plenum, New York.

Redmer, D.A., Rone, J.D., and Goodman, A.L., 1985a, Detection of angiotropic activity from primate dominant follicles. 67th Annu. Meeting Endocr. Soc., p 151, abstr 604.

Redmer, D.A., Rone, J.D., and Goodman, A.L., 1985b, Evidence for a non-steroidal angiotropic factor from the primate corpus luteum: Stimulation of endothelial cell migration in vitro. Proc. Soc. Exp. Biol. Med. 179:136-140.

Redmer, D.A., Kirsch, J.D., and Grazul, A.T., 1987, In vitro production of angiotropic factor by bovine corpus luteum: partial characterization of activities that are chemotactic and mitogenic for endometrial cells. In: Regulation of ovarian and testicular function, edited by Dhindsa, D., Anderson, E., Kalra, S., pp 683-688. Plenum Press, New York.

Redmer, D.A., Grazul, A.T., Kirsch, J.D., and Reynolds, L.P., 1988a, Angiogenic activity of bovine corpora lutea at several stages of luteal development. J. Reprod. Fert. 82:627-634.

Redmer, D.A., Grazul, A.T., and Reynolds, L.P., 1988b, Secretion of angiogenic activity by ovine and bovine corpora lutea throughout pregnancy. J. Reprod. Fert. Abst. Ser. 1:57.

Redmer, D.A., Kirsch, J.D., and Reynolds, L.P., 1991, Production of mitogenic factors by cell types of bovine large estrogen-active and estrogen-inactive follicles. J. Anim. Sci. 69:237-245.

Reynolds, L.P., 1986, Utero-ovarian interactions during early pregnancy: role of conceptus-induced vasodilation. J. Anim. Sci. 62(Suppl. 2, :47-61.

Reynolds, L.P., and Redmer, D.A., 1988a, Secretion of angiogenic activity by placental tissues of cows at several stages of gestation. J. Reprod. Fert. 83:497-502.

Reynolds, L.P., and Redmer, D.A., 1988b, Secretion of angiogenic activity by endometrial tissues of cyclic and ovariectomized, steroid-treated ewes. J. Reprod. Fert. Abstr. Ser. 1:43.

Reynolds, L.P., and Redmer, D.A., 1992, Growth and microvascular development of the uterus during early pregnancy in ewes. Biol. Reprod. 47:(In press, .

Reynolds, L.P. and D.A. Redmer. 1995. Utero-placental vascular development and placental function. *Review Article*. J. Anim. Sci. 73:1839-1851.

Reynolds, L.P., Magness, R.R., and Ford, S.P., 1984, Uterine blood flow during early pregnancy in ewes: interaction between the conceptus and the ovary bearing the corpus luteum. J. Anim. Sci. 58:423-429.

Reynolds, L.P., Ferrell, C.L., Robertson, D.A., and Ford, S.P., 1986, Metabolism of the gravid uterus, foetus and uteroplacenta at several stages of gestation in cows. J. Agric. Sci., Cambridge 106:437-444.

Reynolds, L.P., Millaway, D.S., Kirsch, J.D., Infeld, J.E., and Redmer, D.A., 1987, Angiogenic activity of placental tissues of cows. J. Reprod. Fert. 81:233-240.

Reynolds, L.P., Millaway, D.S., Kirsch, J.D., Infeld, J.E., and Redmer, D.A., 1990, Growth and in vitro metabolism of placental tissues of cows from day 100 to day 250 of gestation. J. Reprod. Fert. 89:213-222.

Reynolds, L.P., Killilea, S.D., and Redmer, D.A., 1992, Angiogenesis in the female reproductive system. FASEB J. 6:886-892

Reynolds, L.P., A.T. Grazul-Bilska, S.D. Killilea and D.A. Redmer, 1993a, Angiogenesis in the female reproductive system: Patterns and mediators. In: *Local Systems in Reproduction*, R.R. Magness and F. Naftolin, eds., , Serono Symp. Publ. Vol. 96. Raven Press, NY, pp 189-211.

Reynolds, L.P., Killilea, S.D., and Redmer, D.A., 1993b, Endometrial growth and vascular development: Patterns and mediators. In: Exogenous hormones and dysfunctional uterine bleeding, edited by Alexander, N.J., D'Arcangues, C., AAAS Press, New York, NY, pp 35-48..

Reynolds, L.P., S.D. Killilea, A.T. Grazul-Bilska and D.A. Redmer. 1994. Mitogenic factors of corpora lutea. *Review Article*. Progr. Growth Factor Res. 5:159-175.

Reynolds, L.P., J.D. Kirsch, K.C. Kraft and D.A. Redmer. 1995. Time-course of the uterine response to estradiol-17β, E2, in ovariectomized, OVX, ewes: Uterine growth and cell proliferation. Biol. Reprod. 52(Suppl.): :119.

Ricke, W.A., D.A. Redmer and L.P. Reynolds. 1995a. Cellular growth of porcine corpora lutea throughout the estrous cycle. J. Anim. Sci. 73, Suppl. 1, :(In press, .

Ricke, W.A., D.A. Redmer and L.P. Reynolds. 1995b. Initial characterization of mitogenic factors produced by porcine corpora lutea throughout the estrous cycle. Biol. Reprod. 52(Suppl. 1, :(Accepted) .

Shepro, D., and D'Amore, P.A., 1984, Physiology and biochemistry of the vascular wall endothelium. In: Handbook of physiology, edited by Renkin, E.M., Michel, C.C., Vol. IV, pp 103-164. Waverly Press, Baltimore.

138

Stirling, D., Magness, R.R., Stone, R., Waterman, M.R., and Simpson, E.R., 1990, Angiotensin II inhibits luteinizing hormone-stimulated cholesterol side chain cleavage expression and stimulates basic fibroblast growth factor expression in bovine luteal cells in primary culture. J. Biol. Chem. 265:5-8.

Stirling, D., Waterman, M.R., and Simpson, E.R., 1991, Expression of mRNA encoding basic fibroblast growth factor, bFGF, in bovine corpora lutea and cultured luteal cells. J. Reprod. Fert. 91:1-8.

Taraska, T., Reynolds, L.P., and Redmer, D.A., 1989, In vitro secretion of angiogenic activity by ovine follicles. In: Growth factors and the ovary, edited by Hirshfield, A., pp 267-272. Plenum, New York.

Teasdale, F., 1976, Numerical density of nuclei in the sheep placenta. Anat. Rec. 185:187-196.

Tiollier, J., Uhlrich, S., Chirouze, V., Tardy, M., and Tayot, J.-L., 1990, Biochemical and biological characterization of a crude growth factor extract, EAP, from placental tissue. In: Placental Communications: Biochemical, Morphological and Cellular Aspects. INSERM, Natl. Inst. of Health and Med. Res., France, Colloquium, 199:98-100, John Libbey Eurotext, London.

Vlodavsky, I., G. Korner, and R. Ishaimichaeli, et al., 1990, Extracellular matrix-resident growth factors and enzymes: possible involvement in tumor metastasis and angiogenesis. Cancer Metastasis Rev 9:203-26.

Weidner, N., Semple, J.P., Welch, W.R., and Folkman, J., 1991, Tumor angiogenesis and metastasis--correlation in invasive breast carcinoma. New England J. Med. 324:1-8.

Wimsatt, W.A., 1950, New histological observations on the placenta of the sheep. Am J Anat 87:391-457. Yan, Z.,

Weich, H.A., Bernart, W., Breckwoldt, M., and Neulen, J., 1993, Vascular endothelial growth factor (VEGF) messenger ribonucleic acid, mRNA, expression in luteinized human granulosa cells in vitro, J. Clin. Endocrinol. Metab. 77: 1723-1725.

Yasuda, T., Grinspan, J., Stern, J., Franceschini, B., Bannerman, P., and Pleasure, D., 1995, Apoptosis occurs in the oligodendroglial lineage, and is prevented by basic fibroblast growth factor, J. Neurosci. Res. 40: 306-317.

Zeleznik, A.J., Schiler, H.M., and Reichert, L.E., 1981, Gonadotropin-binding sites in the Rhesus monkey ovary: Role of the vasculature in the selective distribution of human chorionic gonadotropin to the preovulatory follicle. Endocrinology 109:356-361.

Zheng, J., 1995, Studies of uterine growth during early pregnancy in ewes, Ph.D. Dissertation, North Dakota State University, Fargo, ND, USA.

Zheng, J., D.A. Redmer and L.P. Reynolds. 1993. Vascular development and heparin-binding growth factors in the bovine corpus luteum at several stages of the estrous cycle. Biol. Reprod. 49:1177-1189.

Zheng, J., P.M. Fricke, L.P. Reynolds and D.A. Redmer. 1994. Evaluation of growth, cell proliferation, and cell death in bovine corpora lutea throughout the estrous cycle. Biol. Reprod. 51:623-632.

Zheng, J., A.T. Grazul-Bilska, V. Doraiswamy, D.A. Redmer and L.P. Reynolds. 1995a. Ovine endometrium of early pregnancy secretes an angiogenic factor that stimulates endothelial cell migration. Biol. Reprod. 52(Suppl. 1):190.

Zheng, J., R.R. Magness, D.A. Redmer and L.P. Reynolds. 1995b. Angiogenic activity of ovine placental tissues: Immunoneutralization with FGF-2 and VEGF antisera. J. Soc. Gynecol. Invest. 2(2):289 (Abstr. P146).

ANGIOGENESIS IN SKELETAL MUSCLE

O. Hudlicka[1], M.D.Brown[2] & S. Egginton[1]

Departments of Physiology[1] and Sport and Exercise Sciences[2]
University of Birmingham
Birmingham B15 2TT, UK

ABSTRACT

Growth of vessels in normal adult skeletal muscles occurs during development, cold exposure, increased activity, administration of certain hormones, and increased physical activity (such as endurance exercise or chronic electrical stimulation). It always starts as growth of capillaries, with growth of larger vessels following later, or sometimes not at all: in endurance training growth of capillaries is not accompanied by growth of larger vessels,while a long-term increase in activity due to electrical stimulation leads to growth of the whole vascular bed (demonstrated by increased capillarization, corrosion casts, number of arterioles and maximal conductance measurements. One factor involved in capillary growth in stimulated muscles is the greater shear stress accompanying an increased velocity of flow, as similar growth was found in animals where long-term increase in blood flow was induced by the alpha$_1$ blocker prazosin. Increased shear stress damaged the luminal glycocalyx and also caused a release of prostaglandins. These appear to mediate capillary growth as simultaneous administration of indomethacin decreased incorporation of bromodeoxyuridine into capillary-linked nuclei and attenuated capillary growth. In addition, distortion of the capillary basement membrane by increased capillary wall tension, and by continuous stretching and relaxation of surrounding muscle fibres,may also involved. Long-term muscle stretch due to extirpation of agonists induced capillary growth, but without an increase in blood flow. Disturbance of the basement membrane may lead to release of growth factors. While the evidence for the involvement of bFGF was negative, a low molecular weight angiogenic factor (ESAF) was demonstrated in both stimulated and stretched muscles. Capillary growth in stimulated muscles may also be enhanced by pericyte withdrawal as there was significantly less of capillary perimeter covered by pericytes in muscles with demonstrated capillary growth. Thus mechanical factors acting both from luminal and abluminal side can initiate capillary growth in skeletal muscle by activating either prostaglandins, ESAF or possibly other growth factors, but not bFGF.

Molecular, Cellular, and Clinical Aspects of Angiogenesis
Edited by Michael E. Maragoudakis, Plenum Press, New York, 1996

INTRODUCTION

Growth of vessels in skeletal muscles occurs under physiological circumstances during development and stretch-induced hypertrophy, during exposure to cold and perhaps to chronic hypoxia, during increased levels of certain hormones (thyroxine, anabolic steroids) and mainly in connection with increased muscle contractile activity (see Hudlicka et al, 1992). Under pathological circumstances vessel growth is essential for growth of solid tumours (Folkman & Cotran, 1976) and also occurs in regenerating muscles (eg Hansen-Smith et al, 1980), in wound healing (eg Knighton et al, 1981), and even in some types of muscle diseases linked with muscle dystrophy (Burch et al, 1981; Atherton et al, 1982).

Growth of the vascular bed starts with growth of capillaries (vessels composed only of endothelium and its basement membrane often associated with pericytes apposed to the endothelium on the abluminal side). Larger vessels, such as arterioles or venules, usually develop from capillary networks by apposition of fibroblasts or other mesenchymal cells which are transformed into smooth muscle cells (Clark & Clark, 1940). This is particularly well documented during pre- and postnatal development (Wolff et al, 1975; Stingl & Rhodin, 1994) where arterioles only start developing postnatally, and the expansion of the whole vascular bed is to a great extent governed by the mechanical stretch imposed on the vessels by growing muscle fibres. The extent to which capillary growth under other circumstances is accompanied by growth of larger vessels is much less clear. There is little evidence for it in muscles with intensive capillary growth occurring during endurance training, where maximal blood flow (indicative of growth of larger vessels) is no greater than in normal muscles (Clausen & Trap-Jensen, 1970; Saltin et al, 1986). Thus, any study on vessel growth in skeletal muscle must first deal with capillary growth.

The purpose of this presentation is to describe under what physiological circumstances capillary growth occurs and what factors may be involved in it.

GROWTH OF CAPILLARIES INDUCED BY METABOLIC OR HORMONAL FACTORS

It has been assumed for a long time that one of the causes of capillary growth is a long-term exposure to hypoxia (Valdivia, 1958). However, more recent findings (Hoppeler et al, 1990; Hoppeler & Desplanches, 1992) demonstrated that the higher capillary density in such circumstances was due to atrophy of muscle fibres without real capillary growth. Hypoxia is believed to increase the activity of oxidative enzymes and the volume density of mitochondria, and there is a correlation between these two parameters in cases of capillary growth induced by endurance exercise (Andersen & Henriksson, 1977; Zumstein et al, 1983). Although it is still believed that hypoxia is an important trigger for endothelial cell proliferation in vitro (Smith, 1989) or capillary growth in vivo (Adair et al, 1990), with VEGF (vascular endothelial growth factor) as an important factor initiating the growth, there are numerous examples showing that capillary growth occurred without any change in oxidative metabolism and/or oxygen tension indicative of hypoxia. Hudlicka et al (1988) described increased capillary supply in rat fast skeletal muscles subjected to one form of increased activity without concomitant changes in the activity of oxidative enzymes, while another type of activity increased oxidative metabolism but did not change capillary supply. Hudlicka & Egginton (1989) subsequently demonstrated that a pattern of electrical stimulation which activates only fast glycolytic fibres induced capillary growth without changes in the activity of cytochrome oxidase.

Another example of a dissociation between oxidative capacity and capillary growth was presented by Capo & Sillau (1983) who demonstrated capillary growth in conjunction with increased activity of oxidative enzymes in one type of rat muscle (soleus) under the influence of thyroxine, while anatomically closely related gastrocnemius had increased capillary supply without any signs of increased oxidative enzymes activity. They explained these results as being due to increased blood flow by thyroxine. However, long-term treatment with an anabolic hormone, durabolin (Egginton 1987a,b), led to a modest capillary growth without evidence of increased blood flow or activity of oxidative enzymes. Hormonal changes might also be involved in increases in capillary density during cold acclimation (Fairney & Egginton, 1994).

GROWTH OF CAPILLARIES INDUCED BY INCREASED MUSCLE ACTIVITY

Vanotti & Magiday (1934) were the first to demonstrate increased size of the capillary bed in skeletal muscles of endurance trained animals. However, their observations were based on capillaries perfused with India ink and therefore might not have represented true capillary growth due to incomplete filling. More recent methods, using identification of capillaries either on the basis of electron microscopy or specific staining of capillary endothelium, confirmed their findings. Hermansen & Wachtlova (1971), Brodal et al (1977) and Andersen & Henriksson (1977) showed around a 20% increase in capillary supply in human muscles after about 8 weeks of endurance training, and other authors have reported similar results (see Hudlicka et al, 1992). Ingjer (1979) described a similar increase in capillary supply, with a significantly greater increase in number of capillaries surrounding fast oxidative than any other muscle fibres. This selective increase around one particular fibre type can be explained by the fact that blood flow during endurance type of exercise is increased considerably more in the parts of muscles composed of fast oxidative than of other fibre types (Laughlin & Armstrong, 1982). While the relatively modest increase in capillary growth required a long duration of endurance training, a similar increase could be achieved within 7 days when muscle activity was increased by chronic electrical stimulation (Myrhage & Hudlicka, 1978; Hudlicka et al, 1992; Hudlicka, 1994). Within 28 days, the size of the capillary bed doubled (Brown et al, 1976), but no further increase was observed during subsequent stimulation for further 2 months. What mechanism can explain such an intensive capillary growth ?

ROLE OF BLOOD FLOW AND MECHANICAL FACTORS IN CAPILLARY GROWTH

Capillary growth started in the vicinity of glycolytic fibres during the early stages of chronic electrical stimulation (Hudlicka et al, 1982). These fibres are not activated in endurance trained muscles, but are subjected to long-lasting activity for which they are not metabolically equipped by electrical stimulation. It could be thus assumed that activity induced-hypoxia could trigger capillary growth. However, direct measurement of oxygen tension at a time when capillaries started to grow and prior to it did not reveal lower PO_2 values than in control muscles (Hudlicka et al, 1984). There was no other indication of hypoxia such as increase in muscle content of lactate or adenosine (Hudlicka, 1991).

Several authors described increase in capillary supply as a consequence of long-term increase in blood flow (see Hudlicka et al, 1992). Blood flow is obviously considerably increased in contracting muscles, but the increase is always greater in predominantly glycolytic than in predominantly oxidative muscles (Hudlicka, 1975; Laughlin & Armstrong, 1982). The first appearance of new capillaries in their vicinity of glycolytic fibres could then be explained by a greater increase in blood flow in capillaries supplying these fibres. When we measured velocity of red blood cells (Vrbc) in capillaries supplying the surface of tibialis anterior (composed of 80% of glycolytic fibres) and soleus (100% oxidative), we found a greater increase in Vrbc during muscle contractions in the former than in the latter (Dawson et al, 1987). Other experiments have demonstrated that increased blood flow per se can stimulate capillary growth. Two weeks administration of the alpha$_1$ blocker prazosin doubled and increased flow during contractions by 35% while capillary/fibre ratio increased by 25% (Fulgenzi & Hudlicka, 1994). Further evidence for the importance of capillary blood flow and perfusion in triggering capillary growth is suggested by the fact that increased capillary perfusion precedes capillary growth in chronically stimulated muscles. Hudlicka et al (1984) demonstrated that the percentage of perfused capillaries (studied by timed India ink infusion) was higher in stimulated (84.4±1.4%) than in control (70.8±6.4%) muscles prior to appearance of an increase in the total number of capillaries. Moreover, capillary growth was absent in stimulated muscles where blood supply and hence any increase in flow was limited by ligation of the main supplying (iliac) artery (Hudlicka, 1991).

However, increased blood flow is not the only possible factor inducing capillary growth in chronically stimulated muscles. Although the increase in flow in stimulated muscles is smaller than that induced by prazosin and similar to that induced by long-lasting administration of adenosine (Ziada et al, 1984), the increase in C/F is much greater (Fig 1). It is therefore assumed that mechanical distortion of muscle fibres during contraction and subsequent relaxation may be also important. This was supported by the finding of increased C/F ratio in muscles subjected to long-term stretch after extirpation of an agonist muscle (Tibialis anterior, TA) which resulted in hypertrophy of the extensor digitorum longus (EDL). C/F ratio increased by 36%, although blood flow during contractions was not significantly higher than in control muscles (Egginton & Hudlicka, 1992).

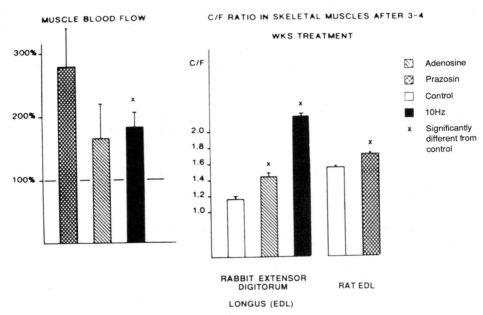

Fig 1.   Changes in blood flow and C/F ratio in muscles of animals receiving adenosine, prazosin or chronic electrical stimulation
Muscle blood flow during acute infusion (3 h) or 42 mmol/hour adenosine, 30 mg/h prazosin and contractions at 10 Hz.

MECHANISMS INVOLVED IN CAPILLARY GROWTH IN SKELETAL MUSCLE UNDER PHYSIOLOGICAL CIRCUMSTANCES

Mechanical factors occurring in skeletal muscles as a result of increased activity, long-term administration of vasodilators or increased stretch affect both the luminal and abluminal sides of capillaries, and may cause a release of a variety of substances which could start angiogenesis by disrupting the capillary endothelial cell basement membrane, thus enabling migration, mitosis and formation of sprouts (Fig 2). Factors acting from the luminal side include increased shear stress, as a consequence of increased red blood cell velocity, and increased capillary wall tension, as a consequence of increased capillary pressure during dilatation of arterioles and/or increased capillary diameter. Factors acting from the abluminal side include stretch of the basement membrane resulting from stretch of surrounding muscle fibres (or alternative stretch and relaxation during muscle contractions, which would cause a longitudinal distortion of capillaries), and interaction with pericytes which are supposed to inhibit proliferation of endothelial cells (D'Amore and Orlidge, 1988).

Factors initiating capillary growth

ACTING FROM THE LUMINAL SIDE

Increased blood flow: vasodilators
                     muscle contractions
leading to
    increased shear stress (damaged glycocalyx, release
                            of serine proteases)
    increased capillary diameter)   {
    increased capillary pressure)   (increased capillary
            wall tension (stretch of basement membrane,
            inhibition of pericytes?)

ACTING FROM ABLUMINAL SIDE

Stretch of muscle fibres: rhythmic contractions stretch
                          (extirpation of agonists)
leading to
    stretch of the basement membrane (inhibition of
        pericytes?)
    activation of metalloproteinases
    release of growth factors

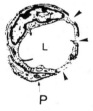

migration of endothelial cells outside capillary
formation of sprouts

Fig 2   Mechanical factors initiating angiogenesis by acting from luminal or abluminal side.

High shear stress stimulated proliferation of endothelial cells in tissue cultures (Ando et al, 1987). It also activates metalloproteinases (Brown et al, 1995) - proteases involved in the breakage of the endothelial cell basement membrane - and it may damage the glycocalyx lining of endothelial cells, and thus make the cells more prone to the action of proteases present in the blood. Calculation of shear stress in capillaries in prazosin-treated and chronically stimulated muscles, on the basis of direct measurements of Vrbc and capillary diameters revealed that shear stress was almost doubled in the former and increased by 50% in the latter (Dawson & Hudlicka, 1993). Electron microscopical studies demonstrated that the glycocalyx, which forms a continuous layer in the luminal side in over 60% of normal capillaries,was discontinuous or absent in about 75% of capillaries in chronically stimulated muscles (Brown et al, 1992a). Further electron microscopical studies demonstrated distortion of the luminal side of capillaries with many irregularities, activation of endothelial cells (demonstrated by increased number of mitochondria and ribosomes) and occasional sites with amorphous basement membrane (Hansen-Smith & Hudlicka, 1993).

The disturbance of the basement membrane may be linked with increased capillary wall tension (T), calculated on the basis of capillary diameters (d) and capillary pressure measurement (P) as $T=Pxd/2$. Myrhage & Hudlicka (1978) demonstrated an increased capillary diameter in chronically stimulated muscles. Using these data and the capillary pressure estimated in resting and contracting muscles (Mellander & Bjornberg, 1992), calculated capillary wall tension was twice as high in chronically stimulated muscles during contractions, but was not altered in prazosin treated animals.

Kuwabara & Cogan (1963) demonstrated withdrawal of pericyte processes from endothelial cells in the retina prior to development of diabetic retinopathy and proliferation of endothelial cells. Orlidge & D'Amore (1987) demonstrated that pericytes inhibit proliferation of endothelial cells in co-cultures. It can thus be assumed that fewer pericyte processes may be found on capillaries in skeletal muscle where angiogenesis was induced by some of the methods mentioned above. Analysis of electron microphotographs of capillaries revealed indeed that pericyte-endothelial cell contact decreased from about 30% in control to about 21% in chronically stimulated muscles (Egginton et al, 1993) at the time when capillarization was increased by about 40%. Since pericytes are located within the basement membrane of capillaries, it can be assumed that their withdrawal may weaken the basement membrane and could be the first step in its disturbance.

RELATION BETWEEN MECHANICAL AND METABOLIC/GROWTH FACTORS

Disturbance of the basement membrane is supposed to release growth factors bound to it, such as basic fibroblast growth factor (bFGF) (D'Amore & Orlidge, 1988). However, we did not find increased expression of mRNA for bFGF in chronically stimulated (Walter & Hudlicka, 1992) or stretch muscles (Brown et al, 1992b) in relation to capillary supply either at the time of increased capillarization or prior to it. Immunolocation with a polyclonal antibody to bFGF in stimulated or stretch muscles also showed no significant difference in staining between control and experimental muscles (Brown et al, 1993). In contrast, there was a direct relationship between increased capillarization and levels of a small molecular weight endothelial cell stimulating angiogenic factor (ESAF) in stimulated (Brown et al, 1993), stretched and prazosin treated muscles (Weiss et al, unpublished results).

Shear stress is known to release prostaglandins (Koller & Kaley, 1990) which then cause dilatation of arterioles and consequently and increased perfusion of capillaries and greater capillary wall tension. A direct effect of prostaglandins on growth of vessels was demonstrated in the rabbit cornea (Ziche et al, 1982) and on CAM (Form & Auerbach, 1983) while indomethacin, a cyclooxygense inhibitor, suppressed lymphocyte-induced angiogenesis (Davel et al, 1985). While studying microcirculation in chronically stimulated muscles, we discovered that arterioles are dilated even in muscles at rest after only 2 days of stimulation, prior to increased capillary counts and this dilatation was eliminated by indomethacin (Pearce & Hudlicka, 1994). Indomethacin also significantly decreased the labelling index for bromodeoxyuridine (BrdU), which was almost trebled in capillary-associated nuclei after only two days of stimulation (Pearce et al, 1995) and significantly attenuated capillary growth in muscles stimulated for 7 days (Pearce et al, 1995). Thus, increased shear stress may affect capillary growth either by release of prostaglandins from endothelial cells or by direct activation of metalloproteinases.

Once the capillary basement membrane is disturbed, endothelial cells start to migrate into the interstitial space where components of the connective tissue surrounding capillaries must also be disrupted to make space for sprouts, while at the same time reorganisation of the extracellular matrix is essential to facilitate further endothelial cell movement. Very little is known about this process in vivo. However, it has been demonstrated in vitro that tube formation by endothelial cells requires disruption of the normally tight adhesion of cells to the extracellular matrix, which at the same time converts cells from proliferative (mitotic) to differentiated (migrating, tube forming) type (Bischoff, 1995). Integrins are essential for binding of endothelial cells to collagen (Ingber & Folkman, 1989) and antibodies against alphavbeta3 integrins inhibited angiogenesis in CAM (Bischoff, 1995). Antibodies against E selectins (which are expressed on the luminal side of capillary endothelium) also inhibited tube formation. Specific inhibitors against various adhesion molecules may thus help to elucidate the factors enabling capillary sprouting.

Maragoudakis et al (1988) described suppression of angiogenesis on CAM by GPA 1734 (8,9-di-hydroxyl-7-methyl-benzo (b) quinolizinium bromide), presuming that this substance inhibits synthesis of collagen type IV and thus prevents the last step in angiogenesis - formation of a new basement membrane. When this substance was tested in our angiogenesis model of stimulated muscles in rats (Dawson & Hudlicka, 1990), it indeed prevented capillary growth, but also diminished the higher Vrbc and high adhesion of leucocytes to postcapillary venules. Although there are no data on other effects of GPA 1734 it might be possible that it also inhibits formation of adhesion molecules and prevents angiogenesis in this way.

CONCLUSIONS

Angiogenesis in normal skeletal muscle elicited either by increased blood flow, increased activity or long-term muscle stretch is initiated by mechanical factors (increased shear stress, increased capillary wall tension, distortion of capillary wall by external forces)

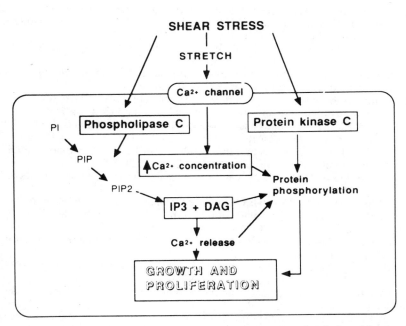

Fig 3   Possible scheme of signal transduction in cell proliferation induced by mechanical factors.

which do not seem to activate bFGF, but do activate ESAF and, in the case of increased muscle activity, release of prostanoids. The transduction signals for cell proliferation may then be mediated by these substances. It is also possible that the mechanical factors would stimulate cell proliferation directly by increasing the influx of calcium via calcium stretch-sensitive channels or by activating phospholipase C and the inositolphosphate cascade with consequent Ca^{2+} release, activation of protein kinase C and protein phosphorylation (see Hudlicka & Brown, 1993) (Fig 3).

REFERENCES

Adair TH, Gay WJ, Montani J-P. Growth regulation of the vascular system; evidence for a metabolic hypothesis. Am.J. Physiol. 259, R393-404,1990.

Andersen P, Henriksson J. Capillary supply of the quadriceps femoris muscle of man: adaptive response to exercise. J. Physiol, Lond. 270: 677-690, 1977.

Ando J, Nomura H, Kamiya A. The effect of fluid shear stress on the migration and proliferation of cultured endothelial cells. Microvasc. Res. 33: 62-70, 1987.

Atherton GW, Cabric M, James NT. Stereological analyses of capillaries in muscles of dystrophic mice. Virchows Arch. A Pathol. Anat. 397: 347-382, 1982.

Bischoff J. Approaches to studying the adhesion molecules in angiogenesis. Trends in Cell. Biol. 5: 69-74, 1995.

Brodal P, Ingjer F, Hermansen L. Capillary supply of skeletal muscles fibres in untrained and endurance-trained men. Am. J. Physiol. 232 (Heart Circ. Physiol. 1): H705-H712, 1977.

Brown LC, Messick FC, Kok MP, Hamilton IG, Girard PR. Fluid flow stimulates metalloproteinase production and deposition into extracellular matrix of endothelial cells. FASEB J. 9: A617, 1995.

Brown MD, Cotter MA, Hudlicka O, Vrbova G. The effects of different patterns of muscle activity on capillary density, mechanical properties and structure of slow and fast rabbit muscles. Pfluegers Arch. 361: 241-250, 1976.

Brown MD, Egginton S, Hudlicka O. Changes in capillary endothelial cell glycocalyx in rat skeletal muscles during chronic electrical stimulation. Int. J. Microcirc: Clin. Exper. 11: 447, 1992a.

Brown MD, Egginton S, Walter HJ, Hudlicka O. Increased capillary supply and localization of basic fibroblast growth factor in rat fast skeletal muscle after stretch-induced overload. Int. J. Microcirc: Clin. Exper. 11: S181, 1992b.

Brown MD, Walter HJ, Weiss JB, Hudlicka O. Growth factors in angiogenesis in skeletal muscle and heart. FASEB J. 7: A884,1993.

Burch TG, Prewitt RL, Law PK. In vivo morphometric analysis of muscle microcirculation in dystrophic mice. Muscle & Nerve, 4: 420-424, 1981.

Capo LA, Sillau AH. The effect of hyperthyroidism on capillarity and oxygen capacity in rat soleus and gastrocnemius muscles. J. Physiol. Lond., 342: 1-14, 1983.

Clark ER, Clark EL. Microscopic observations on the extra endothelial cells of living mammalian blood vessels. Am. J. Anat. 66: 1-49, 1940.

Clausen JP, Trap-Jensen J. Effects of training on the distribution of cardiac output in patients with coronary artery disease. Circulation. 42: 611-624, 1970.

D'Amore P, Orlidge A. Growth factors and pericytes in microangiography. Diabete Metab. 14: 495-504, 1988.

Davel LE, Miguez MM, De Lustig ES. Evidence that indomethacin inhibits lymphocyte-induced angiogenesis. Transplantation Baltimore. 39: 564-565 1985.

Dawson JM, Tyler KR, Hudlicka O. A comparison of the microcirculation in rat fast glycolytic and slow oxidative muscles at rest and during contractions. Microvasc. Res. 33: 167-182, 1987.

Dawson JM, Hudlicka O. Inhibition of capillary growth in skeletal muscle with GPA 1734. Int. J. Microcirc: Clin. Exper. 9: 134, 1990.

Dawson JM, Hudlicka O. Can changes in microcirculation explain capillary growth in skeletal muscle? Int. J. Exp. Path. 74: 65-71, 1993.

Egginton S. Effect of an anabolic hormone on striated muscle growth and performance. Pflugers Arch. 410: 349-355, 1987a.

Egginton S. Effect of anabolic hormone on anaerobic capacity of rat striated muscle. Pflugers Arch. 410: 356-361, 1987b.

Egginton S, Hudlicka O. The effect of long-term activation of glycolytic fibres in rat skeletal muscle on capillary supply and enzyme activities. J. Physiol. 409: 71P, 1989.

Egginton S, Hudlicka O. Effect of long-term muscle overload on capillary supply, blood flow and performance in rat fast muscle. J. Physiol. 452: 9P, 1992.

Egginton S, Hudlicka O, Brown MD. The possible role in angiogenesis of pericytes from chronically stimulated rat skeletal muscles. J. Physiol. 467: 43P, 1993.

Fairney J, Egginton S. The effect of cold acclimation on muscle capillary supply in the Syrian hamster (Mesocricetus auratus). J. Physiol. 475: 61-62P, 1994.

Form DH, Auerbach R. PGE_2 and angiogenesis. Proc. Soc. Exp. Biol. Med. 172: 214-218, 1983.

Folkman J, Cotran R. Relation of vascular proliferation to tumor growth. Int. Rev. Exp. Pathol. 16: 207-248, 1976.

Fulgenzi G, Hudlicka O. The effect of $alpha_1$ blocker prazosin on capillarization, blood flow and performance in ischaemic skeletal muscles. Int. J. Microcirc: Clin. Exper. 14: (S1), 229, 1994.

Hansen-Smith FM, Carlson BM, Irwin KL. Revascularization of the freely grafted extensor digitorum longus muscle in the rat. Am. J. Anat. 158: 65-82, 1980.

Hansen-Smith FM, Hudlicka O. Ultrastructure of capillaries during angiogenesis in electrically stimulated rat extensor digitorum longs (EDL) muscle. FESAB J. 7: 1993.

Hermansen L, Wachtlova M. Capillary density of skeletal muscle in well-trained and untrained men. J. Appl. Physiol. 30: 860-863, 1971.

Hoppeler H, Desplanches D. Muscle structural modifications in hypoxia. Int. J. Sports Med. 13: Suppl 1, S166-168,1992.

Hoppeler H, Kleinert E, Schlegel C, Claassen H, Howald H, Kayar SR, Ceretelli P. Morphological adaptations of human skeletal muscle to chronic hypoxia. Int. J. Sports Med. 11: Suppl 1,53-59, 1990.

Hudlicka O. Uptake of substrates in slow and fast muscles in situ. Microvasc. Res. 10: 17-28, 1975.

Hudlicka O. Review lecture: What makes blood vessels grow? J. Physiol. Lond. 444: 1-24, 1991.

Hudlicka O. Physiological mechanisms of angiogenesis. In Functionality of the endothelium in health and diseased states: A comprehensive review. Ed. G. Pastelin, R. Rubio, G. Ceballos & J. Suarez, Sociedad Mexicana de Cardiologia. Gobierno del Estado de Veracruz, p252-262, 1994.

Hudlicka O, Brown MD. Physical Forces and Angiogenesis. In Mechanoreception by the Vascular Wall. Ed. G.M. Rubanyi, Futura Publishing Co Inc, Mount Kisco, NY, 1993.

Hudlicka O, Brown M, Egginton S. Angiogenesis in skeletal and cardiac muscle. Physiol. Rev. 72: 369-669, 1992.

Hudlicka O, Dodd C, Renkin EM, Gray SD. Early changes in fiber profiles and capillary density in long-term stimulated muscles. Am. J. Physiol. 243 (Heart Circ. Physiol. 12: H528-H535, 1982.

Hudlicka O, Tyler KR, Wright AJA, Ziada AMAR. Growth of capillaries in skeletal muscles. Prog. Appl. Microcirc. 5: 44-61, 1984.

Hudlicka O, Egginton S, Brown MD. Capillary diffusion distances - their importance for cardiac and skeletal muscle performance. News in Physiological Sciences, 3: 134-137, 1988.

Ingber DE, Folkman J. How does extracellular matrix control capillary morphogenesis? Cell. 58: 803-305, 1989.

Ingjer F. Effects of endurance training on muscle fibre ATP-ase activity, capillary supply and mitochondria in man. J. Physiol. Lond. 294: 419-432, 1979.

Knighton DR, Silver IA, Hunt TK. Regulation of wound-healing angiogenesis - effect of oxygen gradients and inspired oxygen concentration. Surgery St Louis, 90: 1981.

Koller A, Kaley G. Prostaglandins mediate arteriolar dilatation to increased blood flow velocity in skeletal muscle microcirculation. Circ. Res. 67: 529-534, 1990.

Kuwabara T, Cogan, DG. Retinal vascular patterns. VI. Mural cells of the retinal capillaries. Arch. Ophthalmol. 69: 492-502, 1963.

Laughlin MH, Armstrong RB. Muscular blood flow distribution patterns as a function of running speed in rats. Am. J. Physiol. 243 (Heart Circ. Physiol. 12): H296-H306, 1982

Maragoudakis ME, Sarmonika M, Panousta-Copoulu M. Inhibition of basement membrane biosynthesis prevents angiogenesis. J. Pharmacol. Exper. Therap. 244: 729-733, 1988.

Mellander S, Bjornberg J. Regulation of vascular smooth muscle tone and capillary pressure. NIPS 7: 113-119, 1992.

Myrhage R, Hudlicka O. Capillary growth in chronically stimulated adult skeletal muscle as studied by intravital microscopy and histological methods in rabbits and rats. Microvasc. Res. 16: 73-90, 1978.

Orlidge A, D'Amore PA. Inhibition of capillary endothelial cell growth by pericytes and smooth muscle cells. J. Cell Biol. 105:1455-1462, 1987.

Pearce SC, Hudlicka O. Are prostaglandins involved in capillary growth in chronically stimulated skeletal muscles? Int. J. Microcirc. 14: 243, 1994.

Pearce S, Hudlicka O, Egginton S. Early in activity-induced angiogenesis in rat skeletal muscles: incorporation of bromodeoxyuridine into cells of the interstitium.J. Physiol. 483P, 1995.

Saltin B, Kiens B, Savard G, Pedersen PK. Role of hemoglobin and capillarization for oxygen delivery and extraction in muscle exercise. Acta Physiol. Scand. 128 Suppl 556: 21-32, 1986.

Smith P. Effect of hypoxia upon the growth and sprouting activity of cultured aorti endothelium from the rat. J. Cell Sci. 92: 505-512, 1989.

Stingl J, Rhodin JAG. Early postnatal growth of skeletal muscle blood vessels of the rat. Cell & Tiss. Res. 275: 419-434, 1994.

Valdivia E. Total capillary bed in striated muscles of guinea pigs native to the Peruvian mountains. Am. J. Physiol. 194: 585-589, 1958.

Vanotti A, Magiday M. Untersuchungen zum Studium des Trainiertsein. V. Uber die Capillarisierung der trainierten Muskulatur. Arbeitsphysiologie 7: 615-622, 1934.

Walter H, Hudlicka O. Expression of basic fibroblast growth factor (bFGF) in chronically stimulated skeletal muscles. Int. J. Microcirc: Clin. Exper. 11: 447, 1992.

Wolff JR, Goerz CH, Bar TH, Gueldner FH. Common morphogenetic aspects of various organotypic microvascular patterns. Microvasc. Res. 10: 373-395, 1975.

Ziada AMAR, Hudlicka O, Tyler KR, Wright AJA. The effect of long-term vasodilatation on capillary growth and performance in rabbit heart and skeletal muscle. Cardiovasc. Res. 18: 724-732, 1984.

Ziche M, Jones J, Gullino PM. Role of prostaglandin E_1 and copper in angiogenesis. J. Natl. Cancer Inst. 69: 475-482, 1982.

Zumstein A, Mathieu O, Howald H, Hoppeler H. Morphometric analysis of the capillary supply in skeletal muscles of trained and untrained subjects - its limitations in muscle biopsies. Pfluegers Arch. 397: 277-283, 1983.

WOUND HEALING ANGIOGENESIS: THE METABOLIC BASIS OF REPAIR

James S. Constant, M.D., David Y. Suh, M.D.,
M. Zamirul Hussain, PhD., Thomas K. Hunt, M.D.

Department of Surgery, University of California, San Francisco

Introduction

Angiogenesis, a critical component of tissue repair, leads to increased oxygenation and endothelial permeability. This improves tissue energetics, enhances the effect of recruited cytokines, and improves protein transport to and from the wound site. We propose metabolic conditions in the wound mediate much of the cellular and cytokine response. We will describe this milieu and further introduce a novel metabolic regulatory mechanism capable of sensing the redox state of tissues and mediating the response. This system acts through Adenosine diphospho ribosylation (ADPR) which mediates hypoxic and lactate induction of critical metabolic events in wound repair including angiogenesis. This is specifically relevant to wound angiogenesis because conditions of hypoxia and elevated lactate coexist in wounds.

Wound Repair

Injury (thermal, mechanical, inflammatory) perturbs the microenvironment by inducing coagulation and activating fibrin deposition and thereby releasing various platelet derived factors. The inflammatory cascade is initiated and auto-amplifies. Inflammatory cells, first leukocytes and then monocytes, enter the area and contribute to two broad functions: defense against infection and the further recruitment and organization of cellular responses which are important in the subsequent phases of healing. This regulation is, in general, accomplished via growth factors and cytokines contributed largely by platelets and macrophages which, together with fibrin- and thrombin- derived growth factors incite angiogenesis, fibroplasia, matrix production and finally, when applicable, epithelization. Long-term remodeling involves continued matrix synthesis and turnover via proteinases and may take years to complete.(Hopf,1992)

Molecular, Cellular, and Clinical Aspects of Angiogenesis
Edited by Michael E. Maragoudakis, Plenum Press, New York, 1996

151

Cellular control of Angiogenesis

Numerous cells participate in wound angiogenesis. Platelets contribute Platelet Derived Growth Factor (particularly PDGF-BB), Transforming Growth Factor (TGF-b), and Insulin-like Growth Factor (IGF-1) and its binding protein IGFBP3. Fibroblasts produce Fibroblast Growth Factor (aFGF or FGF-1 and bFGF or FGF-2) and Insulin-like Growth Factor. Macrophages, a constant feature of wounds, are a continuing source of angiogenic cytokines, and provide Vascular Endothelial Growth Factor (VEGF), Transforming Growth Factor (TGF-b), Tumor Necrosis Factor (TNF-a), and Interleukins (IL-1, IL-6, IL-8).(Sunderkotter, 1994) Neutrophils are not believed to directly signal for angiogenesis, but their metabolic by-products, as we will show, clearly affect the microenvironment and add to the stimuli which drive angiogenesis. Overall, these mechanisms appear complementary(Banda, 1982). As is common in critical cellular systems, the redundant cytokine model allows for rapid paracrine response, and a check and balance regulation of cellular function.

As one of the first blood substances to gain access to wounds, fibrin also contributes to wound healing. It enhances collagen deposition by attracting macrophages and fibroblasts. The lack of fibrin in wounds via experimental defibrinogenation decreases collagen deposition.(Brandstedt, 1980) The fibrin matrix may bind cytokines and growth factors and clearly invites infiltration by inflammatory cells which then find themselves in an acidotic, hypoxic environment. Furthermore, it supports proliferation and migration of endothelial cells, a crucial aspect of angiogenesis. Fibrin clot elicits angiogenesis(Dvorak, 1987) However, it remains unclear whether its matrix properties, one of its degradation products, or it's capacity to stimulate macrophages is most responsible for this angiogenic response. Thrombin has been reported to have angiogenic properties independent of fibrin.(Tsopanoglou, 1993). Fibrinopeptide B and fibrinolytic, fragment E have been reported to have angiogenic activities(Paty, 1987) (Thompson, 1992).

Metabolic control of angiogenesis

We have investigated the broad hypothesis that metabolic aspects of injured and healing tissues drive angiogenesis and collagen synthesis. The wound environment is hypoxic, acidotic and hyperlactated. This represents the sum of three effects: decreased oxygen supply due to vascular damage and coagulation, increased metabolic demand due to the heightened cellular response (anaerobic glycolysis), and aerobic glycolysis by inflammatory cells. Aerobic production of lactate by leukocytes is probably particularly important and occurs because leukocytes contain few mitochondria and develop energy from glucose mainly by production of lactate(Hussain, 1991)(Knighton, 1991).

Macrophages, acting in response to the hypoxia found in wounds, have been identified as a major source of angiogenic factor production (Knighton, 1983). Our current work is directed at identification of macrophage-derived angiogenic factor(s) and the specific conditions responsible for their transcription and secretion. Recently we have directed our attention to VEGF since its been noted that VEGF production is a consequence of tumor cell exposure to hypoxia, and it is likely that macrophages might also respond similarly (Shweiki, 1992)(Finkenzeller, 1995) . VEGF is also *unique* among angiogenic growth factors in that its receptors have only been identified on endothelial cells. In comparison, Fibroblast Growth Factor is angiogenic, but it has no classic secretion sequence, and its receptors are widely distributed. Other cytokines such as IL-1,6, and 8, and TGF-b are also released by certain cells in response to hypoxia, but their effects are not as specific and do not include increasing vascular permeability as is seen in wounds. VEGF has been localized proximate to capillaries in tumors of hepatic, ovarian, glial, and Lewis Lung origin and has been identified in association with hypoxic retinopathy.(Shweiki, 1992)

Intuitively, hypoxia seems a likely instigator of angiogenesis. However, it seems to us a biologically unsatisfactory *sole* agent because hypoxia is dangerously close to death and because angiogenesis occurs in many circumstances in which energy depletion, or lowered redox potential is more prominent than hypoxia. There is at least one known means by which low redox potential can be sensed and converted to biological actions, including collagen deposition and the activities of a number of specific cytoplasmic and nuclear proteins. It involves a set of adenosine diphosphoribose (ADPR) regulatory mechanisms. The presence of these mechanisms was first discovered independently by Hayashi and Chambon in 1964. For many years, the ability of polyADPR (PADPR) to mediate repair of DNA strand breaks was the almost exclusive focus of research in this area. Nevertheless, smaller polymers, monomers, and even a cyclic ADPR have been found to have regulatory roles in the cytosol. These affect a number of cytosolic and membrane-based enzymes as well as mitochondrial and skeletal proteins. Investigators have considered that poly ADP-ribosylation may constitute a link between the metabolic (redox) state and gene regulation.(Loetcher, 1987) Subsequently, Hussain et al have linked poly ADPR to transcription of collagen gene and ADPR to post-translational modification of collagen protein.(Hussain, 1989; Ghani, 1992)

When intracellular lactate is at baseline (non-wounded and normal respiration), the NAD+/NADH ratio favors NAD+ acting as a precursor (after nicotinamide removal) for poly adenosine diphospho ribose (poly ADPR) and ADPR. Poly ADPR suppresses collagen gene transcription and ADPR inhibits post-transcriptional collagen hydroxylation and therefore decreases collagen deposition. Raising lactate (as is often during injury) alters the NAD+/NADH balance and shrinks the NAD+ pool and the subsequent ADPR metabolites. As a consequence, the aforementioned inhibition is released, thereby permitting activation of collagen gene transcription, peptide synthesis, hydroxylation, and deposition (figure 1).(Hussain, 1989; Ghani, 1992) Because macrophages occupy the same wound space and are exposed to the same conditions as fibroblasts, it seems likely that a similar biochemical mechanism involving ADPR may regulate macrophage angiogenesis.

One outstanding feature of the ADPR-mediated mechanism for present purposes is the fact that the several enzymes which remove nicotinamide from

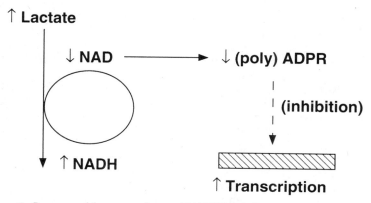

Figure 1: Increased lactate releases PADPR inhibition of Transcription

NAD+, thus converting it to ADPR species, act only on NAD+ and not on the reduced form, NADH. Thus, the pool sizes of ADPR species are directly linked to the redox state of the cell, and ADPR levels are relatively high in well nourished cells. The second important aspect is that ADPR-mediated mechanisms are normally suppressive in the context of collagen synthesis and angiogenesis. When redox potential is lowered and the NAD+ pool is diminished, this suppressive role is released. In the case of collagen mRNA, lactate increases hydroxylation.

Clearly, hypoxia alone shrinks the NAD+ pool, but so will many other metabolic events particularly the presence of large concentrations of lactate. This property allows ADPR-mediated mechanisms to be differentiated from primarily hypoxic regulation acting through a heme protein as has been suggested for regulation of erythropoetin. ADPR-mediated mechanisms can be influenced by increasing lactate concentrations even in the presence of adequate oxygen, and the lactate effect can be abrogated by oxamate which occupies the lactate binding site on LDH. In theory, neither of these would be expected to effect a heme-based mechanism. Exploitation of this property and the finding that VEGF is secreted by macrophages which have been grown in wound-like environments constitute the strategy of this presentation. The basic hypothesis is that certain signals which mediate angiogenesis, VEGF in the instant case, will appear in proportion to the concentrations or pool sizes of certain redox-related substances which change or reflect the availability of NAD+ for conversion to ADPR.

Measurement of Angiogenesis

Angiogenesis is preferably measured by models which allow isolation of individual components of the response. A commonly used in vitro model is the Boyden (dual) chamber. Endothelial cells are placed on one side of a membrane and a chemoattractant or stimulus on the other. Endothelial cell migration across the membrane is quantified. Unfortunately, results from in vitro angiogenic assays do not always correlate reliably with in vivo studies. Even in assay systems such as those incorporating a matrix for cellular migration, in vitro models seem to us to have more value for the study of metastatic invasion rather than for the study of wound healing angiogenesis. This probably reflects the complexity of the wound matrix and the overlapping nature of repair elements.

Wound healing investigators have generally relied on in-vivo assay systems for angiogenesis research. This is relatively easy for us since wounds are an unusual model in which the artifact induced by the observer is the object of the exercise. A simple definition of angiogenic response incorporates three elements: endothelial cell migration, organization, and blood flow. The corneal micropocket assay has been considered the "gold standard". This assay has been performed in multiple laboratories, and provides a reliable, standardized, reproducible, and quantifiable measure of the angiogenic response. It has been difficult to extend this assay to the murine cornea and quantify with similar accuracy. The major disadvantages remain cost, animal cornea use, and the time and skill required to perform the procedures. Other traditional in vivo angiogenesis assays which have been used in wound healing research include the chicken chorioallantoic membrane (CAM) assay and various disc assays (usually a plastic composite with a bio-insert such as collagen gel or fibrin).

An exciting new in vivo angiogenesis assay is the murine matrigel system. Although useful also for in vitro studies, we have found its real strength in wound angiogenesis research as an in vivo assay. Matrigel is a reconstituted basement membrane complex from the murine Engelbreth-Holm-Swarm tumor that contains primarily collagen type IV and laminin. It appears to be an ideal substrate for endothelial cell migration and response, presumably because it closely mimics the biologic basement membrane and matrix. The conditioned test substance is mixed with matrigel as a liquid and injected subcutaneously in the murine dorsum where body temperature changes the solution from liquid to gel. The gel plug is harvested at day 5 or more and H&E slides are prepared. Infiltration of cells, their organization, and blood flow are noted. Non-supplemented plugs (or those from a negative stimulus condition) support minimal infiltration or inflammation except a fibrotic response at the edge. Plugs containing angiogenic substances become infiltrated early by endothelial cells. This is followed by progressive linear organization, and functional vessel development as evidenced by intravasculature erythrocytes (figure 2). Endothelial cell identity is confirmed as necessary with a Factor VIII stain.(Kleinman, 1986)(Passaniti, 1992)

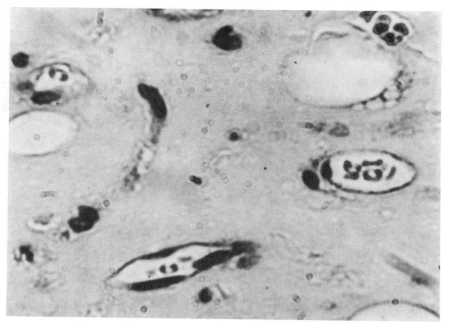

Figure 2: Matrigel section at 11 days X 1000

Analysis of Angiogenic Factors

The experimental design involves the measurement of VEGF from macrophages cultured in the hypoxic and lactated conditions which are characteristic of wounds. PO_2 in human and rodent uninjured subcutaneous tissue is in the range of 55-60 mmHg. The lactate concentration in uninjured, non-exercising tissue is usually less than 1-2 mM. These conditions serve as the reference range for control cultures in set of experiments. In wounds, pO_2 ranges from 10 to 40 mmHg, depending on many conditions including extent of injury and host response. Wound lactate levels can range from 5mM to as high as 20 mM.

We report here results from bone marrow macrophages cultured in wound conditions. We have also successfully used cells gathered from wire mesh Hunt-Schilling wound cylinders (in vivo) or collected from serum, pulmonary alveoli, peritoneum, and established cell lines. An alternative to wound cylinders is the less invasive expanded polytetrafluoroethylene (ePTFE) tissue insert. Surgical wounds can also be studied using immunohistochemical staining for collagen deposition, cytokine presence, and receptor status. Careful control and analysis of specific oxygen and lactate concentrations were performed using environmental control chambers. Tissue, media, and gas phase conditions are monitored with oxygen optodes and analyzed for lactate levels and pH. Cells were analyzed for transcription of angiogenic growth factors by northern blots or PCR. Conditioned supernatant or wound fluid was studied by RIA or ELISA for presence of secreted angiogenic peptides. Putative angiogenic factors were tested for bioactivity in the matrigel and corneal assays. Our prior research of angiogenic cytokine expression in wound conditions of hypoxia and elevated lactate confirms this experimental design for investigation of metabolic control of wound healing angiogenesis.

Experimental Findings

Macrophages in the presence of 15 mM lactate were analyzed for NAD+ and ADPR levels. As described in figure 1, the elevated lactate is expected to decrease NAD+ and ADPR. This is demonstrated in figure 3. Lactate depressed these compounds by approximately 40%.

Treatment	[NAD+] pmol/mg protein	(ADP-ribose)n dpm/mg protein x10E-3
None	243 +/- 93	16.4 +/- 1.8
Lactate, 15mM	142 +/- 39	9.9 +/- 0.8

Figure 3: Effect of lactate on NAD+ level and (ADP-ribose)n synthesis in macrophages

Macrophages were then cultured at a pO2 of 13 mmHg and/or lactate levels of 15 mM and thereby induced to enhance production of Vascular Endothelial Growth Factor (VEGF) (figure 4). Macrophages are the major source of angiogenic cytokines in the wound, and these are common levels of hypoxia and lactate in the wound zones in which macrophages are found.
VEGF production is clearly increased beyond the constitutive level in response to both hypoxia and elevated lactate alone and together. Hypoxia alone is a moderate stimulus. Lactate alone, *even* in the presence of adequate oxygen, is a potent inducer of VEGF protein expression (figure 4). A combination of high lactate concentration and hypoxia elicits the maximal production of VEGF. To our knowledge, the effect of lactate has not been noted previously.

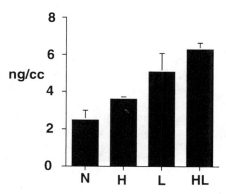

Figure 4: VEGF Protein (ELISA) in conditioned macrophage supernatant (N= normoxic, L= lactate, H= hypoxic). p<0.01.

Comment

We postulate that VEGF secretion represents an effort of the host to correct the metabolic imbalance in the wound by increasing angiogenesis and restoring a favorable energetic state. The ultimate goal of this response is an increased oxygen supply and reduced energy deficit, and a shift toward aerobic glycolysis. Furthermore, the return to a favorable oxidation-reduction balance improves local energetics and attenuates ADPR mediated gene expression because less lactate is produced. In other words, this hypothesis is attractive partly because it contains all the elements necessary for resolution of angiogenesis as well as stimulation, potentially an elegant example of physiologic cellular feedback control.

A considerable portion of in-vivo angiogenic research has been, by necessity, conducted in wound environments. Ironically, wound healing angiogenesis has usually been ignored or abhorred as if it were a special type, or perhaps too mundane a subject in comparison to tumor angiogenesis. One reason for this has been the obvious participation of inflammation in wound repair. In the last NATO conference, the thinking which led inflammation to be a suspect milieu in which to pursue research on tumors was thoroughly discussed and to a large measure dismissed in view of the obvious inflammatory component of many tumors and angiogenesis models. Furthermore, tumor and inflammatory cells share many mechanisms; and some vascularized solid tumors, such as Kaposi's sarcoma, are primarily composed of cells of lymphoid and monocytic lineages. Clearly, cytokines and growth factors exist in both, and for most intents and purposes, the list of these factors is identical. In any case, it would seem tumors which continually grow must, like wounds, expose their cells to stressful, reducing environments.

Arguably, wounds are an ideal model for mechanistic angiogenic research. In particular, there is no doubt that wound angiogenesis is a true angiogenesis because it can be made to occur in spaces which contain no pre-existing tissue. Furthermore, wound angiogenesis can be instigated, augmented, suppressed, and even eliminated by design. Ironically, use of wound models, free as they are of the siren call of "*the* tumor angiogenesis factor," can refocus research on important mechanisms of angiogenesis which precede, or in evolutionary terms "predate," growth factors which probably act well along in the mechanism of blood vessel development just as the appearance of blood vessels occurs well along in the evolution of mammalian life.

Summary

Injury creates a regional energy deficit secondary to decreased oxygen supply and altered local metabolism. Angiogenesis can be conceived as a coordinated response at the wound site aimed at restoring favorable oxygen and energy balance. We propose that these redox gradients are the impetus for the angiogenic response. Macrophages respond to these oxygen and lactate alterations by producing angiogenic cytokines. As a result, we propose that endothelial cells migrate and organize into functional vasculature. Lactate stimulates angiogenic factor production in wound macrophages via ADPR mediated effects on transcription and peptide expression. VEGF is a fundamental angiogenic cytokine secreted by macrophages in wound conditions of hyperlactate and hypoxia. ADPR regulation of wound healing angiogenesis represents feedback control of cellular energy balance and metabolism

Acknowledgments

Support received from:
NIH Program Project Grant
NIH NIGMS Academic Training Grant in Trauma and Burns
Pacific Vascular Research Foundation
Department of Surgery, University of California, San Francisco

References

Banda, M.J., Knighton, D.R. , Hunt, T.K., Werb, Z.,. 1982, Isolation of a nonmitogenic angiogenesis factor from wound fluid. *Proc Natl Acad Sci U S A* **79**(24): 7773-7.

Brandstedt, S., Olson, P.S.,1980., Effect of defibrinogenation on wound strength and collagen formation. A study in the rabbit. *Acta Chir Scand* **146**(7): 483-6.

Dvorak, H.F., Harvey, V.S., Estrella, P., Brown, L.F., McDonagh, J., Dvorak, A.M., 1987, Fibrin containing gels induce angiogenesis. Implications for tumor stroma generation and wound healing. *Lab Invest* **57**(6): 673-86.

Finkenzeller, G., Technau, A., Marme, D., 1995, Hypoxia-induced transcription of the vascular endothelial growth factor gene is independent of functional AP-1 transcription factor. *Biochem Biophys Res Commun* **208**(1): 432-9.

Ghani,Q.P., Hussain, M.Z., Zhang, J., Hunt, T.K., 1992, Control of procollagen gene transcription and prolyl hydroxylase activity by poly(ADP-ribose). In: *ADP-Ribosylation Reactions*, G Poirier and A Moreaer (eds), New York, Springer Verlag, 1992; 111-117.

Hopf, H., Hunt, T.K., 1992, The role of oxygen in wound repair and infection. *Musculoskeletal Infection*. Esterhai, J., Gristina ,A., Poss, R., Park Ridge, IL, American Academy of Orthopaedic Surgeons: 329-39.

Hussain, M.Z., Hunt, T.K., 1991, *Wound Microenvironment*. Philadephia, WB Saunders.

Hussain, M.Z., Ghani, Q.P., Hunt, T.K., 1989, Inhibition of prolyl hydroxylase by poly(ADP-ribose) and phosphoribosyl-AMP. Possible role of ADP-ribosylation in intracellular prolyl hydroxylase regulation. *J Biol Chem* **264**(14): 7850-5.

Kleinman, H.K, McGarvey M.L., Hassel, J.R., Star, V.L., Cannon, F.B., Lauri, G.W., Martin, G.R., 1986, Basement Membrane Complexes with Biological Activity. *Biochemistry* **25**(2): 312-318.

Knighton, D.R., Fiegel, V.D., 1991, Regulation of cutaneous wound healing by growth factors and the microenvironment. *Invest Radiol* **26**(6): 604-11.

Knighton, D.R., Hunt, T.K., Scheuenstuhl, H., Halliday, B.J., 1983, Oxygen tension regulates the expression of angiogenesis factor by macrophages. *Science* **221**(4617): 1283-5.

Loetcher, P., Alvarez-Ganzales, R., Althaus, F.R., 1987, Poly(ADP-ribose) may signal changing metabolic conditions to the chromatin of mammalian cells. *PNAS USA* **84**:1286-1289.

Passaniti, A., Taylor, R.M., Pili, R., Guo, Y., Long, P.V., Haney, J.A., Pauly, R.R., Grant, D.S., Martin, G.R., 1992, A simple, quantitative method for assessing angiogenesis and antiangiogenic agents using reconstituted basement membrane, heparin, and fibroblast growth factor. *Lab Invest* **67**(4): 519-28.

Paty, P., Banda, M.J., Hunt, T.K., 1987, *Fibrin activation of macrophages: One mechanism of angiogenesis in wound healing.* Squibb-Convatic meeting, San Antonio, Texas.

Shweiki, D., Itin, A., Soffer, D., Keshet, E., 1992, Vascular endothelial growth factor induced by hypoxia may mediate hypoxia-initiated angiogenesis. *Nature* **359**(6398): 843-5.

Sunderkotter, C., Steinbrink, K., Goebeler, M., Bhardwaj, R., Sorg, C., 1994, Macrophages and angiogenesis. *J Leukoc Biol* **55**(3): 410-22.

Thompson, W.D., Smith, E.B., Stirk, C.M., Marshall, F.I., Stout, A.J., Kocchar, A., 1992, Angiogenic activity of fibrin degradation products is located in firin fragment E. *J Pathol* (168): 47-53.

Tsopanoglou, N.E., Pipili-Synetos, E., Maragoudakis, M.E., 1993, Thrombin promotes angiogenesis by a mechanism independent of fibrin formation. *Am J Physiol:* C1302-7.

WOUND HEALING, FIBRIN AND ANGIOGENESIS

W. Douglas Thompson, Stephen J. McNally, Naren Ganesalingam,
Deirdre S. E. McCallion, Christina M. Stirk, and William T Melvin*

Departments of Pathology and *Molecular and Cell Biology
University of Aberdeen Medical School
Aberdeen Royal Infirmary
Aberdeen AB9 2ZD
United Kingdom

INTRODUCTION

Fibrin degradation products, specifically fibrin fragment E, have been shown by us to be angiogenic on the chick chorioallantoic membrane (CAM) (Thompson et al., 1985; Thompson et al., 1992). Although we have previously succeeded in demonstrating angiogenic activity in simple extracts of healing skin wounds (Thompson et al., 1991), the rationale for the methodology used has remained uncertain, and the relationship of activity to fibrinolysis has been implied but not demonstrated. Now we show that the angiogenic activity of wound extracts is critically dependant on the methodology applied to the handling of such experimental murine skin wound material. The key factor is the ability to remove fibrinogen from test and control skin extracts whilst preventing the artefactual generation of fibrin degradation products (FDP) (Thompson et al., 1994). Next is the ability to remove or neutralise fibrin fragments with specific antibodies, and first experiments show that angiogenic activity is blocked by prior admixture with a specific polyclonal rat anti human fibrin fragment E antiserum.

Three other aspects of the linkage between angiogenesis and fibrinolysis are being explored, the effect of subcutaneous injection of FDP, the effect of FDP on cells in culture, and the immunohistochemical application to wound healing and chronic inflammation of novel antibodies against fibrin specific epitopes .

MATERIALS AND METHODS

Preparation of Skin Extracts

Male $C_{57}BL_6$ mice were anaesthetised with a mixture of Vetalar (ketamine) (150 mg/kg weight of mice) and Rompun (10 mg/kg), administered intraperitoneally. Two dorsal 2 cm full thickness incisions were created, cutting through the panniculus carnosus muscle

Molecular, Cellular, and Clinical Aspects of Angiogenesis
Edited by Michael E. Maragoudakis, Plenum Press, New York, 1996

161

layer. The incised wounds were immediately sutured. On day three following wounding, the mice were killed and each wound excised with a minimal amount of normal skin, promptly wrapped in tinfoil, and plunged in ice to halt metabolism, clotting and fibrinolysis. If normal skin was being collected, a dorsal area of 3 cm by 3 cm was obtained and promptly placed on ice as with the wound material.

Two different buffer systems were tested for extraction- 3 M EACA buffer (3 M epsilon-amino-caproic acid, 0.005 M Tris, 0.1 M NaCl) at 4x weight per volume, and aprotinin buffer (5mg % aprotinin instead of EACA) also at 4x weight per volume. The samples were subsequently coarsely chopped in Universal containers in 4°C buffer, homogenised, and then ultracentrifuged for 15 min at 15,000 rpm at 4°C, retaining the supernatant. The supernatant was then subjected to repeated ultracentrifugation to remove clotted fibrinogen after temperature adjustment to 18°C, and then 37°C. Carbon tetrachloride was then added at 4 ml for every ml of sample and shaken well, in order to remove lipid and cellular debris. Thereafter the mixture was centrifuged at 2,200 rpm for 10 min and the upper aqueous layer retained. Storage of the filter sterilised material was satisfactory at 4°C. During the extraction process, small sample aliquots were taken immediately following each centrifuge step, and these samples were immunoblotted to assess the effectiveness of the extraction process.

Extract Content of Fibrin(ogen) Related Antigens

Polyacrylamide gel electrophoresis was performed on all wound and skin extracts using the method of Laemmli (1970), followed by immunoblotting to determine the efficiency of the buffer system in preventing formation of artefactual FDP as assessed by band patterns and densities. One dimensional rocket immunoelectrophoresis on samples was performed as described by Laurell (1966) to quantify total fibrinogen and FDP concentrations in wound extracts. Both techniques have been previously applied to wound and atherosclerotic plaque extracts (Thompson et al., 1991; Smith et al., 1990; Thompson et al., 1987)

Effects of Extracts on the Chick Chorioallantoic Membrane

The activity of extracts was assessed by measurement of DNA synthesis as a parameter of angiogenesis in the chick CAM system (Thompson et al., 1992; Thompson et al., 1991; Thompson et al., 1985). The samples were passed through a PD10 buffer exchange column (Pharmacia) to equilibrate in Dulbecco's A buffer before application. Test and control substances were applied in liquid form as 0.3 ml aliquots to the entire "dropped" area of CAM at day 10 of growth. The level of DNA synthesis was measured by methyl - tritiated thymidine incorporation after 18h. Student's t test was used for testing significance on log transformed data.

Admixture of Anti Fibrin E Antiserum

A Lou rat anti human whole fibrin fragment E polyclonal antibody was used which was capable of neutralising active human FDP, as previously found with a rabbit anti fragment E antibody (Thompson et al., 1992). This antibody has now been shown to produce strong specific staining of fibrin on tissue sections, when used for immunohistochemistry. Two aliquots of 2.5 ml wound extract were taken, and 100μl of antibody was added to one. Both aliquots were then incubated at 37°C for one hour before application. Buffer and antibody alone controls were also tested.

Subcutaneous Injection of FDP

Active FDP were prepared from washed fibrin, free of fibrinopeptides A and B, as previously described (Thompson *et al.*, 1985; Thompson *et al.*, 1992), at a concentration range from approximately 100 to 3,200 µg per ml. Two preparations were compared, a 24h digest containing all major degradation products, and a 120h digest containing solely fragment E on immunoblotting. Filter sterilised aliquots of 0.1 ml were injected subcutaneously into the dorsum of Balb/C mice and Sprague Dawley rats. The animals were killed at days 3 and 5, and skin samples fixed in formalin for histological examination of H & E sections.

Effect of FDP in Cell Culture

Fibroblasts derived from 14 day chick embryos were maintained in culture in DMEM plus 10% fetal calf serum. Cells were plated into 6 well culture plates at 3×10^4 per well, in DMEM with 2% fetal calf serum. 15µl of FDP (5.654 mg/ml) were added to half the wells, and the plates maintained in a CO_2 incubator at 37°C. For growth curves, cells from 6 test and 6 control plates were counted.

Immunohistochemical Application of Antibodies against Fibrin Epitopes

A variety of antibodies were tested on paraffin sections of various examples of acute and chronic inflammation, largely on subcutaneous abscess and appendicitis, as suitable test tissues exhibiting major increased vascular permeability, fibrin deposition and lysis amongst a marked inflammatory cell infiltrate. DAKO rabbit polyclonal anti human fibrinogen was taken as a current commercial standard for comparison. These antibodies included NIBn 5F3.C4.B10 (5F3), a mouse monoclonal antibody created by Dr. Patrick Gaffney's group at the National Institute for Biological Standards and Controls, Potters Bar, raised against the disulphide knot of human fragment E (Tymkewycz *et al.*, 1993). The other antibodies were raised either to whole human fibrin fragment E, or to synthetic peptides corresponding to specific regions of the different chains of the fibrin molecule. Rabbit, rat and mouse were used because of previous experience with species similarities and differences attributable to the high degree of conservation of the fibrinogen molecule (Thompson *et al.*, 1993).

Immunostaining was performed using an adaptation of the streptavidin biotin complex method described by Hsu *et al.* (1981). Sections of paraffin embedded material were cut and mounted on APES (α-propyl ethoxy silane) coated glass slides to permit adhesion despite microwave treatment. Standard immunostaining was compared for each antibody with two different antigen retrieval methods, microwave heating and trypsin digestion (Langlois *et al.*, 1994)

RESULTS

Preparation of Skin Extracts

Figure 1 of an antifibrinogen immunoblot shows the progressive diminution of the fibrinogen double bands without increase of FDP bands during the successive stages of extract preparation. This result was obtained consistently using EACA. More limited assessment of aprotinin as an alternative plasmin inhibitor gave comparable findings.

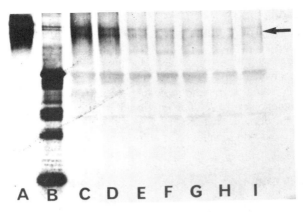

Figure 1. Immunoblot using an antifibrinogen antiserum to show the progressive diminution of the fibrinogen double bands in a wound extract (←), with preservation of density of staining of lower molecular weight fibrin degradation products.

Lanes: A - standard fibrinogen; B - standard fibrin degradation products; C to I - samples of wound extract taken during successive stages of preparation.

Table 1. Effect of processing on FRA content of EACA buffered wound and normal skin samples

| Sample | Total Fibrin Related Antigen (FRA) (μg/ml) | | Percentage change |
	start	end	
normal mouse skin	88	5.8	-93%
day 3 wound	106	86	-19%

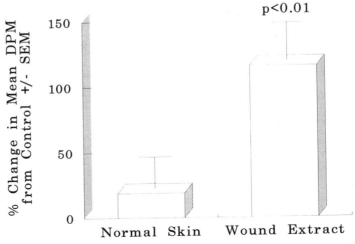

Figure 2. Stimulation of the chick chorioallantoic membrane system by wound but not normal skin extract.

Extract Content of Fibrin(ogen) Related Antigens

The real extent of reduction of fibrinogen is reflected by the quantitative assay of FRA. Table 1 shows an example of 93% reduction after processing of a normal skin extract, and 19% reduction in a day 3 wound extract.

Effects of Extracts on the Chick Chorioallantoic Membrane

Figure 2 demonstrates stimulation of DNA synthesis in the chick CAM by a wound extract, but not a normal skin extract, with reference to a buffer control. Table 2 shows a separate experiment with comparable findings of significant stimulation only by the 3 day wound extract.

Admixture of Anti Fibrin E Antiserum

Table 3 A shows two experiments where there is reduction of stimulatory activity of FDP by the prior admixture of the rat anti human fibrin fragment E antibody, and Table 3 B shows an experiment using a day three wound extract . Figure 3 is a further experiment that demonstrates the blocking of CAM stimulatory effect of wound extract by the antibody.

Table 2. Effect of EACA buffered normal and wounded skin sample extracts on CAM stimulation

Sample	n	mean	SEM	% above control	t-test
control	13	12,735	2,131	-	-
normal skin	9	14,403	3,142	13.1%	p=0.77
control	10	10,448	1,170	-	-
day3 wound	7	26,533	7,044	154%	*p<0.01

Table 3A. Two experiments showing reduction of the mean DNA synthesis in chick CAM stimulated by FDP by prior admixture of anti fibrin E antibody.

Sample	n	mean	SEM	% from control	t-test (p =)
control	14	10,045	1,340	-	-
FDP	13	23,366	4,077	133%	*p < 0.01
FDP plus antibody	11	11,429	2,599	14%	p = 0.90
control	8	8,328	1,321	-	-
FDP	20	21,272	2,633	155%	*p < 0.01
FDP plus antibody	13	12,308	2,160	48%	p = 0.22

Table 3B. The increase in chick CAM DNA synthesis stimulated by a wound extract is neutralised by prior admixture of anti fibrin E antibody.

Sample	n	mean	SEM	% from control	t-test (p =)
control	14	10,045	1,340	-	-
FDP	13	23,366	4,077	133%	*p < 0.01
FDP plus antibody	11	11,429	2,599	14%	p = 0.90
control	8	8,328	1,321	-	-
FDP	20	21,272	2,633	155%	*p < 0.01
FDP plus antibody	13	12,308	2,160	48%	p = 0.22

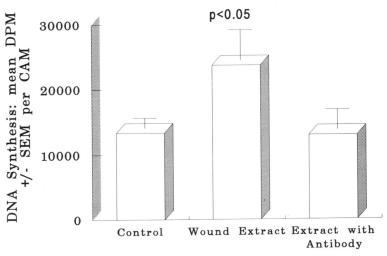

Figure 3. The increase in chick CAM DNA synthesis stimulated by a wound extract is neutralised by prior admixture of anti fibrin E antibody

Subcutaneous Injection of FDP

Subcutaneous injection of FDP into mice led to a fibrovascular response after 3 and 5 days but difficulty was experienced with diffusion of the liquid beneath the mobile skin of the dorsum and in subsequent location of the injection sites. More satisfactory results were obtained from injection into the thicker dermis of the rat. A dose/response relationship was demonstrable, with increased cellularity at day 3 being apparent at doses greater than 500 μg /ml. This was characterised by diffuse macrophage infiltration and patchy foci of polymorphs as well as raised numbers of connective tissue cells. Most inflammatory cells

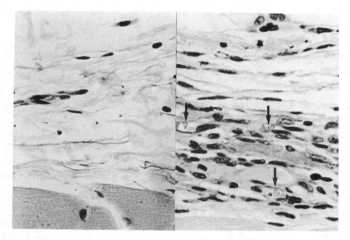

Figure 4. Comparison of skin from normal control rat with rat injected 5 days previously with FDP. (reduced from X 568) Fibrovascular proliferation is evident just above the muscle layer in dermis. [capillaries - ↓]

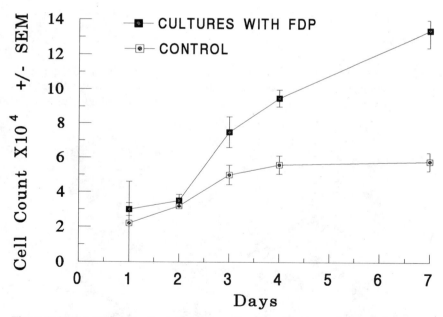

Figure 5. Stimulation of growth of chick embryo fibroblasts in culture by the addition of FDP.

had gone by day 5, when a clear fibrovascular proliferative response was seen, with increased numbers of synthetically active fibroblasts. Figure 4 shows rat tissue demonstrating the pronounced fibrovascular response after 5 days.

Effect of FDP in Cell Culture

Figure 5 shows a typical example of the accelerated growth curve induced by the addition of FDP to chick fibroblast cultures.

Immunohistochemical Application of Antibodies against Fibrin Epitopes

With all antibodies tested which gave any immunostaining, the use of trypsin was found to enhance immunostaining, but the use of microwave pretreatment was found to give by far the best results.

The characteristic meshwork of extracellular fibrin in an inflammatory lesion (abscess cavity in human dermis) is shown to be intensely immunolabelled using the mouse monoclonal antibody 5F3 (Fig 6). Nearly as good results were obtained with our own rat polyclonal anti human fibrin fragment E. DAKO rabbit polyclonal anti human fibrinogen antiserum gave fuzzy immunostaining of fibrin and a large amount of connective tissue background staining likely to be non specific. Of the antibodies directed against the β chain synthetic peptide, (amino acids 15 to 27), the best results came from the mouse antibody, which showed weak fibrin staining. Scanty, mainly single, large inclusion bodies within macrophages were also positively stained (Fig 7). A subtly different pattern of strong immunostaining of multiple inclusion bodies of fibrin in macrophages was obtained with the rat antibody against the γ chain (amino acids 54 to 62) (Fig 8).

DISCUSSION

We have proposed that FDP, and fibrin fragment E in particular, are the major initiator of cell proliferation including angiogenesis, common to all sites of chronic inflammation and tissue damage (Thompson *et al.*, 1985). The rationale for the methodology used previously

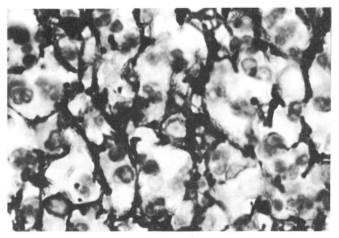

Figure 6. The fibrillar structure of fibrin deposited in a subcutaneous abscess immunostained with the antibody 5F3. (reduced from X 1420).

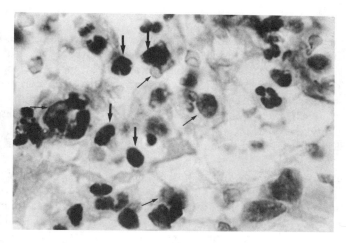

Figure 7. Fibrin stained strongly only within macrophages in the same subcutaneous abscess using the anti fibrinogen β chain antibody for immunoperoxidase immunostaining. Mainly single inclusions (↓) are seen in cells with the characteristic kidney bean shaped nucleus (↗) of the macrophage. (reduced from X 1420)

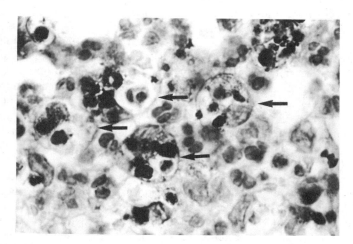

Figure 8. Multiple inclusions are seen within larger macrophages (←)with the γ chain anti fibrin antibody. (reduced from X 1420)

to demonstrate angiogenic activity in simple extracts of healing skin wounds (Thompson *et al.*, 1991) was unclear, and subsequent results were subject to marked operator variation. Part of this variation was eliminated by inclusion of a lipid removal stage to avoid microparticulate lipid and cell debris in tissue extracts coating the surface of the CAM on application. The time of maximum angiogenic activity observed in extracts taken at 3 days, seems plausible as it anticipated maximum vascularity at 5 days in incised mouse skin wounds.

However in order to obtain the biologically meaningful result of stimulatory wound extracts and negative normal skin extracts, these extracts were prepared in serum. Otherwise, normal skin extracts were also stimulatory. Serum itself is non stimulatory on the chick chorioallantoic membrane (CAM), and we have suggested that its real effect was to supply antithrombin III to prevent clotting of endogenous plasma derived fibrinogen on the chick CAM surface. This latter mechanism has been observed by us with other tissue extracts (Stirk *et al.*, 1993), and when whole plasma is applied directly (Stirk and Thompson, 1990), and is consistent with our evidence for the angiogenic activity of fibrin degradation products (FDP), specifically fibrin fragment E (Thompson *et al.*, 1985; Thompson *et al.*, 1992). Serum antithrombin III has been shown to neutralise thrombin in terms of both clotting, and cell growth stimulation in culture by fluid phase thrombin (Hedin *et al.*, 1994). Solid phase, fibrin bound thrombin is protected and may retain mitogenicity (Bar-Shavit *et al.*, 1990).

In the present work, a methodology has been devised that permits the clotting out and removal of fibrinogen derived from the plasma present in the vasculature of both normal and wound skin. This is evident from the loss of intensity of fibrinogen bands on immunoblots of extracts subjected to PAGE (Fig 1), with no accompanying increase in degradation products. This technique is sensitive and, being applied to a surface, does not give more than a crude indication of changes in concentrations. Rocket immunoelectrophoresis indicates the real extent of removal of fibrinogen related antigens (Table 1).

The finding that this methodology results in only the wound extracts being stimulatory to the chick CAM, is in itself evidence for the significance of FDP in extracts. This is strengthened further by very recent work demonstrating blocking of most, if not all of the wound extract activity by prior admixture of a specific rat anti human whole fibrin fragment E (Fig 2, Table 3). This antibody neutralises the activity of FDP prepared in vitro, consistent with previous work with a rabbit anti E (Thompson *et al.*, 1992). Previously we have employed affinity chromatography with specific anti E and D antibodies to demonstrate removal of stimulatory activity from extracts of progressive types of human atherosclerotic plaque only with the anti E antibody (Stirk *et al.*, 1993), but this technique is technically more difficult. It also depends on removal of the of the fragment rather than blocking of the active site.

The demonstration of angiogenic activity in FDP and tissue extracts has until recently depended on the use of the standard angiogenesis model, the chick chorioallantoic membrane. This model has its weaknesses, which we have attempted to overcome (Thompson and Kazmi, 1989). It is an immature, non mammalian system lacking in equivalent inflammatory cell types. Injection of human FDP and semipurified fragment E stimulates angiogenesis in mouse and rat (Fig 4), demonstrating cross species activity between human, chick CAM, mouse and rat. The rat was technically easier to work with, and a fibrovascular response was observed several days after subcutaneous injection of either whole FDP or fragment E rich FDP. Some patchy polymorph leukocyte infiltration was observed despite the absence of fibrinopeptides in the FDP, and this may be attributable to other chemotactic fragments (Skogen *et al.*, 1988). Further work with purified fragments is now required to see whether the accompanying macrophage infiltrate is attributable to fragment D in this system.

FDP have previously been reported to be "toxic" to cells in culture (Ishida *et al.*, 1982), but the concentrations used exceeded those observed by us in tissue extracts (0 to 110 µg/ml). FDP in a concentration range of 10 to 30 µg/ ml are stimulatory to cell growth in culture for chick fibroblasts (Fig 5) and for rabbit aortic smooth muscle cells (Sabally *et al.*, 1994). Our previous observation from autoradiography of the CAM that the FDP effect is not cell type specific (Thompson *et al.*, 1985) is confirmed.

The histological demonstration of fibrin with Lendrum's MSB stain or with antifibrinogen antibodies has never been entirely clear. Dr Patrick Gaffney's 5F3 monoclonal antibody combined with microwave heat exposure for antigen retrieval has allowed crisp demonstration of the characteristic fibrillar structure of fibrin at sites of inflammation in histological sections of human lesions (Fig 6). The demonstration of fibrin as intracellular phagocytosed material by the anti β and γ chain peptides surprised us (Fig 7 & 8). The subtle difference between the two antibodies in terms of multiple inclusions in more macrophages with the anti γ chain antibody may represent different times of epitope exposure during lysosomal degradation. Which comes first is unclear as yet and awaits in vitro studies with fibrin and macrophages. The fact that such intracellular immunostaining is exclusive to macrophages seems most readily attributable to the recently described, non plasmin dependent pathway of degradation of fibrin by the macrophage (Simon *et al.*, 1993), a lysosomal cathepsin D induced phenomenon (Simon *et al.*, 1994). Visualisation would therefore appear to depend on the fortunate combination of concealed epitope exposure during lysis and minor species differences in structure.

Until recently, we have studied FDP in wound extracts in comparison with FDP generated by plasmin at neutral pH. The immunoblot band patterns have always appeared very similar. There may however be minor differences in the FDP from the plasmin and cathepsin pathways, potentially revealed by these new antibodies. The plasmin pathway is optimal at neutral pH, and is extracellular. The cathepsin pathway is optimal at acid pH and operates within lysosomes and probably also in the extracellular microenvironment if at low pH. It is possible that the FDP from each pathway may differ in angiogenic potential. This may be of relevance to the non healing atherosclerotic plaque, where macrophages appear to be present but not fully effective, perhaps due to oxidised lipoprotein interference with lysosomal enzyme function (Hoppe *et al.*, 1994). The possible relation between macrophage function and hypoxia in the low pH non healing wound also awaits exploration.

In conclusion, there are many aspects of fibrinolysis which affect angiogenesis in wound healing, and it must be emphasised that the relative abundance of angiogenic FDP anticipates and accompanies the expression of many other autocrine and paracrine growth factors. It is proposed that the cell proliferative phase of healing is complete when fibrin is removed.

ACKNOWLEDGEMENTS

This work was supported in various aspects by the research funds of the Wellcome Trust, the Medical Research Council, Tenovus Scotland, and the University of Aberdeen. The gift of the monoclonal antibody 5F3 from Dr Patrick Gaffney, National Institute for Biological Standards and Controls, Potter's Bar, UK, is gratefully acknowledged. We wish to acknowledge the technical assistance of Ms Allyson Reid.

REFERENCES

Bar-Shavit, R., Benezra, M., Eldor, A., Hy-Am, E., Fenton, J.W., Wilner, G.D., and Vlodavsky, I., 1990, Thrombin immobilised to extracellular matrix is a potent mitogen for vascular smooth muscle cells: nonenzymatic mode of action, *Cell. Regul.* 1: 453-463.

Hedin, U., Frebelius, S., Sanchez, J., Dryjski, M., and Swedenborg, J., 1994, Antithrombin III inhibits thrombin - induced proliferation in human arterial smooth muscle cells, *Arterioscler Thromb* 14: 254-260.

Hoppe, G., O'Neil, J., and Hoff, H.F., 1994, Inactivation of lysosomal proteases by oxidised low density lipoprotein is partially responsible for its poor degradation by mouse peritoneal macrophages, *J. Clin. Invest.* 94: 1506-1512.

Hsu, S., Raine, L., and Fanger, H., 1981, The use of anti-avidin antibody and avidin-biotin-peroxidase complex in immunoperoxidase techniques, *Am. J. Clin. Pathol.* 75: 816-821.

Ishida, T., and Tanaka, K., 1982, Effects of fibrin and fibrinogen-degradation products on the growth of rabbit aortic smooth muscle cells in culture, *Atherosclerosis* 44: 161-174.

Laemmli, U.K., 1970, Cleavage of structural proteins during the assembly of the head of bacteriophage T4, *Nature* 227: 680-685.

Langlois, N.E.I., King, G., Herriot, R., and Thompson, W.D., 1994, Non-enzymatic retrieval of antigen permits staining of follicle centre cells by the rabbit polyclonal antibody to protein gene product 9.5, *J. Pathol.* 173: 249-253.

Laurell, C.B., 1966, Quantitative estimation of proteins by electrophoresis in agarose gel containing antibodies, *Analyt. Biochem.* 15: 45-52.

Sabally, K., Thompson, W.D., Smith, E.B., and Benjamin, N., 1994, Stimulation of smooth muscle cell and fibroblast proliferation in culture by fibrin degradation products and by human atherosclerotic plaque extracts, *J. Pathol.* 172s: 143A.

Simon, D.I., Ezratty, A.M., Francis, S.A., Rennke, H., and Loscalzo, J., 1993, Fibrin(ogen) is internalised and degraded by activated human monocytoid cells via Mac-1 (CD 11b/ CD 18): a nonplasmin fibrinolytic pathway, *Blood* 82: 2414-2422.

Simon, D.I., Ezratty, A.M., and Loscalzo, J., 1994, The fibrin(ogen)olytic properties of cathepsin D, *Biochemistry* 33: 6555-6563.

Skogen, W.F., Senior, R.M., Griffin, G.L., and Wilner, G.D., 1988, Fibrinogen - derived peptide Bβ 1-42 is a multidomained neutrophil chemoattractant,. *Blood* 71: 1475-1479.

Smith, E.B., Keen, G.A., Grant, A., and Stirk, C., 1990, Fate of fibrinogen in human arterial intima, *Arteriosclerosis* 10: 263-265.

Stirk, C.M., and Thompson, W.D., 1990, Artificial exudate: stimulation of cell proliferation by plasma not serum is associated with fibrinolysis, *Blood Coag. Fibrinol* 1: 537-541.

Stirk, C.M., Kochhar, A., Smith, E.B., and Thompson, W.D., 1993, Presence of growth-stimulating fibrin degradation products containing fragment E in human atherosclerotic plaques, *Atherosclerosis* 103: 159-169.

Thompson, W.D., Campbell, R., and Evans, A.T., 1985, Fibrin degradation and angiogenesis: quantitative analysis of the angiogenic response in the chick chorioallantoic membrane, *J. Pathol.* 145: 27-37.

Thompson, W.D., McGuigan, C.J., Snyder, C., Keen, G.A., and Smith, E.B., 1987, Mitogenic activity in human atherosclerotic lesions, *Atherosclerosis* 66: 85-93.

Thompson, W.D., and Kazmi, M.A., 1989, Angiogenic stimulation compared with angiogenic reaction to injury: distinction by focal and general application of trypsin to the chick chorioallantoic membrane, *Brit. J. Exp. Pathol.* 70:627-635.

Thompson, W.D., Harvey, J.A., Kazmi, M.A., and Stout, A.J., 1991, Fibrinolysis and angiogenesis in wound healing, *J. Pathol.* 165: 311-318.

Thompson, W.D., Smith, E.B., Stirk, C.M., Marshall, F.I., Stout, A.J., and Kocchar, A., 1992, Angiogenic activity of fibrin degradation products is located in fibrin fragment E, *J. Pathol.* 168: 47-53.

Thompson, W.D., Smith, E.B., Stirk, C.M., and Wang J., 1993, Fibrin degradation products in growth stimulatory extracts of pathological lesions, *Blood Coag. Fibrinol.* 4: 113-115.

Thompson, W.D., Ganesalingam, N., and Wang, J.E.H., 1994, Extraction of meaningful angiogenic activity from healing skin wounds by control of artefactual fibrinolysis, *J. Pathol.* 172: 143A.

Tymkewycz, P.M., Creighton Kempsford, L.J., and Gaffney, P.J., 1993, Generation and partial characterization of five monoclonal antibodies with high affinities for fibrin, *Blood Coag. Fibrinol.* 4: 211-221.

TUMOR OXYGENATION AND TUMOR VASCULARITY: EVIDENCE FOR THEIR CLINICAL RELEVANCE IN CANCER OF THE UTERINE CERVIX AND CONSIDERATIONS ON THEIR POTENTIAL BIOLOGICAL ROLE IN TUMOR PROGRESSION

Michael Höckel[1], Karlheinz Schlenger[1], and Margarete Mitze[2]

[1]Department of Obstetrics and Gynecology
[2]Department of Pathology
 University of Mainz
 55101 Mainz, Germany

BIOLOGICAL CHARACTERIZATION OF MALIGNANT TUMORS

Most solid malignancies are thought to be derived from a single neoplastic precursor cell having lost proliferation control and gained the ability to penetrate basement membranes and to invade into the stroma. During the disease course tumors increase their overall cell number by local expansion and the development of regional and distant metastases. Along with the increase in cell number the tumors loose hormonal or other external signal dependencies and acquire resistances towards radio- and chemotherapy. The progressing disease causes symptoms through impaired tissue/organ functions and complications, and finally kills the individual (unless other causes leading to death become manifest earlier).

When producing symptoms most tumors consist already of 10^9 to 10^{11} cells. This number includes tumor cells, stroma cells and blood cells in various relative amounts. The earliest diagnosis of a malignant tumor can be made at 10^7 to 10^8 cells by screening programs or accidentally. Tumor masses in the magnitude of 10^{12} cells are usually no longer compatible with the sustained life of the host. Thus malignant tumors are clinically occult for the longest time of their existence.

Clinically detectable tumors are complex interactive ecosystems of neoplastic cell populations and stromal elements (including blood vessels) with chaotic structural and functional features on one hand and still tremendous abilities for adaptation to changes in their local or systemic environment on the other hand. They exhibit a pronounced heterogeneity in tumor cell phenotypes and stroma composition from microregion to microregion (Heppner, 1964). Tumor cell variants differ in genotype and gene expression despite their lineage from one common ancestor (Nowell, 1976). The heterogeneity of tumor cell

Molecular, Cellular, and Clinical Aspects of Angiogenesis
Edited by Michael E. Maragoudakis, Plenum Press, New York, 1996

173

populations, their ability for adaptation and the development of metastases and therapy resistance might be approached by applying the biological principles of micro- and macroevolution (Wright, 1982). However, although there is strong evidence for an increased genetic instability in tumors and for the accumulation of distinct mutations in tumor cell variants associated with aggressiveness, the molecular biological, cell biological and population biological mechanisms of tumor progression remain obscure at present.

CLINICAL CHARACTERIZATION OF MALIGNANT TUMORS

Clinically manifest malignant tumors have to be characterized with respect to the selection of the most adequate primary and adjuvant therapy and the expected prognosis, i.e. the survival probability of the individual subsequent to standard treatment. Furthermore, clinical tumor characterization should help to investigate new therapeutic modalities and strategies.

One way of characterizing a tumor is the *macroscopic description* of its overall extension. The size of the tumor and its spread to locoregional and distant sites are determined by clinical, pathological and imaging investigation. Several staging systems such as the TNM system or the FIGO (International Federation of Gynecology and Obstetrics) classification for gynecologic malignancies have been established for that purpose. Magnetic resonance imaging is the most promising diagnostic tool at present to measure tumor volume pretherapeutically.

Another approach uses *cell biological, biochemical and molecular biological* methods to investigate various features of the malignant cells as the smallest biological units of the disease. A variety of malignant cell characteristics are meanwhile part of the routine clinical tumor characterization, such as hormonal receptor testing in breast cancer or ploidy measurements in borderline cancer of the ovary.

The third possibility to gain clinically relevant information about a solid tumor is to look at *tumor microregions* sized in the range of 100-1000 µm allowing for the study of clusters of tumor cells interacting with host tissue elements. This is the traditional histopathologist's view which is still obligatory to ascertain the definitive diagnosis of malignancy.

We hypothesize that the pretherapeutical *quantitative* determination of microregional features of the tumor-host ecosystem may predict the biological aggressiveness of a malignant tumor and advocate *Tumor Microregion Histography (TMH)* as a new clinical tool of tumor characterization.

TUMOR MICROREGION HISTOGRAPHY (TMH)

We have suggested TMH as predictive assay for cancer of the uterine cervix, but in essence this method can be applied to any other accessible tumor entity as well (Figure 1). The first step of TMH is adequate tumor tissue sampling. The macroscopic tissue samples have to be representative for the whole tumor in comparison with other tumors of the same clinical entity. Random biopsies obviously do not fulfill these criteria. However, biopsies should be defined with respect to the topography both of the anatomic site of the tumor and within the tumor. For locally advanced cervical cancer we sample two 20 mm long macroscopic tumor tissue biopsies from the intravaginal 12 o'clock and the 6 o'clock positions of the cervix. The next step of TMH is the selection of a sufficiently large number of microregions from the defined macroscopic tumor areas representing intratumoral heterogeneity. According to criteria of systematic random sampling for stereological analysis (Elias *et al.*, 1971; Weibel *et al.*, 1969) a fixed number of microregions is randomly identified within all subareas outlined by a grid superimposed on the whole macroscopic sample. In cervical cancer we select between 50 and 500 microregions per tumor depending on the parameter to be evaluated.

The feature of interest is measured by computerized detector systems or image analysis systems within the microregions and displayed in a histogram.

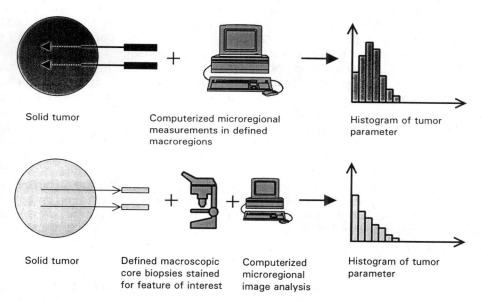

| Solid tumor | Computerized microregional measurements in defined macroregions | Histogram of tumor parameter |

| Solid tumor | Defined macroscopic core biopsies stained for feature of interest | Computerized microregional image analysis | Histogram of tumor parameter |

Fig. 1. Schematic representation of Tumor Microregion Histography (TMH).
Upper panel: Within two tumor macroregions of defined size and location computerized measurements on the feature of interest (e.g. tumor oxygenation) are performed randomly in a fixed number of microregions. The pooled results are displayed in a histogram.
Lower panel: Two microregions of defined size and localization are biopsied from the tumor and processed for histologic work up. The feature of interest is highlighted by specific staining in one histologic section per macroregion and quantified in a fixed number of randomly selected microregions by computerized image analysis. The pooled results are displayed in a histogram.

The histograms for each macroregion are pooled to represent the individual tumor. The information of the whole histogram can be further reduced to statistical parameters like the mean, median, range, etc.

Finally, it has to be shown that *intertumoral variation* is significantly larger than *intratumoral* and *intra- and interobserver* variation to justify the use of the tumor parameters obtained by TMH for clinical evaluation.

Applying TMH in advanced cervical cancer we found two features representing microregional tumor-host interactions - *tumor oxygenation* and *tumor vascularity* - to be of significant clinical relevance.

THE CLINICAL RELEVANCE OF TUMOR OXYGENATION IN CANCER OF THE UTERINE CERVIX

We have developed, evaluated and standardized the first clinically applicable method to measure tumor oxygenation in locally advanced cancer of the uterine cervix by use of the Eppendorf computerized polarographic needle electrode pO_2 histograph (Höckel et al., 1993a,b; Höckel et al., 1991). Applying this procedure according to the TMH methodology in patients with primary cervical cancers of >3 cm diameter at the 12 o'clock and 6 o'clock tumor positions we have been studying the clinical relevance of tumor oxygenation in an open prospective trial. 92 patients with advanced cancers of the uterine cervix, FIGO stages Ib bulky (n = 14), IIa,b (n = 38), IIIa,b (n = 37), and IVa (n = 3) entered the study. Tumor oxygenation exhibited marked intraindividual heterogeneity from microregion to microregion, however, interindividual heterogeneity from tumor to tumor was more pronounced. Forty-nine percent of the patients had carcinomas with median pO_2 readings < 10 mmHg defined as *low pO_2 tumors*. Tumor oxygenation was found to be independent from various patient demographics and gross as well as microscopic tumor features, such as stage, size, histologic type, and differentiation. However, *low pO_2 tumors* showed a significantly higher lymphatic vascular space involvement compared to better oxygenated tumors. Thirty-nine patients receiving complete radiation therapy (± chemotherapy) and 42 patients who underwent radical surgery (± chemotherapy) were analyzed for treatment outcome after a median observation period of 20 months (range 4 to 68 months). In the whole group as well as in the radiation and surgery subgroups patients with low pO_2 tumors had significantly worse disease-free and overall survival probabilities compared to the patients with better oxygenated tumors. Cox regression analysis revealed tumor oxygenation and FIGO stage as most significant independent prognostic factors in patients treated with surgery or radiation for local control.

THE CLINICAL RELEVANCE OF TUMOR VASCULARITY IN CANCER OF THE UTERINE CERVIX

We established a computerized image analysis system to quantify tumor microvascularity in advanced cervical cancer by using the *closest-individual method*, which determines the distribution of distances from random points within the tumor to the closest microvessel (DTCMV). Tumor microvascularity was assessed according to the TMH methodology in paraffin sections of two cylindrical 2 x 20 mm core biopsies obtained transvaginally from the 12 and 6 o'clock positions of each tumor, then immunohistochemically stained for Factor VIII related antigen. The oncologic relevance of tumor vascularity has been studied in an open prospective trial (Schlenger et al., 1995). Tumor vascularity was quantified in 42 patients with cervical cancers of >3 cm in largest diameter, FIGO stages Ib bulky to IVa. Tumor vascularity was independent of various other patient and tumor characteristics, including age, FIGO stage, tumor size, lymph node metastases and lymphatic space involvement. Thirty-nine patients were treated with curative intent either by primary surgery (n = 22) or radiation (n = 17). After a median observation time of 18 months (range 4 to 41 months) the patients with *highly vascularized tumors* (mean DTCMV < 83 μm) had significantly shorter disease-free (p = 0.025) and overall (p = 0.032) survival probabilities than patients with lower tumor vascularity (mean

DTCMV $\geq$ 83 μm). Cox regression analysis identified tumor vascularity as the strongest independent prognostic factor in this group of patients.

THE POTENTIAL ROLE OF TUMOR OXYGENATION AND VASCULARITY IN CANCER PROGRESSION

By applying the TMH methodology in advanced carcinoma of the uterine cervix we identified tumor oxygenation and tumor vascularity as new powerful prognostic parameters which proved to be independent from the classic prognostic factors such as tumor stage and size. Both, *low pO_2 (hypoxic) tumors* and *highly vascularized tumors* have a significantly higher probability to recur or progress than better oxygenated or less vascularized tumors of the same stage or size. Tumor oxygenation and tumor vascularity were independent from each other. Multivariate Cox regression analysis identified cervical cancers which were both hypoxic and exhibited a high vascularity as tumors with the worst prognosis (Table 1). Almost no patient with a cancer of >3 cm diameter showing these features could be salvaged by standard treatment, whereas the 5 year survival probabilities of all other patients with tumors of a similar size and stage distribution is more than 50%.

Table 1. Multivariate Cox regression analysis demonstrating the impact of tumor oxygenation and tumor vascularity on the disease-free survival probability of 39 patients with advanced cancer of the uterine cervix treated with surgery or radiation for cure.

Variable	p-value
Tumor oxygenation Median pO_2< 10mmHg vs $\geq$ 10 mmHg	0.014
Tumor vascularity Mean DTCMV < 83μm vs $\geq$ 83μm	0.036

Tumor oxygenation <u>and</u> vascularity
    Median pO_2 < 10mmHg and mean DTCMV < 83 μm vs all
    others
	0.001

Follow up: Median 18 months (range 4 - 41 months);
20 events; DTCMV, distance to closest microvessel.

There are several observations in experimental tumors and in vitro systems with tumor cells which may explain the prognostic significance of oxygenation and vascularity in terms of tumor biology. For those patients treated with radiation a reduced *oxygen enhancement effect* due to the lack of molecular oxygen may account for the lower tumor control probability (Drescher and Gray, 1959; Gray *et al.*, 1953). However, this may not be the only or even the most probable explanation. Tumor cells, like proliferating normal cells, when stressed by hypoxia can respond by the expression of growth arrest genes (such as p53) to induce a G1/S check point (Giaccia and Graeber, 1995). Similarly, genes coding for stress proteins can also be activated by hypoxia (Hall, 1994).

Both cellular phenomena at the transcriptional level may contribute to the increase of (intrinsic) radioresistance under hypoxia (O'Dwyer *et al.*, 1995; Zhou *et al.*,1995).

Microregional tumor hypoxia appears to be a major stimulus for tumor *angiogenesis*. VEGF, a strong mediator substance for neovascularization, has been immunohistochemically highlighted at areas of hypoxia in histologic tumor sections (Hlatky *et al.*, 1994; Brown *et al.*, 1993; Plate *et al.*, 1992; Shweiki *et al.*, 1992). As a consequence of the formation of new microvessels, reoxygenation of former hypoxic microregions may occur. Theoretically, hypoxic tumor cells can also be reoxygenated by change of their location through migration, lymphatic and hematogenous spread to better oxygenated microregions. Moreover, microregional changes in nutritive blood flow may lead to reoxygenation as well. Cells that were in the S phase during the time of hypoxia can react upon reoxygenation by DNA overreplication, which may result in the amplification of further resistance genes (Rice *et al.*, 1986).

DNA overreplication-recombination events may also lead to chromosomal aberrations and rearrangements increasing cellular heterogeneity and malignant progression (Schimke, 1984). *Hypoxia-reoxygenation* induced enhancement of the metastatic potential of tumor cells has been demonstrated in vitro (Young *et al.*, 1988). In addition, we propose another hypothesis: Hypoxia-reoxygenation events in a solid tumor might be regarded as microregional counterparts of the ischemia-reperfusion syndrome with the production of *oxygen-derived free radicals* (e.g. by the xanthine oxidase pathway) as the key event (Russell *et al.*, 1989; McCord, 1985). The generation of hydroxyl radicals may increase genetic instability through mutations and facilitate clonal evolution to more aggressive cellular phenotypes (Malins *et al.*, 1995; Malins *et al.*, 1993). The mutagenic effect of hydroxyl radicals is highly increased in an oxidative redox state, which should be expected for reoxygenated tumor cells. In tumors which are hypoxic *and* highly vascular hypoxia-reoxygenation events might be particularly frequent.

Likewise, the hypothesis that tumor hypoxia-reoxygenation may accelerate tumor progression into metastatic, radio- and chemoresistant variants is supported by our observation that hypoxic cervical cancers treated by surgery (i) also relapsed significantly more frequently than well oxygenated tumors of the same size and stage, and (ii) the majority of these relapses occurred in the pelvis and only responded poorly to salvage radiation or chemotherapy.

Further studies are planned to elucidate the importance of tumor hypoxia and vascularity for malignant disease progression and may help to develop new strategies for cancer treatment and prevention.

ACKNOWLEDGEMENTS

This work has been supported by a grant from the Deutsche Krebshilfe (M40/91/Va1). M.H. thanks LTS Lohmann Therapie Systeme and Else-Kröner-Fresenius Stiftung for financial support.

REFERENCES

Brown, L.F., Berse, B., Jackman, R.W., Tognazzi, K., Manseau, E.J., Dvorak, H.F., and Senger, D.R., 1993, Increased expression of vascular permeability factor (VEGF) and its receptor in kidney and bladder carcinomas, *Am. J. Pathol.* 143:1255.

Drescher, E.E., and Gray, L.H., 1959, Influence of oxygen tension on X-ray induced damage in Ehrlich ascites tumor cells irradiated in vitro and in vivo, *Radiol. Res.* 11:115.

Elias, H., Henning, A., and Schwartz, D.E., 1971, Stereology: Applications to biomedical research, *Histological Reviews* 51:158.

Giaccia, A.J., and Graeber, T.G., 1995, Regulation of cell proliferation by hypoxia, 43rd Annual Meeting of the Radiation Research Society, USA (1. April - 6. April 1995).

Gray, L.H., Conger, A.D., Ebert, M., Hornsey, S., and Scott, O.C.A., 1953, The concentration of oxygen dissolved in tissues at the time of irradiation as a factor in radiotherapy, *Br. J. Radiol.* 26:638.

Hall, E., 1994, Radiosensitivity and cell age in the mitotic cycle, *in: Radiobiology for the Radiologist,* J.B. Lippincott Company, Philadelphia, Pennsylvania, USA.

Heppner, G.H., 1984, Tumor heterogeneity, *Cancer Res.* 44:2259.

Hlatky, L., Tsionou, C., Hahnfeldt, P., and Coleman, N., 1994, Mammary fibroblasts may influence breast tumor angiogenesis via hypoxia-induced vascular endothelial growth factor up-regulation and protein expression, *Cancer Res.* 54:6083.

Höckel, M., Schlenger, K., Knoop, C., and Vaupel, P., 1991, Oxygenation of carcinomas of the uterine cervix: Evaluation by computerized O_2 tension measurements, *Cancer Res.* 51:6098.

Höckel, M., Knoop, C., Schlenger, K., Vorndran, B., Mitze, M., Knapstein, P.G., and Vaupel, P., 1993, Intratumoral pO_2 predicts survival in advanced cancer of the uterine cervix, *Radiother. Oncol.* 26:45.

Höckel, M., Vorndran, B., Schlenger, K., Baußmann, E., Knapstein, P.G., and Vaupel, P., 1993, Tumor oxygenation: A new predictive parameter in locally advanced cancer of the uterine cervix, *Gynecol. Oncol.* 51:141.

Malins, D.C., Holmes, E.H., Polissar, N.L., and Gunselman, S.J., 1993, The etiology of breast cancer: Characteristic alterations in hydroxyl radical-induced DNA base lesions during oncogenesis with potential for evaluating incidence risk, *Cancer* 71:3036.

Malins, D.C., Polissar, N.L., Nishikida, K., Holmes, E.H., Gardner, H.S., and Gunselman, S.J., 1995, The etiology and prediction of breast cancer, *Cancer* 75:503.

McCord, J.M., 1985, Oxygen-derived free radicals in post-ischemic tissue injury, *N. Engl. J. Med.* 312:159.

Nowell, P.C., 1976, The clonal evolution of tumor cell populations, *Science* 194:23.

O'Dwyer, P.J., Filali, M., Hamilton, T.C., and Yao, K-S, 1995, Mechanisms of altered gene expression under hypoxic conditions, 43rd Annual Meeting of the Radiation Research Society, USA (1. April - 6. April 1995).

Plate, K.H., Breier, G., Weich, H.A., and Risau, W., 1992, Vascular endothelial growth factor is a potential tumor angiogenesis factor in human gliomas in vivo, *Nature* 359:845.

Rice, G.C., Hoy, C., and Schimke, R.T., 1986, Transient hypoxia enhances the frequency of dihydrofolate reductase gene amplification in chinese hamster ovary cells, *Proc. Natl. Acad. Sci. USA* 83:5978.

Russel, R.C., Roth, A.C., Kucan, J.O., and Zook, E.G., 1989, Reperfusion injury and oxygen free radicals, a review, *J. Reconstr. Microsurg.* 5:79.

Schimke, R.T., 1984, Gene amplification, drug resistance, and cancer, *Cancer Res.* 44:1735.

Schlenger, K., Höckel, M., Mitze, M., Schäffer, U., Weikel, W., Knapstein, P.G., and Lambert, A., 1995, Tumor vascularity - A novel prognostic factor in advanced cervical carcinoma, *Gyn. Oncol.* in press:

Shweiki, D., Itin, A., Soffer, D., and Keshet, E., 1992, Vascular endothelial growth factor induced by hypoxia may mediate hypoxia-initiated angiogenesis, *Nature* 359:843.

Weibel, E.R., 1969, Stereological principles for morphometry in electron microscopic cytology, *Int. Rev. Cytol.* 26:235.

Wright, S., 1982, The shifting balance theory and macroevolution., *Annu. Rev. Genet.* 16:1.

Young, S.D., Marshall, R.S., and Hill, R.P., 1988, Hypoxia induces DNA overreplication and enhances metastatic potential of murine tumor cells, *Proc. Natl. Acad. Sci. USA* 85:9533.

Young, S.D., Marshall, R.S., and Hill, R.P., 1988, Hypoxia induces DNA overreplication and enhances metastatic potential of murine tumor cells, *Proc. Natl. Acad. Sci. USA* 85:9533.

Zhou, S., Hill, H.Z., and Hill, G.J., 1995, Further characterization and purification of a multi therapy resistance factor (MTRF), 43rd Annual Meeting of the Radiation Research Society, USA (1. April - 6. April 1995).

ROLE OF SCATTER FACTOR IN PATHOGENESIS OF AIDS-RELATED KAPOSI SARCOMA

Eliot M. Rosen[1], Peter J. Polverini[2], Brian J. Nickoloff[3], Itzhak D. Goldberg[1]

[1]Department of Radiation Oncology
Long Island Jewish Medical Center
The Long Island Campus for Campus for Albert Einstein College of Medicine
270-05 76th Avenue
New Hyde Park, New York 11040
[2]Department of Oral Pathology
University of Michigan School of Dentistry
Ann Arbor, MI 48109-1078
[3]Department of Pathology
University of Michigan Medical School
Ann Arbor, MI 48109-0602

INTRODUCTION

Kaposi's sarcoma (KS) is an enigmatic neoplasm that occurs commonly in several distinct clinical settings. The classic (*endemic*) form of KS most frequently occurs in certain Mediterranean Basin populations (especially Sephardic Jews) and African tribes (eg., Bantus) (Kaposi, 1872). An acquired form of KS occurs in patients treated with immunosuppressive agents, such as organ transplant recipients (Zibrob et al., 1980; Penn, 1987). Finally, the epidemic form of KS is found in association with Acquired Immunodeficiency Syndrome (AIDS) (Havercos et al, 1985; Safai et al., 1985; Mitsuyasu, 1988). AIDS-related KS tumors do not occur uniformly in all AIDS patients, but are usually restricted to one particular subset, homosexual males. This finding has given rise to speculation that additional factors besides the AIDS virus [HIV-1 (human immunodeficiency virus-type 1)] may be involved in the pathogenesis of this tumor. These factors may include infection with a second virus, including members of the human papilloma virus (HPV) and herpesvirus families (Huang et al., 1992; Nickoloff et al., 1992; Chang et al., 1994). Several studies suggest AIDS-related KS differs biologically from other forms of KS. While classic KS usually behaves as a indolent low grade disease with a protracted clinical course, AIDS KS is often clinically aggressive. The latter may be associated with significant morbidity and, in its most aggressive form, lethality (Bacchetti et al., 1988; Azon et al., 1990; Payne et al., 1990; Errante et al., 1991).

KS is not regarded as metastasizing cancer with a single clonally derived tumor cell population, but rather as a multifocal neoplastic lesion composed of multiple cellular constituents. These include proliferating endothelial cells (ECs), an expanded population of

Molecular, Cellular, and Clinical Aspects of Angiogenesis
Edited by Michael E. Maragoudakis, Plenum Press, New York, 1996

181

dermal dendrocytes (macrophage-like cells) that express factor XIIIa, CD4+ lymphocytes, and a population of interstitially located spindle-shaped cells thought to be the KS tumor cells (Nickoloff and Griffiths, 1989a,b). KS lesions are highly vascular tumors characterized by a large component of angiogenesis. The KS tumor cells are very difficult to propagate *in vitro*, and require the use of a special KS cell growth medium (designated "KSGM") for long-term culture. An essential component of KSGM is filtered conditioned medium from type II human T cell leukemia virus-infected T cells (designated "HTLV-II CM") (Nakamura et al., 1988; Salahuddin et al., 1988). HTLV-II, a lymphotrophic retrovirus, is closely related to HIV-1.

The cellular mechanism(s) leading to development of the KS tumor and the histogenesis of the KS tumor cell are not well understood. Various investigators have proposed that the KS tumor cell is originally derived from vascular ECs, dermal dendrocytes, and vascular smooth muscle cells (Dorfman, 1984; Weich et al., 1991; Huang et al., 1993). During studies to define the relationships between the different cellular components of the AIDS-related KS tumor, we observed that KSGM could induce the conversion of normal human ECs to a KS tumor cell-like phenotype. In further studies, we identified the cytokine scatter factor (SF) [also known as hepatocyte growth factor (HGF) (Nakamura et al., 1989)] in HTLV-II CM and showed that SF may mediate some of the observed phenotypic changes (Naidu et al., 1994). SF is a potent motogenic (Stoker et al., 1987; Gherardi et al., 1989; Weidner et al., 1990; Rosen et al., 1989, 1990, 1991a; Li et al., 1994) and angiogenic (Rosen et al., 1991b; Bussolino et al., 1992; Grant et al., 1993; Rosen and Goldberg, *In Press*) cytokine that acts through a tyrosine kinase receptor encoded by a proto-oncogene (*c-met*) (Bottaro et al., 1991; Gonzatti-Haces et al., 1988). More detailed information on the properties and angiogenic activity of SF can be found elsewhere in this volume in the chapter entitled "Scatter Factor as a Potential Tumor Angiogenesis Factor" (Rosen and Goldberg) and in a recent review (Rosen et al., 1994b). Below is a description of our studies of the role of SF and *c-met* in the pathogenesis of the AIDS-KS tumor and the development of the phenotype of this tumor.

CONVERSION OF ENDOTHELIAL CELLS TO A KS TUMOR CELL PHENOTYPE

KS tumor cell lines grew with a predominantly spindle-shaped morphology and expressed strong cytoplasmic immunoreactivity for factor XIIIa (a transglutamase) and for VCAM-1 (vascular cell adhesion molecule-1), which are both expressed by KS tumor cells *in vivo* (Huang et al., 1993). KS tumor cells did *not* express any of the usual immunologic markers of ECs, including Factor VIII antigen, the endothelial adhesion molecule ELAM-1 (endothelial leukocyte adhesion molecule-1), and angiotensin converting enzyme (ACE). On the other hand, normal human umbilical vein endothelial cells (HUVECs) cultured in their usual growth medium (designated "HUVEC-GM") [Ham's F-12:Iscove's modified DMEM (1:1) plus 20% (v/v) fetal calf serum, EC growth factor (30 μg/ml), and heparin (20 units/ml)] grew with a cobblestone epithelioid morphology, expressed immunoreactive Factor VIII antigen, and did *not* express immunoreactive Factor XIIIa or VCAM-1.

We were surprised to find that when HUVECs were cultured for only 18-24 hr in the same culture medium required for long-term *in vitro* growth of KS tumor cells (KSGM), they underwent a remarkable "phenotypic conversion" to a KS cell-like phenotype. KSGM is composed of RPMI 1640 plus 15% (v/v) fetal calf serum, 5% Nutrodoma-Hu (a commercial supplement), and 20% HTLV-II CM. HUVECs switched to KSGM acquired a spindle-shaped morphology as well as strong immunoreactivity for both Factor XIIIa and VCAM-1 (Naidu et al., 1994). These changes specifically required the presence of HTLV-II CM but not of the other components of KSGM.

In agreement with the immunologic results, RT-PCR analysis revealed that: (1) multiple KS tumor cell lines constitutively expressed Factor XIIIa mRNA; (2) HUVECs grown as usual in HUVEC-GM did *not* express Factor XIIIa mRNA; and (3) HUVECs switched to KSGM acquired the expression of Factor XIIIa. PCR analysis was performed using a nested set of PCR primers specific for Factor XIIIa mRNA, control reactions lacking various reagents and cDNA, and ß-actin as a control for loading (see Naidu et al. 1994, for details). Phenotypically converted HUVECs could be passaged multiple times in KSGM without loss of their KS-like phenotype. Another vascular wall cell type, human arterial smooth muscle cells (HASMCs) did *not* undergo this phenotypic conversion process. These cells did not constitutively express Factor XIIIa antigen or mRNA, and their expression was not induced by KSGM. Constitutive expression by HASMCs of the smooth muscle marker α-actin was not altered by culture in KSGM. Thus, the phenotypic conversion reaction appears to be specific for ECs. However, the phenotypic conversion of ECs by KSGM was not 100% complete, since converted HUVECs retained their expression of Factor VIII antigen, which is not expressed by KS tumor cells.

While cytokine networks may contribute to the pathogenesis of epithelial malignancies (Leek et al., 1994), the KS lesion may be even more dependent upon a disordered network of cytokines (Nakamura et al., 1988; Vogel et al., 1988; Miles et al., 1990, 1992; Naidu et al., 1994). We compared cytokine mRNA expression of cultured KS tumor cells with that of HUVECs before and after phenotypic conversion by KSGM. RT-PCR analysis revealed that four mRNA transcripts constitutively expressed by KS tumor cells [IL-1ß (interleukin-1ß), IL-6, IL-8, and ICAM-1 (intercellular adhesion molecule-1)] were expressed by KSGM-converted HUVECs but *not* by control HUVECs (Naidu et al., 1994). None of these cell types expressed mRNA transcripts for IL-2, IL-3, IL-4, IL-5, TNFα (tumor necrosis factor-α), or interferon-gamma; while all three cell types expressed TGFß (transforming growth factor-ß) mRNA transcript. However, discrepancies among the cell types were observed in the cytokine expression profiles for IL-10, granulocyte-macrophage colony-stimulating factor, and TGFα. The cytokine expression profile of KSGM-converted HUVECs clearly resembles but is not identical to that of KS tumor cells. Similar to the morphologic/immunologic results, these data suggest that KSGM partially converts ECs into KS-like cells and that an additional event(s) is required for full "transformation" of ECs into KS tumor cells.

ROLE OF SF IN THE PHENOTYPIC CONVERSION OF ENDOTHELIAL CELLS

Since SF induces epithelial cells to become fibroblastic in morphology (Stoker et al., 1987; Rosen et al., 1989), we assayed KSGM and its components for SF. Several different batches of HTLV-II CM, which were generated from the HTLV-II infected human T lymphocyte cell line 38-10, contained large quantities of immunoreactive SF (measured by ELISA) and bioactive SF (measured by the MDCK serial dilution scatter assay) (Naidu et al., 1994) (Table 1); while the other components of KSGM (RPMI, fetal calf serum, Nutrodoma-Hu) did not contain any detectable SF. In contrast, neither the parent cell line (HUT 78) nor purified resting human T cells produced any detectable SF. In preliminary studies, HIV-1 infected HUT 78 T cell line and antigenically-stimulated normal human T cells (eg., by superantigen or MLC culture) produced detectable but quantitatively lower levels of SF. The production of SF by 38-10 cells was equal to or greater than that of human lung fibroblasts, a major SF-producing cell type (Stoker et al., 1987). These findings suggest that activated T cells, especially retrovirally activated cells, are capable of secreting significant titers of SF.

Based on these findings, we tested the ability of SF alone (ie., in the absence of other components of HTLV-II) to induce the phenotypic conversion of HUVECs to KS tumor-like cells. We found that conversion of HUVECs to Factor XIIIa and VCAM-1 positive spindle-shaped cells could be induced by 12-18 hr of treatment with purified native mouse SF (50 ng/ml) or recombinant human SF (50 ng/ml). SF-treated HUVECs also acquired the ability to express Factor XIIIa mRNA, as demonstrated by RT-PCR analysis. Moreover, rabbit polyclonal antiserum against human SF (Grant et al., 1993) blocked the phenotypic conversion of HUVECs that was induced by HTLV-II CM, suggesting that SF is an essential requirement for this conversion process.

Table 1. SF Content of Conditioned Media from Various Cell Types[1]

	SF Content of Sample	
Cell Type Tested	ng/ml	MDCK units/ml
HUT 78 human lymphocyte cell line	0	0
38-10 HTLV-II infected T lymphocyte line	10.5	186
HIV-1 infected HUT78 T cells	1-2	8
Purified resting human T cells	0	0
Antigen-stimulated human T cells	NT	8-32
MRC5 human lung fibroblasts	5	64

	SF Production Rate (units/10^6 cells/48 hr)
38-10 HTLV-II infected T cell line	120
Human lung fibroblasts (N=6 lines)	20-80

[1]Conditioned media from cultures containing about 10^6 cells/ml were collected after 48-72 hr. Aliquots of cell-free media were assayed for immunoreactive SF protein using a specific double-antibody ELISA (Rosen et al., 1994a) and for SF bioactivity using the MDCK serial dilution scatter assay (Rosen et al., 1989). The MDCK scatter assay is a sensitive and specific assay of SF. NT = Not Tested.

DETECTION OF SF AND ITS RECEPTOR *IN VIVO* IN KS LESIONS

In order to determine if SF may be involved in the pathogenesis of the *in vivo* KS lesion, we performed immunostaining of KS lesions from five patients. Obvious positive staining for both SF and *c-met* antigens were detected in each lesion (Naidu et al., 1994). Microscopic examination of the stained slides revealed that multiple cell types in the KS lesions stained positively for either SF and/or *c-met* protein (see Table 2). As reported previously for psoriatic plaques (Grant et al., 1993), cells of the vascular wall (endothelium and pericytes) stained positively for *c-met* protein but not for SF. Populations of perivascular dendritic cells (dermal dendrocytes) and of interstitial spindle-shaped cells (the presumed KS tumor cells), stained positively for both SF and *c-met*. Similarly, perivascular dendrocyte-like cells in psoriatic plaques were previously found to stain positively for SF (Grant et al., 1993).

Immunostaining and RNA analysis of purified populations of various cultured cell types were performed to determine their correlation with the *in vivo* immunostaining results. In agreement with the findings in Table 2: cytospin preparations of endothelium (HUVECs) expressed *c-met* immunoreactive protein; HUVECs and bovine retinal pericytes expressed *c-met* mRNA by RT-PCR analysis; and three of three KS tumor cell lines expressed *c-met* immunoreactive protein and mRNA. Three different KS tumor cell lines also expressed SF mRNA by RT-PCR analysis. [Purified populations of cultured dermal dendrocytes were not available for analysis at the time of the study.] These findings suggest that SF might be an autocrine or paracrine growth factor for KS tumor cells.

STIMULATION OF KS TUMOR CELL PROLIFERATION BY SF

To test the hypothesis that SF is mitogenic for KS tumor cells, three different KS cell lines were treated with SF and/or other agents, and the cells were counted after 80 hr. Representative results for one KS cell line are shown in Table 3. SF stimulated the proliferation of KS tumor cells by at least as much as two known KS tumor cell mitogens, oncostatin-M (Miles et al., 1990) and IL-6 (Miles et al., 1992). All factors were tested at saturating doses; but none of these factors, alone or in combination, stimulated KS cell proliferation as much as KSGM. In dose response studies, 2 ng/ml of SF induced no proliferative response, while 20 and 50 ng/ml of SF induced near maximal and maximal responses, respectively. Rabbit antiserum against human SF (Grant et al., 1993) blocked all of the KS cell proliferation induced by human recombinant SF, but blocked only 30-50% of the proliferation induced by KSGM. These findings suggest that SF is a significant KS cell mitogen present in KSGM, but that additional, as yet unidentified, mitogenic factors for KS tumor cells are also present in KSGM.

Table 2. In Vivo Immunostaining of AIDS KS Lesions[1]

SF Positive Cell Types	c-Met Positive Cell Types
1. Interstitial spindle-shaped cells (KS tumor cells)	1. Interstitial spindle-shaped tumor cells (KS tumor cells)
2. Perivascular dendritic cells	2. Perivascular dendritic cells
3. Round lymphoid cells	3. Microvascular endothelium
	4. Pericytes
	5. Pili-erector muscle bundles

[1]Five-μm thick cryostat sections were stained using an avidin-biotin immunoperoxidase technique (Vector Laboratories). The primary antibodies were: mouse monoclonal antibody (10C11) against human placental SF (1:1000 of ascites) (Bhargava et al., 1992); a rabbit polyclonal antiserum (Ab 978) against human placental SF (1:1000) (Grant et al., 1993); and rabbit polyclonal antiserum to a C-terminal peptide of the *c-met* protein (Ab C28) (1:1000) (Gonzatti-Haces et al., 1988). Normal rabbit serum (1:1000) (negative control) gave no staining.

Table 3. Stimulation of Proliferation of KS Tumor Cells by SF[1]

Agent(s) or Medium Tested	Increased No. of Cells/Well (X10^{-3})		
RPMI + 10% fetal calf serum	3	±	2
KS cell growth medium (KSGM)	42	±	6
Oncostatin-M (100 ng/ml)	20	±	3
Interleukin-6 (500 units/ml)	15	±	2
Human SF (50 ng/ml)	25	±	2
Mouse SF (50 ng/ml)	26	±	5
KSGM + anti-SF Ab 978 (1:40)	23	±	4
Human SF (50) + anti-SF (1:40)	0	±	0

[1]KS tumor cells were seeded into 48-well plates at 25 X 10^3 cells/well; allowed to attached overnight; washed three times; incubated for 80 hr in 0.5 ml of RPMI 1640 containing 10% fetal calf serum plus the agents listed; and counted. Rabbit polyclonal antiserum against human SF (anti-SF Ab 978) was used to neutralize the effect of SF. Values are mean ± range of duplicate assays, and are typical of responses of three different KS cell lines.

STIMULATION OF ANGIOGENESIS BY KSGM

We tested the *in vivo* angiogenic activity of HTLV-II CM, the active component of KSGM, using the well-established rat cornea angiogenesis assay (Polverini and Leibovich, 1984). We found that 5 μl of concentrated HTLV-II CM gave four out of four strongly positive angiogenic responses (Naidu et al., 1994). These responses were similar in intensity to those induced by the positive controls, human recombinant basic fibroblast growth factor (FGF) (150 ng/ml) and SF (100 ng/ml), which are known to be potently angiogenic (Grant et al., 1993). The angiogenic activity in concentrated HTLV-II CM was significantly inhibited by anti-SF antibodies: only 1/4 (25%) positive responses were obtained in the presence of rabbit antiserum against SF (1:200), and only 2/5 (40%) positive responses were obtained in the presence of chicken egg yolk antibody against SF (1:20). These antibody preparations (see Grant et al., 1993) were not angiogenic when tested by themselves, nor did they block the angiogenic activity of basic FGF. These findings suggest that SF is a major angiogenic factor present in KSGM, accounting for over 50% of the angiogenic activity detected in that medium.

ROLE OF SF IN KS TUMOR DEVELOPMENT: HYPOTHESIS AND MODELS

Hypothesis. Based on the findings described above, we hypothesize that three SF-mediated events are important in the development and maintenance of the AIDS-KS tumor:
1. phenotypic conversion of ECs that have been "initiated" by another carcinogenic event.
2. stimulation of proliferation of initiated/phenotypically converted ECs.
3. stimulation of growth of established KS tumor cells and of tumor angiogenesis.

Model of AIDS KS Tumor Development. In the setting of AIDS, ECs undergo a multi-stage "transformation" process that is initiated by a carcinogenic event, such as acquisition of a transforming viral DNA sequence(s) (Fig. 1). Candidate viruses include HIV-1 itself and certain strains of human papilloma virus and herpesvirus (Vogel et al., 1988; Huang et al., 1992; Nickoloff et al., 1992; Chang et al., 1994). These "initiated" ECs then undergo stable irreversible phenotypic conversion to KS tumor cells under the influence of SF (and possibly other cytokines) produced by lymphotrophic retrovirus (HIV-1) infected T cells. SF derived from infected T cells, an expanding population of KS tumor cells, and, perhaps, other cell types within the tumor (see below), induces further phenotypic conversion of ECs, further paracrine and autocrine growth stimulation of KS tumor cells, and chronic persistent tumor angiogenesis. The latter two events ensure sustained growth of the KS lesion.

Several observations suggest that additional cell types may contribute to the accumulation of SF in KS lesions. Perivascular dendritic cells (dermal dendrocytes) in both psoriatic plaques and AIDS-KS tumors express immunoreactive SF (Grant et al., 1993; Naidu et al., 1994), although we do not yet know if these cells produce SF. Macrophages, another KS-associated cell type, express immunoreactive SF, SF mRNA, and bioactive SF protein (Noji et al., 1990; Wolf et al., 1991; Yanagita et al., 1992; Inaba et al., 1993; Rosen and Goldberg, *In Press*). Recruitment of circulating monocytes, which differentiate into macrophages within the tumor, might be induced by monocyte chemoattractants, such as colony-stimulating factors, which are produced by carcinoma cells (Ramakrishnan et al., 1989) and by KS tumor cells (see above).

In our model, the presence of SF-producing HIV-1 infected T lymphocytes functions to "promote" the conversion of carcinogen-initiated ECs into established KS tumor cells. The presence of these activated SF-secretory T cells may serve to increase both the frequency of tumor formation and the rate of tumor growth in susceptible populations of AIDS patients. This feature may contribute to the high incidence and aggressive clinical behavior of AIDS-related KS as compared with classical KS.

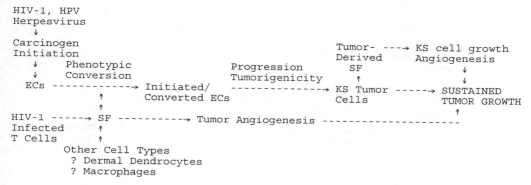

```
HIV-1, HPV
Herpesvirus
    ↓
Carcinogen                                    Tumor- ----→ KS cell growth
Initiation                                    Derived      Angiogenesis
    ↓      Phenotypic        Progression        SF             ↓
    ↓      Conversion        Tumorigenicity     ↑              ↓
  ECs ------------→ Initiated/    -------------→ KS Tumor -----→ SUSTAINED
            ↑       Converted ECs               Cells           TUMOR GROWTH
            ↑                                                      ↑
HIV-1 -----→ SF ----------→ Tumor Angiogenesis --------------------
Infected    ↑
T Cells     ↑
      Other Cell Types
      ? Dermal Dendrocytes
      ? Macrophages
```

Fig. 1. Model of the putative role of SF in the pathogenesis of AIDS-related KS tumor

SUMMARY AND CONCLUSIONS

The major findings of this study can be summarized as follows:

1. HTLV-II CM, which is required for *in vitro* growth of KS tumor cells, induced conversion of normal human endothelial cells to a KS tumor cell-like phenotype, characterized by acquisition of: (a) spindle-shaped morphology; (b) KS tumor cell immunologic markers; and (c) a KS tumor cell-like cytokine expression profile.

2. Many of these phenotypic changes were due to the angiogenic cytokine SF, which was produced by activated T lymphocytes and present in large quantities in HTLV-II CM.

3. SF and its receptor (*c-met*) were detected in different cell populations *in vivo* in human KS lesions and *in vitro* in the corresponding cultured cell types.

4. SF accounted for a significant fraction of the mitogenic activity for cultured KS tumor cells and the *in vivo* angiogenic activity found in HTLV-II CM.

We conclude that SF may play a central mechanistic role in the development and progression of the KS tumor (see proposed hypothesis and model). An accumulating body of evidence indicates that the SF:*c-met* ligand:receptor pair is involved in the interconversion of epithelioid and mesenchymal cells (reviewed in Rosen et al., 1994b). If further studies verify the importance of SF and *c-met* in the pathogenesis of the AIDS-related KS tumor, it may be possible to devise a clinical treatment program for KS tumors based on specific inhibition of the SF-*c-met* signalling pathway.

ACKNOWLEDGEMENTS

This work was supported in part by United States Public Health Service research grants CA64869 and CA64416. Dr. Rosen is an Established Investigator of the American Heart Association (AHA Award No. 90-195).

REFERENCES

Azon-Masoliver A, Mallolas J, Miro JM, Gatell JM, Castel T, Iranzo P, Lecha M, Mascaro JM. Kaposi's sarcoma associated with the acquired immunodeficiency syndrome. An analysis of 67 cases with a study of the prognostic factors. *Med Clin Barcelona* 95: 361-365, 1990.

Bacchetti P, Osmond D, Chaisson RE. Survival patterns of the first 500 AIDS patients in San Francisco. *J Infect Dis* 157: 1040-1047, 1988.

Bhargava M, Joseph A, Knesel J, Halaban R, Li Y, Pang S, Goldberg I, Setter E, Donovan MA, Zarnegar R, Michalopoulos GA, Nakamura T, Faletto D, Rosen EM. Scatter factor and hepatocyte growth factor: Activities, properties, and mechanism. *Cell Growth & Differen* 3:11-20, 1992.

Bottaro DP, Rubin JS, Faletto DL, Chan AM-L, Kmiecik TE, Vande Woude GF, Aaronson SA. Identification of the hepatocyte growth factor receptor as the *c-met* proto-oncogene product. *Science* 251: 802-804, 1991.

Bussolino F, DiRenzo MF, Ziche M, Bocchieto E, Olivero M, Naldini L, Gaudino G, Tamagnone L, Coffer A, Comoglio PM. Hepatocyte growth factor is a potent angiogenic factor which stimulates endothelial cell motility and growth. *J Cell Biol* 119: 629-641, 1992.

Chang Y, Cesarman E, Pessin MS, Lee F, Culpepper J, Knowles DM, Moore PS. Identification of herpesvirus-like DNA sequences in AIDS-associated Kaposi's sarcoma. *Science* 266: 1865-1869, 1994.

Dorfman RF. Kaposi's sarcoma revisited. *Hum Pathol* 15: 1013-1017, 1984.

Errante D, Tirelli U, Tumolo S. Management of AIDS and its neoplastic complications. *Eur J Cancer* 27:380-389, 1991.

Gherardi E, Gray J, Stoker M, Perryman M, Furlong R. Purification of scatter factor, a fibroblast-derived basic protein which modulates epithelial interactions and movement. *PNAS USA* 86: 5844-5848, 1989.

Gonzatti-Haces M, Seth A, Park M, Copeland T, Oroszlan S, Vande Woude GF. Characterization of the TPR-MET oncogene p65 and the MET protooncogene p140 protein tyrosine kinases. *PNAS USA* 85: 21-25, 1988.

Grant DS, Kleinman HK, Goldberg ID, Bhargava M, Nickoloff BJ, Polverini P, Rosen EM. Scatter factor induces blood vessel formation *in vivo*. *PNAS USA* 90: 1937-1941, 1993.

Havercos HW, Drotman DP, Morgan M. Prevalence of Kaposi's sarcoma among patients with AIDS. *N Engl J Med* 321: 1518, 1985.

Huang YQ, Li JJ, Rush MG, Poiesz BJ, Nicolaides A, Jacobson M, Zhang WG, Coutavas E, Abbott MA, Friedman-Kien AE. HPV-16 related DNA sequences in Kaposi's sarcoma. *Lancet* 399: 515-518, 1992.

Huang Y, Friedman-Kien, AE, Li JJ, Nickoloff, BJ. Cultured Kaposi's sarcoma cell lines express factor XIIIa, CD14, and VCAM-1, but not factor VIII or ELAM-1. *Arch Dermatol* 129: 1291-1296, 1993.

Inaba M, Koyama H, Hino M, Okuno S, Terada M, Nishizawa Y, Nishino T, Morii H. Regulation of release of hepatocyte growth factor from human promyelocytic leukemia cells, HL-60, by 1,25-dihydroxyvitamin D3, 12-O-tetradecanoylphorbol 13-acetate, and dibutyryl cyclic adenosine monophosphate. *Blood* 82: 53-59, 1993.

Kaposi M. Idiopathic multiple pigmented sarcoma of the skin. *Arch Dermatol Syphil* 4: 265-274, 1872. (English Translation, *CA - A Cancer J Clin* 32: 342-347, 1982).

Leek RD, Harris AL, Lewis CE. Cytokine networks in solid human tumors: regulation of angiogenesis. *J Leuk Biol* 56: 423-435, 1994.

Li Y, Bhargava MM, Joseph A, Jin L, Rosen EM, Goldberg ID. The effect of scatter factor and hepatocyte growth factor on motility and morphology of non-tumorigenic and tumor cells. *In Vitro Cell Dev Biol* 30A: 105-110, 1994.

Miles SA, Rezai, AR, Salazar-Gonzalez JF, Vander Meyden M, Stevens RH, Logan DM, Mitsuyasu RT, Taga T, Hirano T, Kishimoto T, Martinez-Maza O. AIDS Kaposi sarcoma-derived cells produce and respond to interleukin 6. *Proc Natl Acad Sci USA* 87: 4068-4072, 1990.

Miles SA, Martinez-Maza O, Rezai A, Magpantay L, Kishimoto T, Nakamura S, Radka SF, Linsley PS. Oncostatin M as a potent mitogen for AIDS-Kaposi's sarcoma-derived cells. *Science* 255: 1432-1434, 1992.

Mitsuyasu RT. AIDS-related Kaposi's sarcoma: a review of its pathogenesis and treatment. *Blood Rev* 2: 222-231, 1988.

Naidu YM, Rosen EM, Zitnik R, Goldberg I, Park M, Naujokas M, Polverini PJ, Nickoloff BJ. Role of scatter factor in the pathogenesis of AIDS-related Kaposi's sarcoma. *PNAS USA* 91: 5281-5285, 1994.

Nakamura S, Salahuddin, SZ, Biberfeld P, Ensoli B, Markham PD, Wong-Staal F, Gallo RC. Kaposi's sarcoma cells: Long-term culture with growth factor from retrovirus-infected CD4$^+$ cells. *Science* 242: 425-430, 1988.

Nakamura T, Nishizawa T, Hagiya M, Seki T, Shimonishi M, Sugimura A, Shimizu S. Molecular cloning and expression of human hepatocyte growth factor. *Nature* 342: 440-443, 1989.

Nickoloff BJ, Griffiths CEM. Factor XIIIa-expressing dermal dendrocytes in AIDS-associated cutaneous Kaposi's sarcomas. *Science* 243: 1736-1737, 1989a.

Nickoloff BJ, Griffiths CEM. The spindle-shaped cells in cutaneous Kaposi's sarcoma. *Am J Pathol* 135: 793-800, 1989b.

Nickoloff BJ, Huang Y, Li J, Friedman-Kien AE. Immunohistochemical detection of papillomavirus antigens in Kaposi's sarcoma. *Lancet* 339: 548-549, 1992.

Noji S, Tashiro K, Koyama E, Nohno T, Ohyama K, Taniguchi S, Nakamura T. Expression of hepatocyte growth factor gene in endothelial and Kupffer's cells of damaged rat livers as revealed by *in situ* hybridization. *Biochem Biophys Res Commun* 173: 42-47, 1990.

Payne SF, Lemp GF, Rutherford GW. Survival following diagnosis of Kaposi's sarcoma for AIDS patients in San Francisco. *J Acquir Immune Defic Syndr 3 Suppl* 1: S14-S17, 1990.

Penn I. Neoplastic consequences of transplantation and chemotherapy. *Cancer Detect Prev (suppl)* 1: 149-157, 1987.

Polverini PJ, Leibovich SJ. Induction of neovascularization *in vivo* and endothelial proliferation *in vitro* by tumor-associated macrophages. *Lab Invest* 51: 635-642, 1984.

Ramakrishnan S, Xu FJ, Brandt SJ, Niedel JE, Bast RC Jr, Brown EL. Constitutive production of macrophage colony-stimulating factor by human ovarian and breast cancer cell lines. *J Clin Invest* 83: 921-926, 1989.

Rosen EM, Goldberg ID. Scatter factor and angiogenesis. *Adv Cancer Res* 67, In Press.

Rosen EM, Goldberg ID, Kacinski BM, Buckholz T, Vinter DW. Smooth muscle releases an epithelial cell scatter factor which binds to heparin. *In Vitro Cell Dev Biol* 25: 163-173, 1989.

Rosen, EM, Meromsky L, Setter E, Vinter DW, Goldberg ID. Purification and migration-stimulating activities of scatter factor. *Proc Soc Exp Biol Med* 195: 34-43, 1990.

Rosen EM, Goldberg ID, Liu D, Setter E, Donovan MA, Bhargava M, Reiss M, Kacinski BM. Tumor necrosis factor stimulates epithelial tumor cell motility. *Cancer Res* 57: 5315-5321, 1991a.

Rosen EM, Grant D, Kleinman H, Jaken S, Donovan MA, Setter E, Luckett PM, Carley W. Scatter factor stimulates migration of vascular endothelium and capillary-like tube formation. *In:* "Cell Motility Factors", Goldberg ID, Rosen EM, eds., Birkhauser-Verlag, Basel, 1991b, pp 76-88.

Rosen EM, Joseph A, Jin L, Rockwell S, Elias JA, Knesel J, Wines J, McClellan J, Kluger MJ, Goldberg ID, Zitnik R. Regulation of scatter factor production via a soluble inducing factor. *J Cell Biol* 127: 225-234, 1994a.

Rosen EM, Nigam SK, Goldberg ID. Mini-Review: Scatter factor and the c-met receptor: A paradigm for mesenchymal:epithelial interaction. *J Cell Biol* 127: 1783-1787, 1994b.

Safai B, Johnson KG, Myskowski PL. The natural history of Kaposi's sarcoma in the acquired immunodeficiency syndrome. *Ann Intern Med* 103: 744-750, 1985.

Salahuddin SZ, Nakamura S, Biberfield P, Kaplan MK, Markham PD, Larsson L, Gallo RC. Angiogenic properties of Kaposi's sarcoma-derived cells after long-term culture in vitro. *Science* 242: 430-433, 1988.

Stoker M, Gherardi E, Perryman M, Gray J. Scatter factor is a fibroblast-derived modulator of epithelial cell mobility. *Nature* 327: 238-242, 1987.

Tamura M, Arakaki N, Tsoubouchi H, Takada H, Daikuhara Y. Enhancement of human hepatocyte growth factor production by interleukin-1 alpha and -1 beta and tumor necrosis factor-alpha by fibroblasts in culture. *J Biol Chem* 268: 8140-8145, 1993.

Vogel J, Henrichs SH, Reynolds RK, Lucin PA, Jay G. The HIV tat gene induces dermal lesions resembling Kaposi's sarcoma in transgenic mice. *Nature* 335: 606-611, 1988.

Weich HA, Salahuddin SZ, Gill P, Nakamura S, Folkman J. AIDS-associated Kaposi's sarcoma-derived cells in long-term culture express and synthesize smooth muscle alpha-actin. *Am J Pathol* 139, 1251-1258, 1991.

Weidner KM, Behrens J, Vandekerckhove J, Birchmeier W. Scatter factor: Molecular characteristics and effect on invasiveness of epithelial cells. *J Cell Biol* 111: 2097-2108, 1990.

Wolf, H.K., Zarnegar, R., Michalopoulos, G.K. Localization of hepatocyte growth factor in human and rat tissues: an immunohistochemical study. *Hepatology* 14: 488-494, 1991.

Yanagita K, Nagaike M, Ishibashi H, Niho Y, Matsumoto K, Nakamura T. Lung may have endocrine function producing hepatocyte growth factor in response to injury of distal organs. *Biochem Biophys Res Commun* 182: 802-809, 1992.

Zibrob Z, Haimov M, Schanzer H. Kaposi's sarcoma after kidney transplantation. *Transplantation* 30: 383-384, 1980.

ROLE OF THE EARLY RESPONSE GENE CYCLOOXYGENASE (COX)-2 IN ANGIOGENESIS

Timothy Hla, Ari Ristimäki, Kirsi Narko, Pazit Ben-Av, Menq-Jer Lee, Mark Evans, Catherine Liu and Hajime Sano[1]

Department of Molecular Biology, Holland Laboratory, American Red Cross, Rockville, Maryland, USA and [1] First Department of Medicine, Kyoto Prefectural University School of Medicine, Kyoto, Japan

SUMMARY

Cyclooxygenase (Cox)-2 is a recently-identified isoform of prostaglandin synthase, a rate-limiting enzyme in prostaglandin (PG) biosynthesis. PGs are short-lived mediators that regulate a variety of biological processes in the vasculature including angiogenesis. While PGE_2 is a potent inducer of angiogenesis, the mechanisms involved are not well-understood. We have cloned the Cox-2 gene as a differentiation-induced early response gene from human umbilical vein endothelial cells (HUVEC). The expression of the Cox-2 gene is induced in a sustained manner by the inflammatory mediator interleukin-1 (IL-1) and is suppressed by the anti-inflammatory glucocorticoid dexamethasone (dex). Post-transcriptional regulation of Cox-2 mRNA stability is a major mechanism that regulates the expression of the Cox-2 gene. IL-1 stabilizes the Cox-2 transcript turn-over and thereby achieve a sustained induction of the Cox-2-dependent PG synthesis. In contrast to the Cox-1 enzyme, the expression of the Cox-2 enzyme is low in normal quiescent tissues; however, in acute inflammation, Cox-2 in induced transiently and is suppressed as the inflammatory response subsides. In chronic inflammatory diseases such as rheumatoid arthritis (RA), however, the Cox-2 gene is chronically expressed in an exaggerated manner. High levels of Cox-2 expression is correlated with the highly angiogenic state of the RA synovium. Because PGE_2 is a major prostanoid produced and PGE_2 also induces angiogenesis, the mechanisms of Cox-2-dependent angiogenesis was further investigated. While PGE_2 does not directly induce angiogenic behavior in endothelial cells *in vitro*, PGE_2-treated RA synovial fibroblasts expressed high levels of vascular endothelial cell growth factor (VEGF) mRNA. The induction appears to utilize the EP_2 subtype of the PGE receptor which signals via the G_s/ cAMP/ protein kinase A pathway. These data suggest that Cox-2-derived PGE_2 may be an indirect angiogenic factor in RA. In addition to RA, Cox-2 expression is also upregulated in the colorectal cancer. That Cox-2-

Molecular, Cellular, and Clinical Aspects of Angiogenesis
Edited by Michael E. Maragoudakis, Plenum Press, New York, 1996

191

derived PG may play a causative role in colorectal cancer development is suggested by epidemiological data which demonstrated that Cox inhibitors reduced the incidence of and death from colon cancer. It is not known how high-levels of Cox-2 expression contributes to the development of the malignant disease. However, the ability of Cox enzymes to activate mutagens as well as the property of prostanoids to induce cell proliferation, angiogenesis and immunosuppression may be involved. These data suggest that the dysregulated expression of the early response gene Cox-2 is an important component of many proliferative diseases and that induction of angiogenesis may be a critical component. Better understanding of the mechanisms involved in Cox-2 expression and function may yield novel therapeutic approaches for the control of proliferative and angiogenic diseases such as RA and colon cancer.

EARLY RESPONSE GENES

The concept of early response genes (ERGs) was originally derived from the study of DNA tumor viruses such as SV40 (1). The compact viral genomes express the ERGs such as large T and small T antigens immediately upon infection of host cells. These gene products are critical for the productive replication of virions because they regulate critical events involved in viral gene expression and reproduction (1). Pioneering work of Nathans and colleagues extended these concepts to mammalian cells in the 1980s by the isolation of growth factor-inducible ERGs from growth-arrested 3T3 fibroblasts (2). Critical genes such as c-fos, c-myc, EGR-1 were isolated by these differential cloning efforts (3). Independent of these studies, the discovery of cellular protooncogenes indicated that the growth factor-inducible ERGs are indeed important for normal growth control of cells and that dysregulation of these gene products leads to aberrant growth control *in vitro* and *in vivo* (3). Since these studies, many investigators have characterized ERGs induced by a variety of extracellular stimuli in many differentiated cell types (3). While the early studies suggests that nuclear regulatory proteins such as c-fos, c-jun and c-myc are major members of the ERG superfamily, recent studies indicate that an equally complex repertoire of secreted extracellular factors, intracellular signalling molecules, receptors and enzymes are induced (3). These data point out that cells invest significant energy to precisely regulate ERGs as an adaptive response to stimuli. Thus, dysregulation of ERGs usually results in defective cellular function and ultimately leads to pathological conditions in the organism.

Endothelial cells have been used widely to study different phases of angiogenesis, namely, migration, proliferation, growth arrest and phenotypic differentiation (4). The distinct phases of angiogenesis can be induced by discrete extracellular factors; for example, fibroblast growth factor (FGF) induces proliferation and migration whereas TGFß arrests the growth of endothelial cells (4). We and others have isolated several ERGs from human umbilical vein endothelial cells (HUVEC) (5-9). Our approach focussed on the ERGs induced by the tumor promoter phorbol myristic acetate (PMA), an inducer of growth arrest and tube formation *in vitro*. We isolated and characterized edg-1, an orphan G-protein-coupled receptor (6), edg-2, a putative transcription factor (7), edg-3, the IκBα protein (10) and Cox-2, an inducible cyclooxygenase (11). Dixit *et. al.* have characterized A20, a zinc finger transcription factor involved in apoptosis, B61, a secreted angiogenic cytokine implicated as a mediator of TNF-induced angiogenesis, among others (8). Recently, Introna *et al.* have reported the characterization of an IL-1 induced ERG that encoded a protease-inhibitor molecule (9). Thus, while the repertoire of ERGs in endothelial cells appear to be different from other differentiated cell types, many of these gene products such as A20, B61 and Cox-2 have been shown to be involved in critical regulatory events of angiogenesis.

COX-2: BIOCHEMISTRY, CELL BIOLOGY AND FUNCTION

Prostaglandins, thromboxanes and hydroxyeicosatetraenoic acids, collectively known as prostanoids, are produced by biological oxidation of polyunsaturated fatty acids by the cyclooxygenase (Cox) pathway (12). The oxygenase and peroxidase activities of the Cox enzyme convert free arachidonic acid (AA) to PGH_2. The endoperoxide PGH is then converted to biologically active prostanoids by isomerases that are expressed in a tissue-specific manner (13). For example, platelets express the thromboxane synthase enzyme and thus produce the pro-aggregatory prostanoid thromboxane as a major arachidonate metabolite (12,13).

The Cox enzyme catalyzes the rate-limiting step in the formation of the prostanoids (12,13). This is due to the phenomenon of irreversible self-inactivating property of the enzyme; it is estimated that after 1500 catalytic turnovers, the enzyme molecule becomes irreversibly inactivated. The self-inactivated enzyme is rapidly degraded *in vitro* and thus Cox protein is presumed to turn-over with a rapid rate *in vivo* (13). The level of Cox enzyme protein is regulated by a variety of extracellular factors including cytokines, hormones, growth factors in many differentiated cell-types. Indeed, the Cox enzyme appears to be encoded by two separate genes, Cox-1 and -2 (also known as PGHS-1 and -2). The Cox-2 enzyme is generally inducible by extracellular mediators in a variety of cells whereas the Cox-1 enzyme is normally expressed in quiescent cells in most tissues (14,15). The cDNA for the Cox-1 enzyme was first isolated from ovine seminal vesicles, a tissue that is highly enriched in the Cox enzyme. The enzyme of 576 amino acids is encoded by transcripts of 3 and 5 kb. The smaller transcript was sequenced completely and was shown to possess a small 5'-untranslated region (UTR), 1.8 kb open reading frame and a 1.1 kb 3'-UTR (16). The human and murine genomic sequences encoding the Cox-1 cDNA has been characterized and the open reading frame is encoded by 11 exons spanning approximately 22 kb (14,15). Site-directed mutagenesis analysis of the ovine Cox-1 cDNA has been conducted and the mutant enzymes have been expressed in Cos cells for enzymatic analysis. Using this approach, axial and distal heme-ligated histidine residues, the active-site tyrosine and the N-linked glycosylation sites were identified (17). Furthermore, the role of the aspirin-modified serine residue in the competitive inhibition of the substrate binding was confirmed (17). Rosen et al., using low-stringency hybridization with Cox-1 probes, were the first to predict the existence of the Cox-2 gene (18). However, investigators that were in the process of cloning novel ERGs isolated the Cox-2 cDNAs from chicken, mouse and human tissues (19-21). This is consistent with the observation that an early response of cellular activation by growth factors, cytokines, tumor promoters and viruses is the enhanced productions of PGs (12-14). Thus, the Cox-2 gene appears to encode an early, activation-specific isoform of the Cox enzyme. At the amino acid level, the Cox-1 and -2 polypeptides share approximately 60 % sequence identity (19-21). Both Cox-1 and -2 isoenzymes are membrane proteins that are localized in the lumenal compartment of the endoplasmic reticulum (23). Recent studies, however, have suggested that during the specific stages of the cell-cycle, the Cox-2 enzyme translocates to the inner leaflet of the nuclear membrane (24). The nuclear function of Cox-2 enzyme is not known. Prostanoids such as PGE_2, $PGF_{2\alpha}$, PGI_2 and TXA_2 are rapidly secreted into extracellular milieu (12).

Recent studies on the enzymology and the crystal structure of the Cox enzyme have given unique insights into the differential enzymatic properties of the Cox-1 and -2 isoenzymes (22). Both Cox-1 and -2 isoenzymes oxidize the 1,4-*cis*-dienyl bonds of AA, first at C_{13} and then at C_{10}. The bis-oxygenated AA is immediately cyclized to form the peroxyl-endoperoxide PGG_2. The peroxidase activity of the enzyme then reduces PGG_2 to the hydroxyl-endoperoxide PGH_2. The enzyme requires a single heme molecule as a co-factor for both the bis-oxygenase and peroxidase activities. The peroxidase activity of the Cox enzyme can utilize numerous reducing equivalents such as glutathione, tryptophan, uric acid, etc. as reducing equivalents (12,13). . The crystal structure predicted the existence of the 8 A°x 25 A° 'cyclooxygenase channel' which connects the active site in the interior of the enzyme to the membrane. Since the substrates are derived from the membrane phospholipids, it is assumed that the substrates are transported to the active site via the channel (22).

While the Cox-1 and -2 isoenzymes catalyze generally similar enzymatic reactions, they exhibit differential pharmacological properties with respect to NSAID inhibition (24,25). Studies conducted with the recombinant murine Cox-1 and -2 isoenzymes indicated that several commonly-used NSAIDs, including indomethacin, sulindac sulfide and piroxicam preferentially inhibited Cox-1. In contrast, nabumetone inhibited murine Cox-2 preferentially (26). Surprisingly, studies on recombinant human enzymes did not confirm the Cox-2-selective nature of nabumetone, suggesting that there are species-specific differences in the pharmacological properties of the Cox-1 and -2 isoenzymes (27). Pharmacological selectivity is currently thought to be extremely important in the anti-inflammatory therapeutics area. While Cox-1 is ubiquitously expressed, including in the stomach and the kidney, the Cox-2 enzyme is induced only in inflamed and/or activated tissues (11,15). Basal synthesis of Cox-1-derived prostanoids is important in maintaining the normal physiology of various tissues. Thus, inhibition of Cox-1 results in the serious side-effects associated with NSAID administration, such as gastric ulceration and impaired renal function eventually leading to tubular necrosis and renal failure (12). The side-effect profile of NSAIDs correlate roughly with the Cox-1 selectivity. For example, indomethacin is the most selective drug against the inhibition of human Cox-1 (> 70-fold selective) and also produces the most serious gastric and renal side-effects (27). In contrast, NSAIDs such as ibuprofen which are equipotent at inhibiting both human Cox-1 and -2 are less damaging to the stomach and the kidneys (26,27). Therefore, the search for the Cox-2-selective inhibitors have intensified recently and the experimental drugs NS-398 and DuP-697 appear to be efficacious prototypical agents (24-28). Indeed, data from experimental models of inflammation have indicated that Cox-2-selective inhibitors possess potent anti-inflammatory property and markedly reduced side-effect profile (24-28).

Prostanoids are well-known regulators of acute symptoms of inflammation. For example PGE_2 and PGI_2 induces the dilitation of various vascular beds (12). Prostanoids can also potentiate the vascular permeability action of other vasoactive agents such as histamine. In addition, PGE_2 and PGE_1 can induce pain by their action on peripheral nociceptors (12). Much of the rapid actions of the prostanoids have been characterized pharmacologically and the signalling mechanisms are well understood. Prostanoid receptors belong to the G-protein-coupled receptor superfamily and interact with many second messenger systems including cAMP and phosphoinositide hydrolysis (29). PGE (ep1,ep2,ep3,ep4), PGF, PGI and thromboxane receptors have been cloned recently and many pharmacological agonists and antagonists exist (29 and references therein). While much work has been conducted correlating acute biological effects of prostanoids with receptors and second messenger systems activated, less is known about the chronic effects of prostanoids. It has been known for two decades that PGE_1 and PGE_2, but not $PGF_{2\alpha}$ are capable of inducing an angiogenic response in animal models (30). In addition, tumor-induced angiogenesis in some animal models is inhibited by Cox inhibitors (31). Furthermore, $PGF_{2\alpha}$ and TXA_2 are potent stimulators of vascular smooth muscle growth (32) whereas PGI_2 inhibits the proliferation *in vitro* (33). Overall, much needs to be learned about mechanisms involved in the long-term pathophysiological effects of prostanoids.

REGULATION OF COX-2 EXPRESSION: mRNA STABILITY MAY BE A CRITICAL REGULATORY STEP

The Cox-2 cDNA was initially cloned as an immediate-early gene (19-21). Many extracellular mediators such as hormones, cytokines and growth factors induce the expression of the Cox-2 gene and thus stimulate PG synthesis (15). Thus Cox-2 is considered as major contributor of PG synthesis in activated conditions (15). While many stimuli such as fibroblast growth factor (FGF)-1 induce the Cox-2 mRNA in human endothelial cells transiently, inflammatory cytokines such as interleukin-1 (IL-1) induce it with a sustained kinetics (21). IL-1 induces the Cox-2 gene transcription transiently; however, it independently stabilizes the Cox-2 mRNA resulting in sustained induction of Cox-2 mRNA, protein and prostanoid synthesis (35). In contrast to the proinflammatory cytokine IL-1, the antiinflammatory glucocorticoids down-regulates the Cox-2 mRNA in a variety of cells (36). The ability of dex to down-regulate Cox-2 mRNA can occur even if the steroid is added 16 h after the addition of IL-1. The half-life of the Cox-2 mRNA in the presence of IL-1 and dex is less then 1 h, suggesting that dex antagonizes the IL-1-induced mRNA stability mechanisms (37,38). Endogenous glucocorticoids were shown to be a key factor in the expression of Cox-2 *in vivo* (39). Thus understanding of IL-1 and dex regulation of Cox-2 mRNA stability may yield novel insights into the physiological and pathophysiological mechanisms of inflammation. Furthermore, pharmacological agents that mimic the actions of glucocorticoids to down-regulate Cox-2 without other side effects may prove to be highly efficacious anti-inflammatory agents.

EXPRESSION OF COX-2 IN RA AND COLON CANCER

RA is an aggressive chronic inflammatory disease with a strong angiogenic component (40). In contrast osteoarthritis (OA) is a less severe form of joint inflammatory disease (40). High levels of Cox-2 were detected by immunohistochemical staining of the RA synovium but not in the OA synovium (41). Immunostaining was most prominent in angiogenic endothelial cells, tissue macrophages and synovial fibroblasts (41). While the Cox-1 antigen is also expressed, the Cox-2 polypeptide was strongly induced by *ex vivo* treatment with PMA and IL-1 and was potently suppressed by dex (36). These data suggests that exaggerated and sustained expression of the Cox-2 polypeptide in RA may be an important component of disease progression into a highly angiogenic and destructive synovial pannus. The major prostanoid produced by the RA synovium is PGE_2 (40). It is known that PGE_2 induces bone resorption and induces angiogenesis (30). Thus, Cox-2-derived PGE_2 may function as an angiogenic factor and inducer of bone resorption in RA. The function of other prostanoids produced by Cox-2, i.e., $PGF_{2\alpha}$ and 15-HETE as well as the peroxidase activity of Cox-2 in the context of RA is not known.

The potential role of prostanoids in colon cancer was recently established by large-scale epidemiological studies which demonstrated the protective role of Cox enzyme inhibitors in the prevention of cancer incidence (42,43). Immunohistochemical analysis indicated the enhanced expression of Cox-2 polypeptide is observed in tumor tissues but not in the adjacent normal colon tissues. Highest expression of immunoreactive Cox-2 was found in well-differentiated tumor cells of the colon. Significant immunostaining was also observed in inflammatory cells and vascular endothelial cells. The immunostaining of the Cox-1 antigen was low in both normal and tumor tissues (44). These data indicate that selective up-regulation of the Cox-2 gene is associated with colon cancer. These data are also consistent with, but do not prove that the mechanism of action of Cox inhibitors as preventative agents

195

for colon cancer is via the inhibition of the Cox-2 activity. It is not known if and how enhanced Cox-2 expression contributes to the development of the colon cancer. Because Cox enzymes possess oxygenase as well as peroxidase activities, they have been shown to activate a number of mutagens and carcinogens (45). Thus sustained activation of the Cox-2 pathway may induce potentially carcinogenic mutations within the epithelial cells. In addition, production of prostanoids, hydroperoxy and hydroxy fatty acids may contribute, in a paracrine or autocrine manner, to the development of a neoplastic lesion.

INDUCTION OF VEGF BY PGE$_2$

A potential mechanism via which prostanoids may regulate proliferative diseases is via the induction of the angiogenic response *in situ*. While PGE$_1$ and PGE$_2$ are potent inducers of angiogenesis *in vivo*, mechanism of action is not known (30). Because high levels of Cox-2 are associated with RA (40) and because PGE$_2$ is a major product of the RA tissue (41), we examined the effect of these prostanoids on the angiogenic growth factor gene expression by the RA synovial fibroblasts. We initially focussed our attention on the expression of VEGF, a potent and specific angiogenic factor (46). PGE$_1$ and PGE$_2$ but not PGF$_{2\alpha}$ induced VEGF mRNA expression in RA synovial fibroblasts. The kinetics of induction was rapid; VEGF mRNA levels started to increase within 30 min. of PGE$_1$ addition, maximal levels were achieved by 60-200 min. and VEGF mRNA levels declined by 20 h. Furthermore, treatment of the RA synovial fibroblasts with IL-1α induced a delayed but sustained up-regulation of the VEGF expression. Thus, in response to an inflammatory cytokine and prostanoid action, the RA synovial fibroblast is capable of expressing an angiogenic signal by the up-regulation of VEGF expression (47). The ep1 and ep2 but not ep3 and ep4 subtypes of PGE receptors were expressed in synovial fibroblasts, as determined by RT-PCR analysis. It is known that the ep1 receptor activates the phosphoinositide hydrolysis whereas the ep2 receptor regulates the cAMP pathway (29). Elevation of cAMP by forskolin and cholera toxin treatment also up-regulated the VEGF mRNA levels. In addition, down regulation of protein kinase C by chronic phorbol ester treatment did not block the ability of PGE$_2$ to induce the VEGF mRNA levels. These data suggest that PGE acting on the ep2 receptor/ G$_{\alpha s}$ /cAMP signalling pathway regulates the expression of the VEGF gene transcription in RA synovial fibroblasts.

FUTURE APPROACHES AND SPECULATIONS

Efforts to characterize early response genes in endothelial cells have resulted in the molecular characterization of the Cox-2 gene. While the regulated induction of the Cox-2 gene may be a normal protective mechanism, dysregulated expression is associated with several proliferative and angiogenic pathologies. Because endogenous antiinflammatory glucocorticoids inhibit the Cox-2 expression by enhancing mRNA degradation, understanding the molecular basis of this phenomenon may yield novel approaches to long-acting anti-inflammatory drugs. PGE$_2$ that is released induces the expression of VEGF in fibroblasts by the ep2-coupled pathway. Pharmacological modulation of this pathway may be useful in the control of exaggerated angiogenesis in many proliferative diseases in which enhanced Cox-2 expression is implicated. Finally, better understanding of the function of Cox-2 in the development of colon cancer is warranted in light of the recent epidemiological data.

References

1. Nathans D; Lau LF; Christy B; Hartzell S; Nakabeppu Y; Ryder K (1988) 53, 893-900.
2. Lau, L. and Nathans, D. (1987) *Proc. Natl. Acad. Sci., USA* 84, 1182-1186.
3. Herschman, R. (1991) *Ann. Rev. Biochem.* 60, 281-319.
4. Folkman, J. and C. Haudenschild (1980) *Nature* 288,551-556.
5. Hla, T. and Maciag, T. (1990) *Biochem. Biophys. Res. Comm.* 167, 637-643.
6. Hla, T. and Maciag, T. (1990) *J. Biol. Chem.* 265, 9308-9313.
7. Hla,T., Jackson, A.Q., Appleby, S.B. and Maciag, T. (1994) *Biochim. Biophys. Acta* 1260, 227-229.
8. Dixit, V., Green, J., Sharma, V., Holzman, L.B., Wolf, F.W., O'Rourke, K., Ward, P.A., Prochownik, E.V. and Marks, R.M. (1990) *J. Biol. Chem.* 265, 2973-2978.
9. Breviario, F., d'Aniello, E.M., Golay, J., Peri, G., Bottazzi, B., Bairoch, A., Saccone, S., Marzella, R., Predazzi, V., Rocchi, M., Della Valle, G., Dejana, E., Montovani, A. and Introna, M. (1992) *J. Biol. Chem.* 267, 22190-22197.
10. Hla, T., Zimrin, A., Evans, M., Ballas, K. and Maciag, T. (1995) (submitted)
11. Hla, T. and Neilson, K. (1992) *Proc. Natl. Acad. Sci., USA* 89, 7384-7388.
12. Needleman, P., J. Turk, B.A. Jacshick, A.R. Morrison, and J.B. Lefkowith (1986) *Ann. Rev. Biochem.* 55, 69-102.
13. Smith, W.L., Eling, T., Kulmacz, R.J., Marnett, L.J. and Tsai, A. (1992) *Biochemistry* 31, 2-7.
14. DeWitt, D. (1991) *Biochim. Biophys. Acta* 1083, 121-134.
15. Herschman, H.R. (1994) *Canc. Met. Rev.* 13, 241-256.
16. DeWitt, D. and Smith, W.L. (1988) *Proc. Natl. Acad. Sci. USA* 85, 1412-1416.
17. Shimokawa, T. and Smith, W.L. (1992) *J. Biol. Chem.* 267,12387-12392.
18. Rosen, G.D., T.M. Birkenmeier, A. Raz, and M.J. Holtzman (1989) *Biochem. Biophys. Res. Comm.* 164, 1358-1365.
19. Xie, W., Chipman, J.G., Robertson, D.L., Erikson, R.L. and Simmons, D.L. (1991) *Proc. Natl. Acad. Sci. USA* 88, 2692-2696.
20. Kujubu, D.A., Fletcher, B.S., Varnum, B.C., Lim, R.W. and Herschman, H.R. (1991) *J. Biol. Chem.* 266, 12866-12872.
21. Hla, T., Ristimäki, A., Appleby, S. and Barriocanal, J. (1993)*Ann. New York Acad. Sci.* 696, 197-204.
22. Picot, D., Loll, P.J., Garavito, R.M. (1994) *Nature* 367, 243-249.
23. Morita, I., Schindler, M., Regier, M., Otto, J.C., Hori, T., DeWitt, D.L. and Smith, W.L. (1995) *J. Biol. Chem.* 270, 10902-10908.
24. Masferrer, J., Zweifel, B., Manning, P., Hauser, S., Leahy, K., Smith, W., Isakson, P. and Seibert, K. (1994) *Proc. Natl. Acad. Sci. USA.* 91, 3228-3232.
25. Gans, K., Galbraith, W., Roman, R., Haber, S., Kerr, J., Schmidt, W., Smith, C., Hewes, W., Ackerman, N. (1990) *J. Pharm. Exp. Ther.* 254, 180-187.
26. Meade, E.A., W.L.Smith, D.L.DeWitt (1993) *J.Biol. Chem.* 268,6610-6614.
27. Laneuville, O., Breuer, D.K., DeWitt, D.L., Hla, T., Funk, C.D. and Smith, W.L. (1994) *J.Pharm.Exper.Ther.* 271, 927-934.
28. Futaki, N., Yoshikawa, K., Hamasaka, Y., Arai, I., Higuchi, S., Iizuka, H. and Otomo, S. (1993) *Gen. Pharm.* 24, 105-110.
29. Coleman, R.A., Smith, W.L. and Narumiya, S. (1994) *Pharm. Rev.* 46, 205-229.
30. Ziche, M., Jones, J. and Gullino, P.M. (1982) *J. Natl. Canc. Inst.* 69, 475-482.
31. Peterson, H. (1983) *Invasion Metastasis* 3, 151-159.

32. Watanabe, T., Nakao, A., Horie, Y. and Kurokawa, K. (1994) *J.Biol. Chem.* 269,17169-17625.
33. Libby P, Warner SJ, Friedman GB (1988) *J Clin Invest* 81(2):487-98.
35. Ristimäki, A., Garfinkel, S., Wessendorf, J., Maciag, T. and Hla, T. (1994) *J. Biol. Chem.* 269, 11769-11775.
36. Crofford, L.J., Wilder, R.L., Ristimäki, A., Remmers, E., Epps, H.R. and Hla, T. (1994) *J. Clin. Invest.* 93, 1095-1101.
37. Ristimäki, A., Narko, K. and Hla, T. (1995) submitted
38. Evett, G.E., Xie, W., Chipman, J., Robertson, D., and Simmons, D.L. (1993) *Archives in Biochem. Biophys.* 306, 169-177.
39. Masferrer, J., Seibert, K., Zweifel, B.S., and Needleman, P. (1992) *Proc Natl Acad Sci U S A* 89, 3917-3921.
40. Harris, E.D. (1990) Rheumatoid arthritis. *New Engl. Jrl. Med.* 322, 1277-1289.
41. Sano, H., Hla, T., Maier, J., Crofford, L., Case, J., Maciag, T. and Wilder, R. (1992) *J. Clin. Invest.* 89:97-108.
42. Thun, M.J., Namboodiri, M.M. and Health, C.W., Jr. (1991) *New Eng. J. Med.* 325, 1593-1596.
43. Giardiello, F.M., Hamilton, SR, Krush, AJ, Piantadosi, S., Hylind, LM, Celano, P., Bocker, SV, Robinson, CR and Offerhaus, GJ (1993) *New Eng. J. Med.*, 328, 1313-1316.
44. Sano, H., Kawahito, Y., Wilder, R.L., Hashiramoto, A., Mukai, S., Asai, K., Kimura, S., Kato, H., Kondo, M. and Hla, T. (1995) *Cancer Res.* 55, 3785-3789.
45. Marnett, L.J. (1981) *Life Sci.* 29, 531-546.
46. Koch, A., Harlow, L., Haines, G., Amento, E., Unemori, E., Wong, W., Pope, R., Ferrara, N. (1994) *J. Immunology* 152, 4149-4156.
47. Ben-Av, P., Crofford, L.J., Wilder, R.L. and Hla, T. (1995) *FEBS J.* 372, 83-87.

A HEPARANASE-INHIBITORY, bFGF-BINDING SULFATED OLIGOSACCHARIDE THAT INHIBITS ANGIOGENESIS *EX OVO* HAS POTENT ANTITUMOR AND ANTIMETASTATIC ACTIVITY *IN VIVO*

Robert J. Tressler,[1] J. Wee,[1] N. Storm,[1] P. Fugedi,[1] C. Peto,[1] R.J. Stack,[1] D.J. Tyrrell,[1] and J.J. Killion[2]

[1]Glycomed Incorporated, Alameda, CA
[2]Dept. of Cell Biology, M.D. Anderson Cancer Center, Houston, TX

ABSTRACT

Angiogenesis is essential for solid tumor progression, growth and metastasis. This process requires mitogenic signals from growth factors for endothelial cell proliferation (eg. bFGF), as well as hydrolases for extracellular matrix degradation to permit endothelial cell migration and tumor cell invasion (eg. matrix metalloproteases and heparanase). We have evaluated the antitumor and antimetastatic activity of a series of sulfated oligosaccharides in syngeneic murine and human tumor xenograft models. One of these compounds, GM1474 is a novel polysulfated oligosaccharide that has multiple biological activities, and binds tightly to bFGF as well as inhibiting tumor cell-derived heparanase. GM1474 inhibited bFGF-dependent proliferation of endothelial cells *in vitro* and angiogenesis *ex ovo* in a chick chorioallantoic membrane (CAM) assay. GM1474 did not inhibit the *in vitro* proliferation of human or murine tumor cells. Systemic administration of GM1474 significantly inhibited subcutaneous tumor growth of human mammary (MDA-231), prostatic (PC-3) and hepatocellular (SKHep-1) carcinomas in nude mice. Administration of GM1474, on days 0-4 after iv. injection of B16-F10 melanoma cells into C57/BL6 mice, resulted in a dose-dependent increase in survival of treated animals. Treatment with GM1474 resulted in a >90% reduction in the number of spontaneous lung metastases that occurred in the SN12PM6 orthotopic model of human renal cell carcinoma. In the KM12L4a orthotopic human colorectal carcinoma model, GM1474 significantly inhibited the formation of spontaneous liver metastasis in nude mice. This compound also significantly inhibited the orthotopic growth of the human 253J-BV bladder cell carcinoma in nude mice. These data demonstrate that a sulfated oligosaccharide that binds bFGF and inhibits heparanase is angiostatic, and is a potent antitumor and antimetastatic agent *in vivo*.

INTRODUCTION

The major cause of death from cancer is due to the growth and metastatic spread of tumor cells from the primary site to distant organ sites. Current therapy relies on

Molecular, Cellular, and Clinical Aspects of Angiogenesis
Edited by Michael E. Maragoudakis, Plenum Press, New York, 1996

199

surgical resection, in combination with chemo- and/or radiation therapy. The success of these approaches against metastatic disease has been poor. Some of the reasons for this are the outgrowth of resistant tumor cell subpopulations, patient toxicity to therapy and poor bioavailability of the therapeutic agent (Nicolson, 1984).

Solid tumor progression, growth and metastasis is a dynamic process that is angiogenesis dependent. Angiogenesis is a complex process maintained through a regulated balance of angiogenic and angiosuppressive factors (Holmgren, et al., 1995). When this equilibrium is altered in favor of the former, angiogenesis proceeds. Aberrant angiogenesis is associated with pathological conditions such as tumor growth and metastasis, retinopathies and rheumatoid arthritis (Folkman and Klagsbrun, 1987). Increased growth of an occult avascular tumor mass is dependent on angiogenesis for more efficient nutrient and waste transport. Initially, a small avascular tumor's increase in volume is a linear function, but as the tumor becomes vascularized, its volume increases exponentially (Folkman, 1993).

A cancer cell's utilization of tumor neovasculature as a route of access to the host's circulation is essential for metastatic spread. Tumor vascular density has been shown to correlate with malignancy for a variety of human cancers (Weidner, 1993). The growing tumor consists of a variety of cell types, including proliferating tumor cells, inflammatory cells and fibroblasts. This mileu of normal host and tumor cells elicits an array of angiogenic factors such as VEGF, TGFalpha, TGFbeta, TNFalpha, IL-8, PDGF and FGF (Zagzag, 1995). Several of these factors interact with heparin and heparan sulfate proteoglycans (HSPG), which can modulate their biological effects (Spillman and Lindahl, 1994).

Of the various HSPG-binding angiogeneic factors, bFGF is the primary angiogenic stimulus in most normal and pathological angiogenic processes. bFGF can serve not only as a mitogen, but also is a key inducer of a number of other aspects of the angiogenic phenotype, such as urokinase-type plasminogen activator and collagenase expression, cell locomotion, integrin expression, and gap-junctional intercellular interactions (Gualandris and Presta, 1995). Soluble heparin and cell surface-associated heparan sulfate (HS) are thought to serve as obligatory bFGF-binding cofactors that stabilize bFGF binding to, and dimerization of, the high affinity transmembrane cell surface receptor, all of which is required for signal transduction (Spivak-Kroizman, et al., 1994).

Evaluation of the ability of low molecular weight heparin-derived compounds and analogs to interfere with bFGF-dependent endothelial cell proliferation and angiogenesis provides the basis of a strategy for identifying anti-angiogenic compounds. Towards this end, researchers have demonstrated that heparin and heparin-derived fragments are potent inhibitors of angiogenesis ex ovo and in vivo (Folkman and Klagsbrun, 1987). Subsequently investigators have found heparin-steroid conjugates to have antitumor activity in mice (Thorpe, et al., 1993).

Another requirement for vessel formation is the activity of a variety of hydrolases, such as matrix metalloproteases and heparanase, which degrade the extracellular matrix, allowing endothelial cell migration and tumor cell invasion (Vlodavsky, et al., 1988). Heparanase inhibitors have been shown to prevent metastasis in a murine melanoma model (Irimura, et al., 1986), as well as block the extravasation of inflammatory cells, and inhibit angiogenesis induced by the release of angiogenic heparin-binding growth factors bound to the ECM (Vlodavsky, et al, 1991). These findings indicate that heparanase inhibitors may have potential use as antitumor agents by inhibiting angiogenesis and tumor cell invasion.

The implications of these observations are that a compound which is an antagonist of bFGF and heparanase inhibits angiogenesis and will inhibit the growth and recurrence of the primary tumor, as well as metastasis formation and subsequent outgrowth of occult metastases. Therefore it is reasonable to assume that noncytotoxic agents that inhibit tumor angiogenesis will be useful as adjunct therapies with conventional treatment regimens for the treatment of metastatic cancer and inoperable primary tumors (Teicher, *et al*, 1992). We have synthesized a noncytotoxic polysulfated oligosaccharide, GM1474, with a high affinity for bFGF which also inhibits heparanase activity. This report describes the bFGF and heparanase inhibitory activity of GM1474 *in vitro*, its angiostatic activity in the CAM assay, and the potent antitumor activity of this multifunctional carbohydrate in a variety of *in vivo* tumor models.

MATERIALS AND METHODS

Tumor cells

The PC-3 human prostatic adenocarcinoma and MDA231 human mammary adenocarcinoma cell lines were obtained from the American Type Tissue Culture Collection, and routinely propagated in DMEM (GIBCO, Grand Island, NY) + 10% fetal calf serum (FCS, GIBCO), or in a 1:3 mixture (vol/vol) of Leibovitz's L15 medium + Keratinocyte medium (GIBCO) + 10% FCS, respectively. The B16-F10 murine melanoma sell lime was obtained from Dr. Brian Brandley, Glycomed, Incorporated and was propagated in DMEM + 10% FCS. The human renal cell carcinoma, SN12pm6, is a metastatic variant of the SN12 tumor cell line that was derived by *in vivo* selection for spontaneous lung metastasis (Naito, *et al.*, 1989). The SN12pm6 cell line was cultured in EMEM (M.A. Bioproducts, Walkersville, MD) supplemented with 5% FCS, sodium pyruvate, non-essential amino acids, L-glutamine and vitamins (GIBCO). The spontaneously metastasizing KM12L4a human colorectal carcinoma line was developed in the Department of Cell Biology, M.D. Anderson Cancer Center, Houston Texas as previously described (Morikawa, *et al.*, 1988), and was propagated under the same conditions as the SN12pm6 tumor cell line. All cell lines were cultured in plastic tissue culture flasks (Corning), and maintained in a humidified 5% CO_2 incubator at 37°C. The 253J-BV human bladder carcinoma line was established in the Department of Cell Biology, M.D. Anderson Cancer Center, Houston, TX, from a metastasis of a patient with transitional cell carcinoma and has been maintained by orthotopic passage in the bladder wall of nude mice. The 253J-BV line had been passaged 5 times *in vivo* prior to use in the study.

Animals

Female C57 BL/6 mice, 4-6 weeks old (15-20 grams), were purchased from Simonsen Laboratories (Gilroy, CA), and housed in the Glycomed animal facility. Athymic nude mice were also purchased from Simonsen Laboratories or through the animal contract center of the National Cancer Institute and acclimated for at least one week prior to use in studies. All experimentation was reviewed and approved by the Institutional Animal Care and Use Committees of the facilities where studies were conducted.

Chemicals

GM1474 was supplied as a lyophilized powder by the Department of Chemistry, Glycomed, Incorporated. The compound was routinely stored at 4°C in a dessicator prior to use. Solutions of GM1474 were prepared weekly in PBS, filtered sterilized and stored in the dark at 4°C rior to use in assays. All other reagents were purchased from commercial vendors and were of the highest quality available.

bFGF-binding assay

The ability of GM1474 to inhibit the interaction of RO-12 UC cells and bFGF was carried out as described previously (Ishihara, et al., 1992). Briefly, the wells of 96-well tissue culture plates were coated overnight at 4°C with 50μl of 10μg/ml human recombinant bFGF and then blocked by the addition of 5% (v/v) fetal bovine serum in PBS for 1 h at room temperature. One hundred μl of RO-12 UC cell suspension (3 x 10^6 cells/ml in 5% fetal bovine serum in PBS) containing various concentrations of GM1474 were added to each bFGF-coated well and incubated for 5 min at room temperature. Each well was washed three times with PBS, and then 20μl of 5% SDS was added to lyse the bound cells. The protein concentration of the lysate was then determined after the addition of 200μl of Micro BCA protein assay reagent (Pierce Chemical Co.), using bovine serum albumin as a standard.

Quantification of heparanase activity in soluble extracts of rat hepatoma cells using a CPC precipitation assay

A heparanase assay was developed based on the observation that heparanase-cleaved heparan sulfate (HS) chains can be distinguished from uncleaved chains by selectively precipitating the latter with CPC (LaPierre, et al., 1996). Briefly, this assay is done by combining radiolabelled HS with the test agent. Hepatoma soluble extracts are then added to the reaction mixture containing labelled HS and test agent. The samples are incubated and excess heparin is added to terminate the reaction. Sodium acetate, pH 5.5 is then added to each tube followed by a solution of 0.6% CPC w/v in water. The tubes are incubated for 1 h at ambient temperature, and then centrifuged for 10 min at 4,000 x g. The supernatants are assayed for radioactivity by liquid scintillation counting. All assays were run in triplicate with heparin as a positive control.

Chick chorioallantoic membrane (CAM) assay

The CAM assay for evaluating angiostatic activity was described previously (Castellot, et al., 1986). A sling made of plastic wrap was fashioned inside a styrofoam cup and held in place by a rubber band stretched around the outside of the cup. A fertilized chicken egg was placed in the plastic sling and the cups were placed in a humidified incubator. Eight days after fertilization, a methylcellulose pellet containing test substance or buffer control was placed on the CAM. The CAMs were scored at 24 h after placement of pellets, using a dissecting microscope to examine the microvasculature surrounding each pellet. If a completely avascular zone was observed more than 3/4 of the way around the pellet, it was scored as ++. A + score indicates that either the overall vascularity was significantly reduced, or that an avascular zone existed around 1/4-3/4 of the pellet. No effect indicates that no change in vascularity of the area immediately surrounding the pellet was observed.

In vitro proliferation assays

The effect of GM1474 on bFGF-stimulated rat adrenocortical endothelial cell (ACE) proliferation *in vitro* was carried out as described (Ishihara, et al., 1993). Briefly, ACE cells were seeded at a density of 3000 cells/well in 96-well tissue culture plates and were grown for 4 days in DMEM supplemented with 2ng/ml bFGF, 10% fetal bovine serum, 50µg/ml gentamicin, and GM1474 at varying concentrations.

The effect of GM1474 on the *in vitro* proliferation of tumor cells was assessed for all of the cell lines tested *in vivo*. Briefly, the assays were carried out by this general procedure; Tumor cells were seeded at densities ranging from 2000-20,000 cells/tissue culture well containing 1ml of media. GM1474 was dissolved in PBS and added to the wells (0.1ml vol.) to give a final concentration ranging from 10-100µg/ml of compound. PBS was used as vehicle control. Cells were harvested with trypsin-EDTA and then counted electronically at approximately 48, 72 and 96 hours with a Coulter Counter. Replicates of three to five wells per time point were routinely taken for cell count determinations.

In vivo tumor assays

B16-F10 murine experimental metastasis model: B16-F10 tumor cells were harvested with trypsin-EDTA (GIBCO), washed 2x with PBS, resuspended at a concentration of 5×10^5 cells/ml, and held on ice prior to injection. Animals were challenged with 5×10^4 cells in a volume of 0.1 ml intravenously (9-15 animals/treatment group). Animals were randomly distributed prior to assignment to treatment groups and monitored for survival. Animals received GM1474 (100mg/kg, sc., 0.1ml volume) or vehicle on days 0-4 post tumor challenge. Survival times were recordeded for animals and the data was evaluated by Chi-Squared statistical analysis.

MDA231 human mammary adenocarcinoma model: 3×10^6 MDA231 human mammary adenocarcinoma cells were injected sc. in the dorsal flank of nude mice and animals recieved 50mg/kg GM1474 or vehicle, sc., on days 8-73, alternating 5 days treatment and 2 days no treatment. Tumor measurements were determined weekly and statistical analysis carried out using the Mann-Whitney nonparametric test.

PC-3 human prostatic adenocarcinoma model: 5×10^6 PC-3 cells were injected (0.1ml vol.) sc. in the dorsal flank of male Balb/C nude mice and animals recieved 80mg/kg GM1474 or vehicle, sc., on days 1-23. Tumor measurements were determined biweekly and statistical analysis was carried out with an unpaired two-tailed T-test.

SK-Hep-1 human hepatocellular carcinoma model: 1×10^7 cells were injected sc. in the dorsal flank of female Balb/c nude mice. Treatment with GM1474 was started 24 hours after tumor challenge, 50 mg/kg/day, sc., on a schedule of 5 days treatment, 2 days no treatment. Tumor dimensions were measured once weekly and tumor volumes calculated. There were 10-12 animals per treatment group and the Mann-Whitney unpaired test used to analyze the data.

Orthotopic 253J-BV human bladder carcinoma model: The effect of GM1474 on the orthotopic outgrowth of a human bladder carcinoma in nude mice was assessed. 1×10^6 253J-BV bladder carcinoma cells were injected into the bladder wall of nude mice on day 0. On day 7 animals were randomly distributed, and treatment with GM1474 (50mg/kg/day, sc., 0.05ml injection volume) or vehicle started on a schedule of 5 days treatment, 2 days no treatment, for a total of 4 weeks. Animals were then sacrificed, the bladders removed, and tumor-bearing bladder weights determined.

Orthotopic human renal carcinoma metastasis assay: SN12pm6 cells were harvested at 70% confluency with trypsin-EDTA (GIBCO), washed with HBSS and resuspended at a concentration of 20×10^6 cells/ml for injection. Animals were injected with 1×10^6 SN12pm6 cells in the left renal subcapsular space (5-9 animals/treatment group). Animals were randomly distributed prior to assignment to treatment groups. Animals recieved 100mg/kg GM1474 or vehicle (PBS) sc., on days 3-24, Nephrectomy of the tumor-bearing kidney was carried out on day 26. Animals were sacrificed ~21 days after nephrectomy, the lungs were placed in Bouin's fixative and the number of lung metastases counted using a dissecting microscope. Statistical analysis was carried out using the Mann-Whitney µ-test.

Orthotopic human colorectal carcinoma metastasis assay: KM12SM cells were harvested at 70% confluency with trypsin-EDTA (GIBCO), washed with HBSS and resuspended at a concentration of 20×10^6 cells/ml for injection. Animals were challenged with 1×10^6 cells in a volume of 0.1 ml intravenously (9-15 animals/treatment group). Animals were randomly distributed prior to assignment to treatment groups. Animals recieved 50mg/kg GM1474 or vehicle (PBS) sc., 5X/week on days 4-27 and cecectomy was performed on day 28. Animals were sacrificed 49 days after cecectomy, the livers were placed in Bouin's fixative and the incidence of liver metastases scored using a dissecting microscope. Chi-square statistical analysis was carried out.

RESULTS

In Vitro results

Previously we have shown that a heparin-derived hexasaccharide is the smallest fragment of heparin with significant bFGF-modulating activity *in vitro* (Tyrrell, et al., 1993). Based on this structural data, GM1474, a low molecular weight sulfated oligosaccharide was synthesized and evaluated *in vitro* for its bFGF-modulating and heparanase-inhibitory activity.

A panning procedure utilizing RO-12 UC cells for the evaluation of interactions between heparin-related molecules and bFGF has been described previously (Ishihara, et al., 1992). GM1474 was a potent inhibitor ($IC_{50} < 1\mu g/ml$) of RO-12 UC cell binding to immobilized bFGF in this assay (Table 1.). GM1474 also was an effective inhibitor of bFGF-stimulated rat adrenocortical endothelial cell proliferation (Table 1.).

When tested in the chick chorioallantoic membrane assay, GM1474 inhibited angiogenesis in the presence of hydrocortisone (Table 1.).

Table 1. *In Vitro* and *Ex Ovo* activities of GM1474

Assay	Results	
	GM1474	Heparin
	(IC_{50}, $\mu g/ml$)	
bFGF-UC cell binding assay	<1	1
bFGF-stimulated ACE cell proliferation	10	8-15
Heparanase assay	4.5	5.0
Chorioallantoic membrane angiogenesis assay (score)	+/++	+++

Rat hepatoma cells were used to isolate heparanase, which was used to assess the inhibitory activty of GM1474 *in vitro* in a CPC precipitation assay with radiolabelled heparan sulfate as previously described (LaPierre, *et al.*, 1995). Results indicated that GM1474 inhibited heparanase with the same potency as heparin (IC_{50}=4.5μg/ml, Table1.).

Routine testing of GM1474 for direct antiproliferative effects on the various tumor cell lines was carried out *in vitro*. With the exception of the SK Hep-1 and the 253JBV cell lines, GM1474 did not significantly inhibit *in vitro* proliferation of all other cell lines tested (data not shown). In the presence of 10μg/ml of GM1474, the SK-Hep-1 tumor cell line was inhibited ~ 54% at 97 hours. The 253J-BV tumor line was inhibited ~ 35% under similar conditions.

In Vivo results

A variety of human and murine tumor growth and metastasis models were used to assess the activity of GM1474 *in vivo*. In the B16-F10 murine melanoma experimental metastasis assay, GM1474 (100 mg/kg/day, sc, days 0-4) significantly prolonged the survival of C57/Bl6 mice challenged by intravenous injection with 5×10^4 B16-F10 tumor cells versus vehicle control treated animals (Fig. 1).

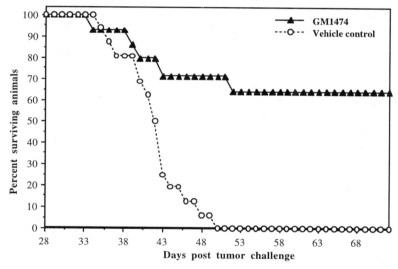

Figure 1. Effect of GM1474 on survival in a B16-F10 murine experimental metastasis assay. Animals recieved GM1474 (100mg/kg/day, sc.) on days 0-4.

The ability of GM1474 to inhibit the sc. growth of the MDA-231 human mammary adenocarcinoma in nude mice was evaluated. MDA-231 sc. tumor growth was significantly inhibited in animals in the GM1474 treatment group versus vehicle control treated animals, with a maximal inhibition of 59% occuring on day 62 post tumor challenge (Fig. 2).

In the PC-3 human prostatic adenocarcinoma tumor growth model, there was moderate inhibition of PC-3 sc. tumor growth in the GM1474 -treated group versus the vehicle control group with maximal inhibition of 37% occurring at day 20 (p<0.05, Fig. 3).

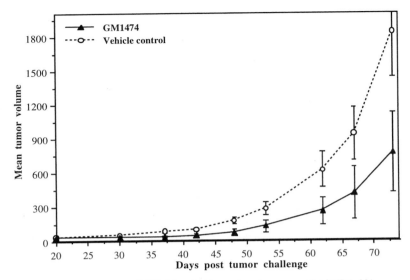

Figure 2. Effect of GM1474 (50mg/kg/day, sc.) on the sc. growth of MDA-231 human mammary carcinomas in nude mice. Animals recieved treatment on days 8-73, alternating 5 days treatment, 2 days no treatment.

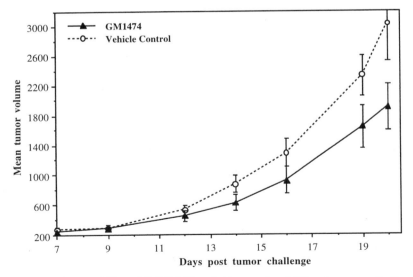

Figure 3. Effect of GM1474 (80mg/kg/day, sc., days 1-23) on the sc. growth of PC-3 human prostatic carcinomas in nude mice.

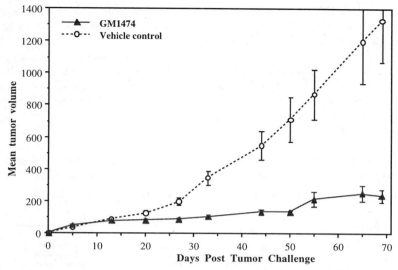

Figure 4.   Effect of GM1474 (50mg/kg/day, sc.) on the sc. growth of SK-Hep-1 carcinomas in nude mice. Treatment was recieved on days 1-33, on a schedule of 5 days treatment, 2 days no treatment.

GM1474 significantly inhibited sc. outgrowth of SK-Hep-1 tumors beginning on day 20, with maximal inhibition of 82% occuring on day 69 (p<0.05, Fig. 4).

The effect of GM 1474 on the orthotopic growth of the 253J-BV human bladder carcinoma that is routinely maintained by *in vivo* passage in nude mice was also assessed. GM1474 (50 mg/kg/day, sc.) inhibited orthotopic growth of the 253J-BV carcinoma in the mouse bladder wall >60% (p<0.001, Fig. 5).

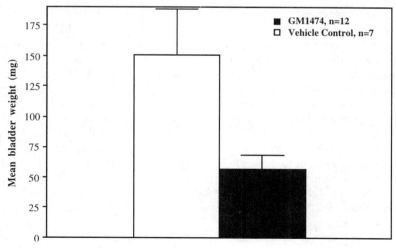

Figure 5.   Effect of GM1474 (50mg/kg/day, sc.) on the orthotopic growth of the 253J-BV human bladder carcinoma in nude mice. Animals recieved compound on days 7-xx, 5 days treatment, 2 days no treatment. The average bladder weight is 20-25 mg.

In the SN12pm6 human renal cell carcinoma orthotopic spontaneous lung metastasis model, GM1474 (100mg/kg/day, sc.) inhibited the incidence of spontaneous lung metastases in nude mice >90% (p<0.001, Table 2).

In the KM12L4a human colorectal carcinoma orthotopic model, which assesses spontaneous liver metastasis formation, GM1474 significantly inhibited the incidence of liver metastases. 6/6 GM1474-treated animals had no evidence of liver metastasis formation versus 8/8 animals with liver metastasis in the vehicle control group (p<0.05, Table 2).

Table 2. Antimetastatic activity of GM1474 in the SN12pm6 human renal and KM12L4a human colorectal orthotopic spontaneous metastasis models in nude mice

Treatment	Tumor Model	
	SN12pm6 (average number of lung metastases, $\pm$ S.E.M.)	KM12L4a (incidence of liver metastasis)
GM1474	5 ± 2[1]	0/6
Vehicle control	81 ± 10[2]	8/8[3]

[1]3/10 animals had no visible lung metastases.
[2]6/10 animals had >100 visible lung metastases.
[3]Metastatic burden ranged from >50% of the liver involved to ~10 small to medium sized metastases evident (<0.5 ~ 2 mm in diameter).

DISCUSSION

With the exception of the SK-Hep-1 and 253J-BV tumor lines, GM1474 did not inhibit tumor cell proliferation *in vitro*. This suggests that GM1474's antitumor activity is not primarily via a cytotoxic mechanism, but involves two distinct mechanisms: a) bFGF growth factor antagonism, preventing bFGF receptor dimerization and signal transduction (Spivak-Kroizman, *et al.*, 1994) and b) significant heparanase-inhibitory activity. These activities would facilitate the angiostatic activity of GM1474 by inhibition of endothelial cell proliferation and migration, while inhibition of heparanase could prevent tumor cell invasion as well as angiogenesis (Vlodavsky, *et al*, 1991). GM1474's ability to prolong survival in the B16-F10 murine melanoma experimental metastasis model suggests that the heparanase inhibitory activity of GM1474 and not its bFGF antagonism prevents tumor cell invasion and subsequent metastasis to the lungs in this assay system (Irimura, *et al.*, 1986).

In the case of the *in vitro* antiproliferative activity noted for GM1474 with the SK-Hep-1 and 253J-BV tumor lines, it is possible that these cells are bFGF responsive. SK-Hep-1 cells have been reported to produce bFGF (M. Presta, personal communication) as does the 253J-BV bladder carcinoma (J.J. Killion, personal communication) and these lines may utilize bFGF as an autocrine growth factor, which may explain the *in vitro* activity of GM1474 with these tumor lines. Nonetheless, GM1474 was potent in both models and not only in tumor xenografts implanted sc., but also in an orthotopic site, which is more biologically relevant to the disease state.

The tumor growth inhibitory activity of GM1474 in the MDA231 and PC-3 models is not due to a cytotoxic effect. The *in vitro* and *ex ovo* data suggest that suppression of tumor growth in these models is due to inhibition of tumor angiogenesis, with the MDA231 model being more responsive to treatment with GM1474 than the PC-3 tumor model. This observation may be explained in part by the more aggressive rate of growth of the PC-3 tumor line versus the rate of tumor outgrowth of the MDA231 mammary adencarcinoma line, which could decrease the therapeutic opportunity for an angiostatic agent. It should be noted that there was extensive central tumor necrosis noted at necropsy in the PC-3 model and this tumor may be less dependent on angiogenesis for its growth.

While GM1474 demonstrated good efficacy in a murine experimental metastasis model, a more biologically relevant setting would be to test the agent for its ability to inhibit spontaneous metastasis in human orthotopic models. Towards this objective, GM1474 was evaluated in two models of this type. The SN12pm6 human renal carcinoma model is an excellent system for this type of evaluation. This tumor line mimics the clinical progression of metastatic renal cancer, which primarily spreads to the lungs in humans. If the initial diagnosis indicates that the patient has metastatic renal cancer, partial or complete nephrectomy of the involved kidney may be carried out with followup adjuvant therapy. To date this treatment regimen has not met with great success in patients with advanced metastatic renal cancer. Similarly, the KM12L4a human colorectal model mimics the pattern of growth and spread of metastatic colorectal cancer in patients. This cancer often colonizes the liver resulting in the death of the patient. Again, the execution of this model has clinical parallels in that after outgrowth of the primary tumor, cecectomy is performed to remove the primary. This procedure does not prevent the occurrence of liver metastases in nude mice and thus allows the evaluation of the antimetastatic potential of a test agent. As has been shown, GM1474 is an effective inhibitor of metastasis in both of these models.

The antimetastatic activity of GM1474 in the renal and colorectal models was not via a direct cytotoxic mechanism, as is indicated by the lack of effect of GM1474 *in vitro* on cellular proliferation. The mechanisms that resulted in decreased lung and liver metastasis could be multiple. GM1474 may have exerted its effect by inhibition of tumor angiogenesis and subsequent escape of metastatic tumor cells from the primary site, or by inhibiting outgrowth of occult metastases that were able to seed the lungs or liver in these models. Another possibility is that the compound inhibited the invasiveness of the tumor cells, preventing the arrest of the tumor cells in the liver or lungs of the colorectal and renal models, respectively. None of these potential mechanisms are mutually exclusive, and the antimetastatic effects observed may have been due to a combination of the abovementioned activities, with varying degrees of dependence on one particular mechanism based on the model tested. An example of this potential variation in dependence is demonstrated by the observation that KM12 tumors that are propagated in the subcutis versus intracecally in nude mice express significantly lower levels of heparanase and that sc. KM12 tumors are nonmetastatic, while intracecal tumors do metastasize to the liver, suggesting that a heparanase antagonist may play a greater role in inhibition of liver metastasis in this model than a bFGF inhibitor (Nakajima, et al., 1990).

This series of studies has built on the biological and biophysical data generated previously on the structure and function of heparin-derived bFGF antagonists to generate a low anticoagulant, low molecular weight carbohydrate structure with multiple biological activities that inhibits angiogenesis *ex ovo*, has antitumor activity *in vivo*, yet is largely noncytotoxic for tumor cell lines *in vitro*. The potent activity of GM1474 *in vivo* and the evidence presented suggest that this molecule's mechanisms of action are to inhibit bFGF-stimulated angiogenesis and heparanase-mediated tumor and possibly endothelial cell migration and invasion. These activities, which can prevent tumor growth and metastasis, are thought to be the reason for the potent antitumor and antimetastatic activity of GM1474. Preliminary toxicology studies to date with this agent do not reflect any major limiting toxicities.

These data suggest that GM1474 may have efficacy in humans for metastatic solid tumors, which to date are generally poor responders to current conventional therapies. It is key that one realize that a noncytotoxic angiostatic agent such as GM1474 alone is not likely to induce regression of clinically significant tumor masses, but that agents of this type have great potential for facilitating disease stabilization, increasing time to recurrence and decreasing occurrence of detectable metastases in the patient when used as a nontoxic adjunct therapy.

REFERENCES

Castellot, J., Kambe, A.M., Dobson, D.E., and Spriegelman, B.M., 1986, Heparin Potentiation of 3T3-Adipocyte Stimulation: Angiogenesis Mechanisms of Action on Endothelial Cells, *J. Cell Physiol.* 127: 323-329.

Folkman, J., 1993, Tumor Angiogenesis, in: *Cancer Medicine*, (J.F. Holland, E. Frei III, R.C. Bast, D.W. Kufe, D.L. Morton, and R.R. Weichselbaum, eds.), Lea & Febiger, Philadelphia, pp. 153-170.

Folkman, J., and Klagsbrun, M., 1987, Angiogenic Factors, *Science* 235: 442-447.

Gualandris, A., and Presta, M., 1995, Transcriptional and Postransscriptional Regulation of Urokinase-Type Plasminogen Activator Expression in Endothelial Cells by Basic Fibroblast Growth Factor, *J. Cell. Physiol.* 162: 400-409.

Holmgren, L., O'Reilly, M.S., and Folkman, J., 1995, Dormancy of micrometastases: Balanced proliferaiton and apoptosis in the presence of angiogenesis suppression, *Nature Medicine* 1: 149-153.

Irimura, T., Nakajima, M., and Nicloson, G.L., 1986, Chemically Modified Heparins as Inhibitors of Heparan Sulfate Specific Endo-beta-glucuronidase (Heparanase) of Metastatic Melanoma Cells, *Biochemistry* 25: 5322-5328.

Ishihara, M., Tyrrell, D.J., Kiefer, M., Barr, P.J., and Swiedler, S.J., 1992, A Cell-Based Assay for Evaluating the Interaction of Heparin-like Molecules and Basic Fibroblast Growth Factor, *Anal. Biochem.* 202: 310-315.

Ishihara, M., Tyrrell, D.J., Stauber, G.B., Brown, S., Cousens, L.S., and Stack, R.J., 1993, Preparation of Affinity-fractionated, Heparin-derived Oligosaccharides and Their Effects on Selected Biological Activities Mediated by Basic Fibroblast Growth Factor, *J. Biol. Chem.* 268: 4675-4683.

LaPierre, F., Holme, K., Lam, L., Tressler, R.J., Storm, N., Wee, J., Stack, R.J., Castellot, J., and Tyrrell, D.J., 1996, Low Anticoagulant Heparin Derivatives which Inhibit Endoglycosidase (Heparanase) Activity Possess Anti-angiogenic, Anti-metastatic and Anti-tumor Properties, submitted, *Biochemistry*.

Morikawa, K., Walker, S.M., Nakajima, M., Pathak, S., Jessup, J.M., and Fidler, I.J., 1988, Influence of Organ Environment on the Growth, Selection, and Metastasis of Human Colon Carcinoma Cells in Nude Mice, *Cancer Res.* 48: 6863-6871.

Naito, S., Walker, S.M., and Fidler, I.J., 1989, In vivo selection of human renal cell carcioma cells with high metastatic potential in nude mice, *Clin. Expl. Metastasis* 7: 381-389.

Nakajima, M., Morikawa, K., Fabra, A., Bicana, C.D., and Fidler, I.J., 1990, Influence of Organ Environment on Extracellular Matrix Degradative Activity and Metastasis of Human Colon Carcinoma Cells, *J. Natl. Cancer Inst.* 82: 1890-1898.

Nicolson, G.L., 1984, An Introduction to Cancer Invasion and Metastasis, in: *Cancer Invasion and Metastasis: Biologic and Therapeutic Aspects*, (G.L. Nicolson and L. Milas, eds.), Raven Press, New York, pp. 1-4.

Spillmann, D., and Lindahl, U., 1994, Glycosiminoglycan-protein interactions: a question of specificity, *Curr. Opin. Struct. Biol.* 4: 677-682.

Spivak-Kroizman, T., Lemmon, M.A., Dikic, I., Ladbury, J.E., Pinchasi, D., Huang, J., Jaye, M., Crumley, G., Schlessinger, J., and Lax, I., 1994, Separin-Induced Oligomerization of FGF Molecules Is Responsible for FGF Receptor Dimerization, Activation, and Cell Proliferation, *Cell* 79: 1015-1024.

Teicher, B.A., Sotomayor, E.A., and Huang, Z.D., 1992, Antiangiogenic Agents Potentiate Cytotoxic Cancer Therapies against Primary and Metastatic Disease, *Cancer Res.* 52: 6702-6704.

Thorpe, P.E., Derbyshire, E.J., Andrade, S.P., Press, N., Knowles, P.P., King, S., Watson, G.J., Yang, Y.-C., and Rao-Bette', M., 1993, Heparin-Steroid Conjugates: New Angiogenesis Inhibitors with Antitumor Activity in Mice, *Cancer Res.* 53: 3000-3007.

Tyrrell, D.J., Ishihara, M., Rao, N., Horne, A., Kiefer, M.C., Stauber, G.B., Lam, L.H., and Stack, R.J., 1993, Sturcture and Biologicval Activities of a Heparin-derived Hexasaccharied with Hith Affinity for Baswic Fibroblast Growth Factor, *J. Biol. Chem.* 268: 4684-4689.

Vlodavsky, I., Fuks, Z., Ishai-Michaeli, R.I., Bashkin, P., Levi, E., Korner, G., Bar-Shavit, R., and Klagsbrun, M., 1991, Extracellular Matrix-Resident Basic Fibroblast Growth Factor: Implication for the Control of Angiogenesis, *J. Cell. Biochem.* 45: 167-176.

Vlodavsky, I., Michaeli, R.I., Bar-Ner, M., Fridman, R., Horowitz, A.T., Fuks, Z., and Biran, S., 1988, Involvement of Heparanase in Tumor Metastasis and Angiogenesis, *Israel J. Med. Sci.* 24: 464-470.

Weidner, N., 1993, Tumor angiogenesis: review of current applications in tumor prognostication, *Sem. Diagn. Pathol.* 10: 302-313.

Zagzag, D., 1995, Angiogenic Growth Factors in Neural Embryogenesis and Neoplasia, *Am. J. Pathol.* 146: 293-309.

INHIBITORS OF ANGIOGENESIS IN HUMAN URINE

Theodore Fotsis[1,6], Michael S. Pepper[2], Erkan Aktas[1], Antonia Joussen[3], Friedrich Kruse[3], Herman Adlercreutz[4], Kristina Wähälä[5], Tapio Hase[5], Roberto Montesano[2], and Lothar Schweigerer[1]

[1]Department of Hematology and Oncology, Children´s University Hospital, University of Heidelberg, INF 150, 69120 Heidelberg, Germany.
[2]Institute of Histology and Embryology, Department of Morphology, University Medical Center, 1211 Geneva 4, Switzerland.
[3]Department of Ophthalmology, University of Heidelberg, 69120 Heidelberg, Germany.
[4]Department of Clinical Chemistry, University of Helsinki, Meilahti Hospital, SF-00290 Helsinki, Finland.
[5] Department of Chemistry, University of Helsinki, Vuorikatu 20, SF-00100 Helsinki, Finland

[6] To whom correspondence should be addressed

INTRODUCTION

Angiogenesis, the formation of new blood vessels, is essential for normal development. However, in the adult organism angiogenesis is essentially absent with the exception of conditions such as the cyclical changes in the female reproductive tract and the response to wounding. In these situations the co-ordinated, sequential cellular events leading to formation of new capillaries are tightly regulated both spatially and temporally (Klagsbrun and Folkman, 1990). In contrast, many diseases are driven by persistent, unregulated formation of new capillaries resulting in destruction of normal tissue function and architecture (Klagsbrun and Folkman, 1990; Folkman and Shing, 1992). In rheumatoid arthritis, a highly vascularized pannus invades the joint and destroys cartilage. In diabetes, new capillaries in the retina invade the vitreous, bleed, and cause blindness. Ocular neovascularization is the most common cause of blindness. Formation of new blood vessels is also critical for the growth of tumors (Klagsbrun and Folkman, 1990; Folkman et al., 1989; Blood and Zetter, 1990). Indeed, avascular tumors do not grow beyond a diameter of 1 to 2 mm (Folkman and Cotran, 1976; Folkman, 1985). Furthermore, new tumor vessels provide a port of exit for tumor cells to metastasize to distant sites (Weidner et al., 1991). These diseases were recently called angiogenic diseases and share at least two characteristics in common, i) an abnormality of capillary blood vessel growth is the principal pathological

Molecular, Cellular, and Clinical Aspects of Angiogenesis
Edited by Michael E. Maragoudakis, Plenum Press, New York, 1996

213

feature and ii) therapeutic control of the abnormal capillary growth would ameliorate or eliminate other manifestations of the disease.

Recognition of the importance of angiogenesis has led to the search for angiogenesis inhibitors, which, by stopping the process of persistent, unregulated, pathological neovascularization, could find important applications in the treatment of angiogenic diseases, including cancer. The human organism contains a number of endogenous angiogenesis inhibitors both of proteinic and lipophilic structure (Blood and Zetter, 1990; Bouck, 1990; Schweigerer and Fotsis, 1992). The latter, because of the small molecular size, are excreted in urine in hydrophilic conjugated forms. Examples of endogenous urinary-excreted angiogenesis inhibtors are cortisone and its dihydro and tetrahydro derivatives (Crum et al., 1985) and vitamin D_3 metabolites (Oikawa et al., 1990; Majewski et al., 1993). Moreover, urine is also the excretory route of the metabolites of the numerous dietary-ingested compounds. These phytochemicals are postulated to be, at least in part, responsible for the long-known preventive effect of plant-based diets on tumorigenesis and other chronic diseases (Adlercreutz, 1990). Epidemiological data have clearly demonstrated that dietary factors contribute to about a third of potentially preventable cancers (Miller, 1990).

To examine the possibility that urine might contain potential antiangiogenic compounds, we screened urine of human subjects consuming a diet rich in plant products for the presence of such compounds using a purification procedure developed by us previously to separate lipophilic compounds of steroid or steroid-like structure (Fotsis and Adlercreutz, 1987; Fotsis, 1987; Adlercreutz et al., 1991a). Six fractions (1-6) were obtained from the purification of urine (Fig 1) and their effect on bFGF-stimulated proliferation of bovine brain-derived capillary endothelial (BBCE) cells was tested (Fotsis et al., 1993). Fractions 1, 2, and 4 were able to inhibit the proliferation of endothelial cells. Fraction 4 was not further examined as it contained neutral steroids (Fotsis and Adlercreutz, 1987), which are known angiogenesis inhibitors (Crum et al., 1985).

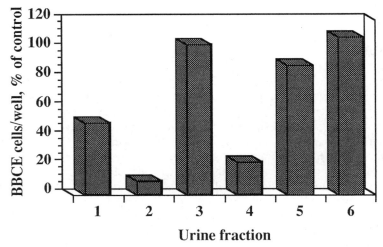

Figure 1. Effect of semi-pure urine preparations on the proliferation of vascular endothelial cells. Urine was purified as described (Fotsis and Adlercreutz, 1987; Fotsis, 1987; Adlercreutz et al., 1991a) and the fractions obtained (1-6) contained the following compound groups: 1. Diphenolic compounds I, 2. Compounds with vicinal-cis hydroxyls, 3. Diphenolic compounds II, 4. Neutral compounds, 5. Non-polar monophenolic compounds and 6. Polar monophenolic compounds. (Fotsis et al., 1995)

In this report, we summarize the results of the investigation of fractions 1 and 2 . This work led to the identification of the isoflavonoid genistein (Fotsis et al., 1993) and the endogenous estrogen metabolite 2-methoxyestradiol as potential antiangiogenic compounds (Fotsis et al., 1994). We also provide preliminary results on the identification of metabolites of flavonoids that show potent inhibitory activities on *in vitro* angiogenesis (Fotsis et al. in preparation).

ISOFLAVONOIDS

Using GC-MS (gas chromatography-mass spectrometry), initial identification was carried out by comparing the mass spectra of the different peaks with those of reference mass spectra collections. Synthetic standards of potential candidate compounds were then prepared and definite identification was established by comparing their mass spectra with those of the unknown compounds. In this way, we were able to demonstrate the presence of the isoflavonoids genistein, daidzein and O-desmethylangolensin in fraction 1 and the lignans enterodiol, enterolactone and matairesinol as well as the isoflavonoid equol in fraction 3 (Fotsis et al., 1993).

Synthetic genistein targets proliferating but not quiescent cells

The synthetic analogs of the identified compounds were then tested with regard to their effect on the bFGF-stimulated proliferation of BBCE cells. Genistein had a potent and dose-dependent inhibitory effect on BBCE cell proliferation with half-maximal and maximal effects at 5 and 50 μM concentrations, respectively. In contrast, the remaining isoflavonoids, some of which are closely related to genistein, were considerably (5-20 times) less potent. Genistein also inhibited the proliferation of vascular endothelial cells derived from bovine adrenal cortex (ACE) or aorta (BAE) (Fig. 2A). However, endothelial cells were not the only target of genistein. The proliferation of low density cultures of various normal (NIH-3T3 mouse embryonic fibroblasts) (Fig 2A) and tumor (SH-EP and Kelly human neuroblastoma, SK-ES1 Ewing's sarcoma, RD and A-204 human rhabdomyosarcoma and Y-79 human retinoblastoma) (Fig. 2B) cells was also inhibited by genistein with half-maximal concentrations varying between 10-45 μM (Schweigerer et al., 1992; Fotsis et al., 1995). Though endothelial cells are slightly more sensitive ($IC_{50} = 6$ μM) to the inhibitory effect of genistein, it appears that genistein has a broader inhibitory effect on proliferating cells.

In low-density cultures of proliferating endothelial cells, genistein induced marked morphological changes: at concentrations up to 25 μM, genistein induced a highly spread morphology compatible with growth arrest ; cell densities always exceeded those determined at seeding (data not shown), indicating that cell death was not involved. When exposed to genistein concentrations above 25 μM, the cells acquired an elongated morphology, and eventually died (Fotsis et al., 1993). These results indicate that genistein is cytostatic up to concentrations of approximately 25 μM and that it becomes cytotoxic above this level. Reversibility experiments, in which low-density cultures of endothelial cells were exposed to bFGF (2.5 ng/ml) and increasing concentrations of genistein and then to medium without genistein, confirmed this assumption (Fotsis et al., 1995). The result was the same on uncoated and gelatin coated substrata. In contrast, confluent, quiescent endothelial cells did not exhibit toxicity signs even at genistein concentrations up to 200 μM (Fotsis et al., 1995). These data clearly suggest that genistein targets only proliferating cells leaving quiescent, non-dividing cells unaffected. This property may be important with respect to possible use of the compound in therapeutic applications, as few side effects are likely to be encountered.

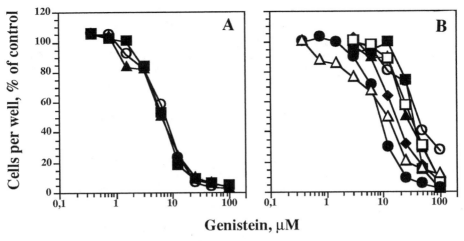

Figure 2. Effect of genistein on the proliferation of various normal and tumor cells. A). ACE (■), BAE (O) cells were seeded at a densitiy of 5,000 cells per well, and every other day received 10 μl aliquots of bFGF (2.5ng/ ml) and of buffer with or without the indicated concentrations of genistein. NIH-3T3 cells (▲)were seeded at a density of 10,000 cells per well, and received every other received day 10 μl of buffer with or without the indicated concentrations of genistein. B) Kelly (■) and SH-EP (●) neuroblastoma , A-204 (♦), A-673 (□) and RD (▲) rhabdomyosarcoma, SK-ES1 Ewing's sarcoma (O), and Y-79 retinoblastoma (△) cells were seeded at densities of 10,000 cells per well, and received every other day received 10 μl of buffer with or without the indicated concentrations of genistein. Cells in A) and B) were counted after 5 to 6 days (Fotsis et al., 1995).

Genistein inhibits angiogenesis *in vitro* and *in vivo*

Having established the inhibitory effect of genistein on the proliferation of endothelial cells, it was interesting to investigate whether genistein had additional effects on other functions of endothelial cells important for angiogenesis. Angiogenesis is a complex process requiring the co-ordinated, sequential involvent of a number of cellular events other than proliferation. Indeed, formation of new capillaries begins with a localized breakdown of the basement membrane of the parent vessel, through the finely-tuned elaboration of proteolytic enzymes and their inhibitors (Pepper and Montesano, 1990), followed by migration of endothelial cells and invasion of the surrounding matrix. The initial sprout further elongates as a result of continued migration and endothelial cell replication. Once a lumen has been formed, the capillary fuses with the tip of another maturing sprout to form a functional capillary loop (Folkman and Klagsbrun, 1987). Proteolytic degradation of the extracellular matrix by endothelial cells is controlled by angiogenic factors, like bFGF, which induce the production of urokinase type plasminogen activator and its physiological inhibitor, plasminogen activator inhibitor-1 (reviewed by Pepper and Montesano, 1990). As demonstrated (Fotsis et al., 1993), genistein markedly reduced both bFGF-stimulated and basal levels of both plasminogen activator and plasminogen activator inhibitor-1 activity in BME (bovine microvascular endothelial) cells. Moreover, genistein inhibited the bFGF-induced migration of endothelial cells in wounded confluent monolayers of endothelial cells (data not shown). Inhibition of proteolytic enzyme production and migration of endothelial cells by genistein, therefore, reflects a more complex interference of the compound with important early events of angiogenesis other than endothelial cell proliferation.

The combined effects of genistein on proliferation, proteolytic enzyme production and migration of endothelial cells prompted us to investigate the effects of the compound in an experimental *in vitro* system which mimics angiogenesis *in vivo*. As previously shown (Montesano et al., 1986), BME cells seeded on the surface of collagen gels (Fig. 3A) invade the gels when exposed to bFGF and form capillary-like tubes beneath the gel surface (Fig.3B). Genistein alone at 200 μM concentrations had no effect on confluent BME cultures (Fig. 3C). However, when added together with bFGF, it inhibited their ability to invade the gels and generate capillary-like structures (Fig. 3D). An initial quantitative analysis revealed that genistein inhibited bFGF-induced invasion of BME cells with a half-maximal effect at approximately 150 μM concentration (Fotsis et al., 1993). In subsequent studies we have observed that the half-maximal concentration is approximately 8 μM, closely correlating with the value of the *in vitro* antiproliferative effect on endothelial cells (6 μM). The higher value reported in our early studies is a consequence of poor solubility obtained with inappropriate diluents (M.S. Pepper, unpublished observation).

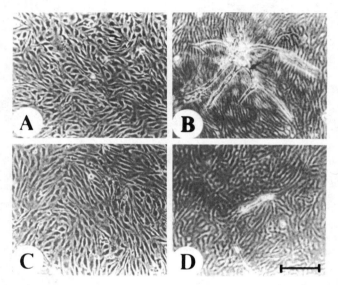

Figure 3. Effect of genistein on *in vitro* angiogenesis. A-D) Three-dimensional collagen gels were prepared in 18-mm tissue culture wells as described (Montesano and Orci, 1985). 5×10^4 BME cells (in a volume of 0.5 ml) were seeded into each well, and medium was changed every 2-3 days until the cells were confluent. After confluence, they received either no additions (A), bFGF (30 ng/ml) only (B), genistein (200 μM) only (C) or genistein (200 μM) 2 h prior to bFGF (30 ng/ml) (D). After 2 days of incubation, medium was changed and cells exposed to the same conditions again. Representative pictures were taken after another 3 days. In (B), the arrow shows the lumen of a capillary-like tube and in (D), genistein inhibits the formation of such tubes (Fotsis et al., 1993)

Next, we investigated whether genistein could inhibit angiogenesis in an *in vivo* system. Assessment of neovascularization in the rabbit cornea is a well established and widely used *in vivo* angiogenesis assay (Rastinejad et al., 1989). In the present study, a modified corneal micropocket assay in NZW rabbits was employed to assess the *in vivo* antiangiogenic effects of genistein. Originating from a central incision, a tunnel was prepared up to 2 mm central to the limbus. Methylcellulose pellets containing 200 ng bFGF were inserted and the incision was closed. Daily subconjuctival injections (0.1 ml) of genistein (0.4 mg/ml in sterile Ringer solution) significantly (p<0.001) reduced the number of vessels at the limbus and 1 mm central to the limbus as well as the total vascularized area (Fig. 4). These results clearly indicate that genistein can inhibit several functions of endothelial cells that play a critical role during angiogenesis. The same experiment was carried out using Hydron pellets with identical results.

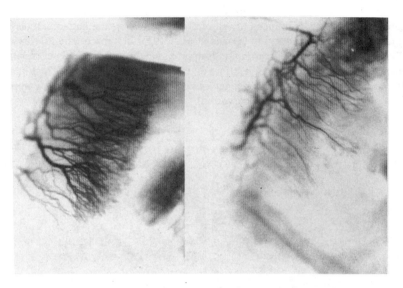

Figure 4. Effect of genistein on the neovascularization assay in NZW rabbit corneas. Daily subconjuctival injections (0.1 ml) of genistein (0.4 mg/ml in sterile Ringer solution significantly (p<0.001) reduced the number of bFGF-induced vessels at the limbus and 1 mm central to the limbus as well as the total vascularized area (right panel) comparing to the other eye of the rabbit treated only with buffer (left panel).

Possible molecular mechanisms of action

The cytostatic and cytotoxic effects of genistein on proliferating, but not quiescent, cells of different origin suggests that this compound interferes with a proliferation-associated event. Though we have not addressed the mechanism of action of genistein , there are several known properties of genistein which point towards possible molecular events affected by it. Genistein could inhibit basal or bFGF-stimulated endothelial cell proliferation and *in vitro* angiogenesis through attenuation of the activity of tyrosine kinases. Indeed, genistein has been shown to be a competitive inhibitor of ATP binding to the catalytic domain of tyrosine kinases (Akiyama et al., 1987) and was found to inhibit EGF receptor (Akiyama et al., 1987) and PDGF receptor (Hill et al., 1990) tyrosine kinase activities both in intact cells and *in vitro*. This appears as an attractive hypothesis, since high affinity FGF receptors are tyrosine kinases (Ullrich and Schlessinger, 1990) and since vanadate, an inhibitor of phosphotyrosine phosphatases, induces angiogenesis in the collagen gel assay Montesano et al., 1988). Alternatively, genistein might elicit its action on proliferating cells by attenuating the activity of S6 kinase(Linassier et al., 1990), an enzyme which is also activated by bFGF (Blackshear et al., 1985; Pelech et al., 1986). As a third possibility, genistein might exert its effects by modulating the activity of topoisomerases I and II (Yamashita et al., 1990; Okura et al., 1988), enzymes which are involved in nuclear events like transcription, replication, recombination and mitotic chromosome segregation. It is, however, possible that genistein could inhibit proliferation via other, hitherto unknown molecular mechanisms which remain to be discovered.

Biological relevance

The effects of genistein on cell proliferation and angiogenesis raise the possibility that this compound may play a physiological role in the organism. Genistein precursors are present in soy products (Setchell and Adlercreutz, 1988) and genistein itself is present at high concentration in urine of individuals consuming a traditional soy-rich Japanese diet (about 6 μmol/24h) (Adlercreutz et al., 1991b), but at more than 30-fold lower concentration in urine of omnivores (0.18 μmol/24h) (Adlercreutz et al., 1991a) (Table 1). It is therefore tempting to speculate that genistein might contribute to the long-known preventive effect of plant-based diet on chronic neovascular diseases including solid tumor growth. In the latter case, genistein may have both direct effects on tumor cell proliferation and indirect effects through inhibition of tumor-induced neovascularization, which is so crucial for the development of a solid tumor. In this context it is interesting that Japanese women and women of Japanese origin in Hawaii consuming a diet similar to the original traditional Japanese diet have low breast cancer incidence and mortality (Muir et al., 1987). Also, men of Japanese origin in Hawaii have low mortality from prostate cancer although the incidence of *in situ* prostate cancer in autopsy studies is similar to that of men in Western societies (Severson et al., 1989). Finally, soy products, which are important constituents of Japanese diet, inhibit mammary tumorigenesis in rat models (Baggot et al., 1990; Troll et al., 1980; Barnes et al., 1990).

ESTROGENS

Fraction 2, which exhibited a strong inhibitory activity on endothelial cell proliferation contained catecholic estrogen metabolites, suggesting their involvement in the estrogen-dependent regulation of growth and angiogenesis. We therefore examined synthetic standards of these and other related estrogen metabolites for their effects on basic fibroblast growth factor (bFGF) -induced proliferation of bovine brain capillary endothelial cells (BBCE) *in vitro* (Table 1). 2-Methoxyestradiol exhibited the strongest inhibitory effect of all metabolites tested, the half-maximal concentration being approx. 0.15 μM (Table 1) The inhibition appeared to be specific to the 2-methoxyestradiol molecule as isomeric and closely related structures were 40 to 250 times less potent, with some estrogen metabolites having no effect at all (Fotsis et al., 1994).

2-Methoxyestradiol inhibits invasion, migration and proliferation of endothelial cells *in vitro*

The inhibitory effect of 2-methoxyestradiol was essentially the same (data not shown) on different types of endothelial cells (bovine aortic, bovine adrenal cortex microvascular (BME), human umbilical vein and human skin) with half-maximal inhibitory concentrations varying between 0.2 and 2.5 μM. 2-Methoxyestradiol reversibly inhibited endothelial cell proliferation at concentrations up to 3-5 μM, being increasingly cytotoxic (reduction of cell number below the initial seeding density) at higher doses (data not shown). Confluent cultures of endothelial cells were, however, virtually unaffected by 2-methoxyestradiol; marginal cytotoxicity was observed at 100 μM. Interestingly, confluent cultures of endothelial cells unable to further divide exhibited cytotoxicity similar to low density cultures, when treated simultaneously with bFGF (data not shown). It appears, therefore, that 2-methoxyestradiol only targets cells stimulated by bFGF or other growth factors. In addition to the effects on proliferation, 2-methoxyestradiol also had effects on other functions of endothelial cells important for angiogenesis. 2-Methoxyestradiol inhibited the *in vitro* migration of BBCE and BME cells (data not shown) and the ability of BME cells to invade collagen gels and form capillary-like structures (Montesano et al., 1986), at concentrations comparable to those that inhibited their *in vitro* proliferation; the half maximal inhibition of *in vitro* angiogenesis was approximately 0.35 µM (Fotsis et al., 1994).

Table 1. Effect of 2-methoxyestradiol and related structures on the proliferation of endothelial cells (Fotsis et al., 1994).

Trivial name	Systematic name	IC 50 (µM)
2-Methoxyestradiol	1,3,5(10)-Estratriene-2,3,17β-triol 2-methyl ether	0.134
2-Hydroxyestrone	1,3,5(10)-Estratriene-2,3-diol-17-one	5.9
4-Methoxyestradiol	1,3,5(10)-Estratriene-3,4,17β-triol 4-methyl ether	7.24
2-Methoxyestradiol 3-methyl ether	1,3,5(10)-Estratriene-2,3,17β-triol 2,3-dimethyl ether	7.78
2-Methoxyestrone	1,3,5(10)-Estratriene-2,3-diol-17-one 2-methyl ether	11.5
2-Hydroxyestradiol	1,3,5(10)-Estratriene-2,3,17β-triol	15.7
2-Methoxyestriol	1,3,5(10)-Estratriene-2,3,16α,17β-tetrol 2-methyl ether	21.3
2-Hydroxyestradiol 3-methyl ether	1,3,5(10)-Estratriene-2,3,17β-triol 3-methyl ether	21.42
Estrone	1,3,5(10)-Estratriene-3-ol-17-one	26
16-Epiestriol	1,3,5(10)-Estratriene-3,16β,17β-triol	31
16α-Hydroxyestrone	1,3,5(10)-Estratriene-2,16α-diol-17-one	31.1
Estriol	1,3,5(10)-Estratriene-3,16α,17β-triol	32
Estradiol-17β	1,3,5(10)-Estratriene-3,17β-diol	34.5
15α–Hydroxyestriol	1,3,5(10)-Estratriene-3,15α,16α,17β-tetrol	n.i
Estriol-16α-glucuronide	1,3,5(10)-Estratriene-3,16α,17β-triol 16α-glucosiduronate	n.i.

n.i : no inhibition

The inhibitory effects of 2-methoxyestradiol were not restricted to endothelial cells. The proliferation of low density cultures of various normal (bovine granulosa, NIH-3T3 mouse embryonic fibroblast and HFK2 human skin fibroblast) and tumor (SH-EP human neuroblastoma, A 204 human rhabdomyosarcoma and Y-79 human retinoblastoma) cells were inhibited with half-maximal concentrations between 0.35 and 2.2 µM (data not shown). Again, quiescent confluent monolayers of HFK2 cells were virtually unaffected at doses up to 100 µM (Fotsis et al., 1994). These results indicate that actively proliferating cells are the target of 2-methoxyestradiol, and that the compound probably interferes with a proliferation-associated cellular event.

2-Methoxyestradiol suppresses angiogenesis and tumor growth *in vivo*

The *in vitro* effects of 2-methoxyestradiol on angiogenesis and proliferation of malignant cells prompted us to investigate the *in vivo* properties of the compound (Fotsis et al., 1994). 2-Methoxyestradiol, administered orally in mice, inhibited the growth of tumors arising from subcutaneously injected Meth A sarcoma and B16 melanoma cells (Fig. 5 A). The suppressive effect of 2-methoxyestradiol on tumor growth appeared to be a consequence of inhibition of tumor-induced angiogenesis rather than a result of direct inhibition of the proliferation of tumor cells. This was assessed by quantitating the vasculature of tumors by means of microspheres injected in the left ventricle shortly prior to sacrificing the mice and excising the tumors. After determination of tumor weights, the tumor tissue was dissolved and the microparticles were counted. There was a significant reduction of the number of microspheres per g tumor tissue in the treated compared to the control tumors (Fig. 5 B). The result was confirmed by visualizing the vessels with indian ink. Apart from their marginally lower weight (approx. 15 %), the treated mice exhibited no apparent signs of toxicity and were all alive after 12 days of daily treatment. We have not observed hair loss, intestinal disturbance or infection, toxic side-effects associated with conventional chemotherapy. The lack of toxicity is in agreement with the *in vitro* results showing no effect on quiescent, non-dividing cells.

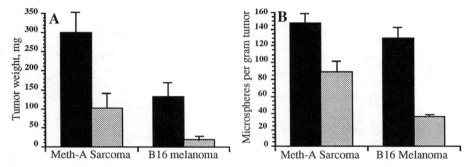

Figure 5. Inhibition of growth (A) and neovascularization (B) of tumors by oral administation of 2-methoxyestradiol in mice. (A) Meth-A sarcoma or B16 melanoma cells (1 x 10[6]) were inoculated subcutaneously in 0.1 ml saline in the dorsal skin of C3H mice (n = 20). On the same day, the mice received orally either 2-methoxyestradiol, at 100 mg per kg in body weight, suspended in 300 μl of olive oil (n = 10) or 300 μl olive oil alone (n = 10). This treatment was carried out every day and the diameter of the tumors was monitored every second day . On day 12 the mice were sacrificed and tumor weights determined. The weight in mg ± SD of both the Meth-A sarcoma and B16 melanoma tumors on day 12 is presented. (□, 2-methoxyestradiol-treated; ■, control. (B) In six control (■) and 2-methoxyestradiol-treated (□) mice the neovascularization of the Meth-A sarcoma and B16 melanoma tumors was quantitated by injecting 5 x 10[5] microspheres (10 μm, E-Z Trac, Interactive Medical Technology Ltd, Los Angeles, CA, USA) into the left ventricle 5 minutes before sacrificing the mice. After determination of tumor weights (see above), the tumor tissue was dissolved with a sodium hydroxide-SDS solution and the microspheres were counted microscopically. The degree of neovascularization of the treated and control tumors is expressed as microspheres per g of tumor tissue ± S (Fotsis et al., 1994).

Possible mechanisms of action

Unlike the angiostatic steroids of corticoid structure (Crum et al., 1985), heparin or sulphated cyclodextrins are not required for antiangiogenic activity of 2-methoxyestradiol indicating a different mechanism of action. The results of Table 1 and the negligible affinity of 2-methoxyestradiol for the estrogen receptor (MacLusky et al., 1983) also exclude the possibility that the effects of this steroid on endothelial cells are mediated through interactions with the estrogen receptor. We have observed, by immunofluorescence, disruption of microtubules, but not actin microfilaments, vimentin intermediate filaments, nor vinculin containing adhesion plaques in endothelial cells treated with 2-methoxyestradiol (data not shown), suggesting that abnormal microtubule assembly might be responsible for the effects of this compound on proliferating cells (Seegers et al., 1989). Indeed, it has been shown that 2-methoxyestradiol inhibits tubulin polymerization by interacting at the colchicine site (D'Amato et al., 1994).

FLAVONOIDS

GC-MS analysis revealed that fraction 2, in addition to catechol estrogens, contained also several flavonoids. Since flavonoids are isomeric compounds to the isoflavonoids, and because they are more abundant in nature compared to the latter (Harborne, 1973), we decided to investigate their potential antiangiogenic activity. In this communication we will present some preliminary results of these studies which are still ongoing (Fotsis et al., in preparation).

Table 2. The chemical formulae and the names of the flavonoids tested.

Chemical formula	Name	Substitution					
		5	7	2'	3'	4'	5'
Flavones							
	FLAVONE	H	H	H	H	H	H
	5-METHOXYFLAVONE	OCH₃	H	H	H	H	H
	CRYSIN	OH	OH	H	H	H	H
	2',3'-DIHYDROXYFLAVONE	H	H	OH	OH	H	H
	3',4'-DIHYDROXYFLAVONE	H	H	H	OH	OH	H
	APIGENIN	OH	OH	H	H	OH	H
	LUTEOLIN	OH	OH	H	OH	OH	H
	LUTEOLIN-7-GLUCOSIDE	OH	OGlc	H	OH	OH	H
Flavanones							
	HESPERETIN	OH	OH	H	OH	OCH₃	H
	ERIODICTYOL	OH	OH	H	OH	OH	H
Flavonols							
	3-HYDROXYFLAVONE	H	OH	H	H	H	H
	FISETIN	H	OH	H	OH	OH	H
	QUERCETIN	OH	OH	H	OH	OH	H
	MYRISETIN	OH	OH	H	OH	OH	OH

Catechin

Coumarin

In order to evaluate this compound class, we have screened a series of flavonoid structures (Table 2), obtained commercially or synthesized by us, with regard to their ability to inhibit the proliferation of endothelial and various other normal and tumor cells. The aim of this screening was to establish i) the antiproliferative effect of flavonoid structures *in vitro*, ii) the selectivity of the antiproliferative effect with regard to endothelial cell replication and iii) the structure-activity correlation of the possible effects. The preliminary results of this evaluation are seen in Table 3. Some of the flavonoids have stronger antiproliferative effects than genistein. Moreover, the antimitotic effect of flavonols exhibited a selectivity towards endothelial cells. In every case, however, it was evident that the heterocyclic ring C and especially the keto group in position 4 plays an important role in the antiproliferative effects of the flavonoids. Removal of this group in catechin resulted in complete abolishment of the antiproliferative effects. Further *in vitro* and *in vivo* experiments are underway to explore the antiangiogenic potential of this interesting compound group.

Table 3. The antiproliferative effects of flavonoids on various normal and tumor cells. Half-maximal concentrations in μmol/l.

	BBCE	HFK2	HaCaT	MCF7	SHEP
Flavones					
Flavone	8.5				
5-Methoxyflavone	12.0				
3´,4´-Dihydroxyflavone	1.8		1.5	0.5	5.0
Chrysin	9.0	25.0	20.0	7.0	5.0
Apigenin	6.5	10.0	6.0	3.5	4.0
Luteolin	2.0	4.5	4.0	6.0	4.0
Luteolin-7-glucoside	18.0		6.0	6.0	4.0
Flavonoles					
3-Hydroxyflavone	4.0				
Fisetin	4.0	20.0	20.0	15.0	7.0
Quercetin	3.5	22.0	>50.0	10.0	8.0
Myricetin	28.0	40.0	40.0	n.e.	
Flavanones					
Eriodictyol	7.0	>50.0	15.0	18.0	10.0
Hesperetin	30.0	n.e.	n.e.	40.0	>50.0
Catechin	>50.0	>50.0	n.e.	n.e.	n.e.
Cumarin	n.e				

n.e. = no effect
BBCE : Bovine brain capillary endothelial cells
HFK2 : Human fibroblasts
HaCaT : Human keratinocytes
MCF7 : Breast cancer cells
SHEP : Neuroblastoma cells

CONCLUDING REMARKS

The results presented in this communication clearly show that human urine contains several endogenous and dietary derived compounds which exhibit potent antiangiogenic and antimitotic properties. Irrespective of whether or not these substances play any physiological role in humans, they might be suitable as pharmacological agents. The lack of toxicity on confluent cells *in vitro* and their potent inhibitory effect on angiogenesis and tumor cell proliferation, suggest that they could be used in the treatment of solid malignant tumors and other angiogenic diseases. Further work should focus on the investigation of more potent synthetic analogs of the compounds and their detailed *in vivo* effects on angiogenesis and tumorigenesis. If one of the original substances or their analogs succesfully passes all the tests, it could then be used either intravenously for immediate therapeutic effects or orally in low quantities as long-term dietary supplements to reduce the incidence of tumors or other angiogenic diseases like rheumatoid arthritis, psoriasis and diabetic retinopathy.

Another approach would be targeting of the compounds to elicit site specific effects, thereby obviating high doses of administration and possible toxic effects in other organs. Indeed, genistein has already been successfully used in the biotherapy of human B-cell precursor (BCP) leukemia in an immunodeficient mouse model (Uckun et al., 1995). In these experiments genistein coupled to the monoclonal antibody B43 was targeted to the B cell-specific receptor CD19. The B43-genistein immunoconjugate bound to BCP leukemia cells with high affinity and triggered rapid apoptotic cell death. At less than one-tenth the maximum tolerated dose more than 99.999 percent of the human BCP leukemia cells were killed, which led to 100 percent event-free survival from an otherwise invariably fatal leukemia. Pharmacokinetic analysis revealed that the B43-genistein immunoconjugate had substantially longer elimination life and slower plasma clearance than unconjugated genistein. It appears, therefore, that genistein can exert its *in vitro* effects also *in vivo* provided that it is targeted appropriately to the desired site of action. The same might be also true for 2-methoxyestradiol and the flavonoids, and this possiblity should be explored. It should be mentioned in this context that recent improvements in our understanding of the role of angiogenesis in the pathogenesis of several important diseases and the mechanisms that control it, have opened the way to the antiangiogenic treatment of these diseases. This approach has so far been applied to the treatment of refractory hemangiomas with alpha-interferon (White et al., 1989; Ezekowitz et al., 1991) and nerve-sheath tumors with AGM-1470 (Takamiya et al., 1993).

ACKNOWLEDGMENTS

We thank C. Di Sanza and M. Quayzin for excellent technical assistance, and P.-A. Ruttiman for photographic work. Work in Heidelberg was supported by grants from Schwerpunkt "Entzündung" (Land Baden-Württemberg), Deutsche Krebshilfe and Tumorzentrum Heidelberg-Mannheim, work in Geneva by grants from the Swiss National Science Foundation (31-34097.92) and work in Helsinki by grants of the Finnish Cancer Foundation and the Sigrid Juselius Foundation.

REFERENCES

Adlercreutz, H. (1990). Western diet and western diseases: some hormonal and biochemical mechanisms and associations. Scand. J. Clin. Lab. Invest. *50, Suppl.201*, 3-23.

Adlercreutz, H., Fotsis, T., Bannwart, C., Wähälä, K., Brunow, G., and Hase, T. (1991a). Isotope dilution gas chromatographic-mass spectrometric method for the determination of lignans and isoflavonoids in human urine, including identification of genistein. Clin. Chim. Acta *199*, 263-278.

Adlercreutz, H., Honjo, H., Higashi, A., Fotsis, T., Hämäläinen, E., Hasegawa, T., and Okada, H. (1991b). Urinary excretion of lignans and isoflavonoid phytoestrogens in Japanese men and women consuming traditional Japanese diet. Am. J. Clin. Nutr. *54*, 1093-1100.

Akiyama, T., Ishida, J., Nakagawa, S., Ogawara, H., Watanabe, S., Itoh, N., Shibuya, M., and Fukami, Y. (1987). Genistein, a specific inhibitor of tyrosine-specific protein kinases. J. Biol. Chem. *262*, 5592-5595.

Baggot, J.E., Ha, T., Vaughn, W.H., Juliana, M.M., Hardin, J.M., and Grubbs, C.J. (1990). Effect of miso (Japanese soybean paste) and NaCl on DMBA-induced rat mammary tumors. Nutr. Res. *14*, 103-109.

Barnes, S., Grubbs, C., and Setchell, K.D.R. (1990). Soybeans inhibit mammary tumors in models of breast cancer. In Mutagenesis and carcinogens in diet. M. Pariza, ed. (New York: Wiley-Liss), pp. 239-253.

Blackshear, P.J.; Witters, L.A., Girard, P.R., Kuo, J.F., and Quamo, S.N. (1985). Growth factor-stimulated phosphorylation in 3T3-L1 cells. Evidence for protein kinase C-dependent and -independent pathways. J. Biol. Chem. *260*, 13304-13315.

Blood, C.H. and Zetter, B.R. (1990). Tumor interactions with the vasculature: Angiogenesis and tumor metastasis. Biochim. Biophys. Acta Rev. Cancer *1032*, 89-118.

Bouck, N.P. (1990). Tumor angiogenesis: The role of oncogenes and tumor suppressor genes. Cancer Cells *2*, 179-185.

Crum, R., Szabo, S., and Folkman, J. (1985). A new class of steroids inhibits angiogenesis in the presence of heparin or a heparin fragment. Science *230*, 1375-1378.

D'Amato, R.J., Lin, C.M., Flynn, E., Folkman, J., and Hamel, E. (1994). 2-Methoxyestradiol, an endogenous mammalian metabolite, inhibits tubulin polymerization by interacting at the colchicine site. Proc. Natl. Acad. Sci. USA *91*, 3964-3968.

Ezekowitz, A., Mulliken, J., and Folkman, J. (1991). Interferon alpha therapy of haemangiomas in newborns and infants. Br. J. Haematol. *79 Suppl. 1*, 67-68.

Folkman, J. (1985). Tumor angiogenesis. Adv. Cancer Res. *43*, 175-203.

Folkman, J. and Cotran, R.S. (1976). Relation of vascular proliferation to tumor growth. Int. Rev. Exp. Pathol. *16*, 207-248.

Folkman, J. and Klagsbrun, M. (1987). Angiogenic factors. Science *235*, 442-447.

Folkman, J. and Shing, Y. (1992). Angiogenesis. J. Biol. Chem. *267*, 10931-10934.

Folkman, J., Watson, K., Ingber, D.E., and Hanahan, D. (1989). Induction of angiogenesis during the transition from hyperplasia to neoplasia. Nature *339*, 58-61.

Fotsis, T. (1987). The multicomponent analysis of estrogens in urine by ion exchange chromatography and GC-MS-II. Fractionation and quantitation of the main group of estrogen conjugates. J. Steroid Biochem. *28*, 215-226.

Fotsis, T. and Adlercreutz, H. (1987). The multicomponent analysis of estrogens in urine by ion exchange chromatography and GC-MS-I. Quantitation of estrogens after initial hydrolysis of conjugates. J. Steroid Biochem. *28*, 203-213.

Fotsis, T., Pepper, M., Adlercreutz, H., Fleischmann, G., Hase, T., Montesano, R., and Schweigerer, L. (1993). Genistein, a dietary-derived inhibitor of *in vitro* angiogenesis. Proc. Natl. Acad. Sci. USA *90*, 2690-2694.

Fotsis, T., Pepper, M., Adlercreutz, H., Hase, T., Montesano, R., and Schweigerer, L. (1995). Genistein, a dietary ingested isoflavonoid, inhibits cell proliferation and in vitro angiogenesis. J. Nutr. *125 Suppl.*, 790S-797S.

Fotsis, T., Zhang, Y., Pepper, M.S., Adlercreutz, H., Montesano, R., Nawroth, P.P., and Schweigerer, L. (1994). The endogenous estrogen metabolite 2-methoxyestradiol inhibits angiogenesis and suppresses tumor growth. Nature *368*, 237-239.

Harborne, J.B. (1973). Flavonoids. In Phytochemistry. L.P. Miller, ed. (New York: Van Nostrand Reinhold Co.), pp. 344-380.

Hill, T.D., Dean, N.M., Mordan, L.J., Lau, A.F., Kanemitsu, M-Y., and Boynton, A.L.(1990). PDGF-induced activation of phospholipase C is not requird for induction of DNA synthesis. Science *248*, 1660-1663.

Klagsbrun, M. and Folkman, J. (1990). Angiogenesis. In Peptide growth factors and their receptors II. M.B. Sporn and A.B. Roberts, eds. (Berlin: Springer Verlag), pp. 549-586.

Linassier, C., Pierre, M., LePecQ, J.-B., and Pierre, J. (1990). Mechanism of action in NIH-3T3 cells of genistein, an inhibitor of EGF receptor tyrosine kinase activity. Biochem. Pharmacol. *39*, 187-193.

Majewski, S., Szmurlo, A., Marczak, M., Jablonska, S., and Bollag, W. (1993). Inhibition of tumor cell-induced angiogenesis by retinoids, 1,25-dihydroxyvitamin D3 and their combination. Cancer Lett. *75*, 35-39.

MacLusky, N.J., Barnea, E.R., Clark, C.R., and Naftolin, F. (1983) Catechol Estrogens. Merriam, G.R. and Lipsett, M.B., eds. (Raven Press, New York), pp. 151-165.

Miller, A.B. (1990). Diet and cancer - A review. Rev. Oncol. *3*, 87-95.

Montesano, R. and Orci, L. (1985). Tumor-promoting phorbol esters induce angiogenesis in vitro. Cell *42*, 469-477.

Montesano,R, Vassalli, J.-D., Baird, A., Guillemin, R., Orci, L. (1986). Basic fibroblast growth factor induces angiogenesis *in vitro*. Proc. Natl. Acad. Sci. USA *83*, 7297-7301.

Montesano, R., Pepper, M.S., Belin, D., Vassalli, J.D., and Orci, L. (1988). Induction of angiogenesis in vitro by vanadate, an inhibitor of phosphotyrosine phosphatases. J. Cell Psysiol. *134*, 460-466.

Muir, C., Waterhouse, J., Powell, M.T., and Whelan, S. (1987). NN. In Cancer Incidence in five continents. N. N, ed. (Lyon: International Agency for Research on Cancer), pp. NN.

Oikawa, T., Hirotani, K., Ogasawara, H., Katayama, T., Nakamura, O., Iwaguchi, T., and Hiragun, A. (1990). Inhibition of angiogenesis by vitamin D3 analogues. Eur. J. Pharmacol. *178*, 247-250.

Okura, A., Arakawa, H., Oka, H., Yoshinari, T., and Monden, Y. (1988). Effect of genistein on topoisomerase activity and on the growth of [Val 12] Ha-ras-transformed NIH 3T3 cells. Biochem. Biophys. Res. Commun. *157*, 183-189.

Pelech, S.L., Olwin, B.B., and Krebs, E.G. (1986). Fibroblast growth factor treatment of swiss 3T3 cells activates a subunit S6 kinase that phosphorylates a synthetic peptide substrate. Proc. Natl. Acad. Sci. USA *83*, 5968-5972.

Pepper, M.S. and Montesano, R. (1990). Proteolytic balance and capillary morphogenesis. Cell Differ. Dev. *32*, 319-328.

Rastinejad, F., Polverini, P., and Bouck, N.P. (1989). Regulation of the activity of a new inhibitor of angiogenesis by a cancer suppressor gene. Cell *56*, 345-355.

Schweigerer, L., Christeleit, K., Fleischmann, G., Adlercreutz, H., Wähälä, K., Hase, T., Schwab, M., Ludwig, R., and Fotsis, T. (1992). Identification in human urine of a natural growth inhibitor for cells derived from solid pediatric tumors. Eur. J. Clin. Invest. *22*, 260-264.

Schweigerer, L. and Fotsis, T. (1992). Angiogenesis and angiogenesis inhibitors in pediatric diseases. Eur. J. Pediatr. *151*, 472-476.

Seegers, J.C., Aveling, M.-L., Van Aswegen, C.H., Cross, M., Koch, F., and Joubert, W.S. (1989). The cytotoxic effects of estradiol-17b, catecholestradiols and methoxyestradiols on dividing MCF-7 and HeLa cells. J. Steroid Biochem. *32*, 797-809.

Setchell, K.D.R. and Adlercreutz, H. (1988). Mammalian lignans and phytoestrogens. Recent studies on their formation, metabolism and biological role in health and disease. In Role of the gut flora in toxicity and cancer. I.R. Rowland, ed. (London: Academic Press), pp. 315-345.

Severson, R.K., Nomura, A.M.Y., Grove, J.S., and Stemmerman, G.N. (1989). A prospective study of demographics and prostate cancer among men of Japanese ancestry in Hawaii. Cancer Res. *49*, 1857-1860.

Takamiya, Y., Friedlander, R.M., Brem, H., Malick, A., and Martuza, R.L. (1993). Inhibition of angiogenesis and growth of human nerve-sheath tumors by AGM-1470. J. Neurosurg. *78*, 470-476.

Troll, W., Wiesner, R., Shellabarger, C.J., Holtzman, S., and Stone, J.P. (1980). Soybean diet lowers breast tumor incidence in irradiated rats. Carcinogenesis *1*, 469-472.

Uckun, F.M., Evans, W.E., Forsyth, C.J., Waddick, K.G., T.-Ahlgren, L., Chelstrom, L.M., Burkhardt, A., Bolen, J., and Myers, D.E. (1995). Biotherapy of B-cell precursor leukemia by targeting genistein to CD19-associated tyrosine kinases. Science *267*, 886-891.

Ullrich, A., and Schlesinger, J. (1990). Signal transduction by receptors with tyrosine kinase activity. Cell *61*, 203-212.

Weidner, N., Semple, J.P., Welch, W.R., and Folkman, J. (1991). Tumor angiogenesis and metastasis--Correlation in invasive breast carcinoma. N. Engl. J. Med. *324*, 1-8.

White, C.W., Sondheimer, H.M., Crouch, E.C., Wilson, H., and Fan, L.L. (1989). Treatment of pulmonary hemangiomatosis with recombinant interferon alpha-2a. N. Engl. J. Med. *320*, 1197-1200.

Yamashita, Y., Kawada, S.-Z., and Nakano, H. (1990). Induction of mammalian topoisomerase II dependent DNA cleavage by nonintercalative flavonoids, genistein and orobol. Biochem. Pharmacol. *39*, 737-744.

FACTITIOUS ANGIOGENESIS III: HOW TO SUCCESSFULLY ENDOTHELIALIZE ARTIFICIAL CARDIOVASCULAR BIOPROSTHESES BY EMPLOYING NATURAL ANGIOGENIC MECHANISMS

Peter I. Lelkes, Victor V. Nikolaychik, Mark M. Samet, Dawn M. Wankowski, and Valerie Chekanov*

Laboratory of Cell Biology and *Section of Cardiology, Department of Medicine, University of Wisconsin Medical School, Milwaukee Clinical Campus, Milwaukee, WI 53201

INTRODUCTION

In this series of NATO ASI meetings on angiogenesis, we introduced the concept of "factitious angiogenesis" to describe a novel approach in tissue engineering aimed at the generation of permanent, hemocompatible blood conduits (31,32). We define as blood conduits a variety of cardiovascular prostheses, such as artificial vascular grafts, ventricular assist devices and total artificial hearts, and skeletal muscle ventricles. Without belaboring the profound technical and surgical problems associated with their manufacture, implantation, and/or long term use, all of these cardiovascular prostheses share a major, common obstacle: the inadequate hemocompatibility of their blood-contacting surfaces, which are made of various types of biopolymers (23,30,52). We hypothesized that the hemocompatibility in these novel blood conduits can be significantly improved by lining their blood-contacting surfaces with a non-thrombogenic monolayer of autologous endothelial cells (ECs).

In the past, we have discussed in some detail attempts made by us and others to improve the hemocompatibility of artificial cardiovascular prostheses by exploiting some aspects of the angiogenic cascade (31,32). In this communication, we will first briefly introduce the notion that the exploitation of natural angiogenic mechanisms has become an integral part of "tissue engineering". We will then summarize some of the progress achieved by us and others over the past two years in endothelializing the blood-contacting surfaces of various types of cardiovascular prostheses. Finally, we will discuss a novel aspect of factitious angiogenesis, namely the use of angiogenic matrices, such as fibrin glue, in dynamic cardiomyoplasty.

NOVEL CONCEPTS IN TISSUE ENGINEERING

Over the past few years, "tissue engineering" has emerged as a novel concept in basic and applied biomedical research. One of its foremost goals is to combine advanced techniques of cell culture, biomaterials and biomedical engineering to re-create *in vitro* viable and functional tissues (organoids) which then can be used to replace or repair failing tissues/organs *in vivo*. For an excellent review of this subject, see reference (28). Obviously,

Molecular, Cellular, and Clinical Aspects of Angiogenesis
Edited by Michael E. Maragoudakis, Plenum Press, New York, 1996

a necessary prerequisite for *in vitro* tissue re-creation is the appropriate, organ-specific differentiation of constituent cells of a given tissue in co-culture (58). Moreover, in order for tissue formation to occur, three-dimensional co-cultures must be established. A recurrent problem with conventional three-dimensional cultures, such as in stirred fermentors or rotatory flasks, is fluid motion which might abrogate organ-specific heterotypic interactions between the various cell types in co-culture. The elevated, flow induced shear stress in these vessels might not only cause physical damage to the cells, but also directly prevent cell-cell aggregation and clustering. In addition, the convectional diffusion in the stirred medium causes rapid dispersion of the local concentration of humoral factors, which are pivotal for tissue-specific differentiation *in vivo* (13). Some of the more esoteric, yet, probably, most successful approaches to three-dimensional co-culture is the use of microgravity, either in space or simulated in the laboratory, in unique, NASA-designed rotating wall cell culture reactors (16,53). Recent studies suggest that cells cultured under (simulated) microgravity undergo an unprecedented level of tissue-specific differentiation (14,17,25).

No tissue is viable without an adequate blood supply, which provides oxygen and nutrients while concomitantly eliminates noxious metabolic products. Therefore, the generation of genuine, EC-lined "blood-vessels" in three-dimensional tissue culture by *in vitro* angiogenesis has become a central focus of modern tissue engineering. Currently, numerous attempts are being made to include vascular ECs in the various organ-specific co-cultures. Some approaches rely on unique, bio-degradable polymeric scaffolds (12) which, amongst other advantages, also facilitate the three-dimensional organization of ECs into capillaries (28,44). In other studies, exogenous angiogenic growth factors, such as bFGF or some of the other factors contained in Matrigel, are added to the culture medium with the assumption that these factors will promote *in vitro* tubular morphogenesis (39). However, given the appropriate three-dimensional culture conditions (such as microgravity), concoction of parenchymal and endothelial cells will secrete just the right mixture of differentiative factors to assure an organ-specific, functional differentiation of each of the individual cell types and will trigger their assembly into macroscopic (up to 2-3 cm diameter) three-dimensional "organoids". Specifically, some of these organoids contain EC-lined neo-"blood" vessels (14). Although this exciting approach is still in its infancy, it appears to have far-reaching clinical potential, e.g., for the repair or replacement of failing organs and/or for gene therapy. However, a prerequisite for the applicability of these organoids will be to demonstrate that these neo-vessels are indeed functional and can anastomose with the vasculature of the host.

VASCULAR GRAFTS

The most common artificial cardiovascular prostheses are the vascular grafts. Of particular interest are the small diameter (< 5 mm) artificial grafts which, as anticipated, will one day supplant "natural" autologous grafts, presently used on a routine basis for, e.g., coronary bypass operations. However, since compliant vascular grafts are made of common biomaterials, such as expanded tetrafluoroethylene (ePTFE, Teflon) or various types of segmented polyurethanes (9), they all share the problem of insufficient hemocompatibility. The tubular geometry of vascular grafts represents the simplest possible form of cardiovascular prothesis. Therefore, these grafts were the first vehicles in which a solution to the problem of hemocompatibility was addressed by endothelial cell seeding (19,62). Recent studies suggest that this approach is clinically feasible, but requires further optimization (60).

Endothelialization of the luminal surface of an small diameter, porous graft can be enhanced by stimulating transmural angiogenesis and ingrowth of fibrous tissue (38). As discussed above, one of the prerequisites for such transmural endothelialization is the availability of appropriate scaffolds, in this case of microporous grafts, which allow for directional tissue ingrowth and neovascularization (15,26). In addition, the process of neovascularization can be accelerated and/or its efficiency enhanced by providing additional stimuli, such as precoating graft surface with angiogenic growth factors, e.g., acidic fibroblast growth factor (18), or with angiogenic adhesive matrices, such as fibrin glue (40), or with a combination of the two. At this time, it is not known whether (or how) the ingrowing blood vessels will contribute to the establishment of the EC monolayer covering the graft surface. It is also intriguing whether these capillaries will serve as a sort of "vasa vasorum" and, thus, stabilize the neointima by supplying oxygen and nutrients.

Another, most promising approach to achieving factitious angiogenesis is the *in vitro* regeneration of the entire vascular wall of a synthetic graft by hierarchically layering smooth muscle cells, fibroblasts and ECs, respectively, into and onto a permissive adhesive matrix made out of collagen and dermatan sulfate (24). Following implantation of this compliant graft into dogs and exposure to pulsatile blood flow, the EC layer remains intact and non-thrombogenic, while the smooth muscle layers orient themselves longitudinally and circumferentially in, respectively, the neomedia and neoadventitia (22). This and other ingenious approaches in vascular tissue engineering (2,27,37,43) hold the promise that by cleverly exploiting basic angiogenic mechanisms, such as EC migration, proliferation, and extracellular matrix remodelling, we will finally be able to overcome some of the most challenging problems in biomaterials research and generate truly hemocompatible artificial vascular grafts.

VENTRICULAR ASSIST DEVICES AND TOTAL ARTIFICIAL HEARTS

Every year, end-stage cardiomyopathy and similar progressive fatal heart diseases claim some 75,000 victims in the USA alone (20). Approximately 50% of these patients could be saved by long term cardiac support. Due to limited donor heart supplies, ca. 35,000 of the patients will die each year, unless a permanent total artificial heart (TAH) or ventricular assist device (VAD) becomes available. Besides the immense technological problems associated with the mechanics and electronics of an implantable TAH or VAD, a major stumbling block for these prostheses is the hemocompatibility of the blood-contacting biomaterials. Once again, a possible remedy is to line the luminal surfaces of these devices with autologous ECs (30,61). As previously described, we successfully endothelialized polyurethane-made blood sacs of ventricular assist devices by applying some of the general principles of "factitious angiogenesis", such as enhancing EC adhesion, migration and proliferation on angiogenic matrices, and by facilitating the remodelling of the extracellular matrix (30-32,42,51).

A particularly promising approach towards endothelialization has been to precoat the luminal surfaces of the ventricles with a heat-stabilized form of autologous cryoprecipitate or with fibrin glue of controlled thickness and composition (42). As noted above, fibrin glue, the reaction product of fibrinogen and thrombin, has become one of the most beneficial extracellular matrices to support both the growth of ECs and instant establishment of a stable, shear-resistant monolayer in vascular grafts (40). Moreover, fibrin glue is a useful "angiogenic" matrix, since activated ECs migrate into the matrix and form networks of capillary-like tubes (10,45).

In exploring feasible matrix supports for successfully endothelializing cardiovascular prostheses, preliminary experiments and preclinical trials are always carried out with animal-derived cells or in animal models, such as rats, rabbits, dogs, sheep and cows. Unfortunately, as past experiments have clearly demonstrated, one should not expect that the results from such pre-clinical trials could be predictably transferred into the clinical setting. For example, we have previously reported that adult human ECs, when plated on biomaterials, exhibit a significantly lower growth/proliferation potential than ECs of animal origin (32). Indeed, as shown in Figure 1, ECs isolated from adult human aortae need significant supplementation of their culture medium with complex nutrients and growth factors to maintain an adequate level of proliferation. By contrast, bovine aortic ECs grow without complex growth supplements, presumably since they produce and secrete their own growth factors, such as basic fibroblast growth factor. Similar findings were made with a number of large-vessel ECs isolated from various animals, including sheep (Figure 1), canines and rodents (data not shown). However, as seen in Figure 1, fibrin glue may be useful as a ubiquitous extracellular matrix which universally supports the growth of ECs, irrespective of their species origin. This finding underscores the potential usefulness of fibrin glue or similar autologous protein complexes in a clinical setting as a provisional matrix for "factitious angiogenesis".

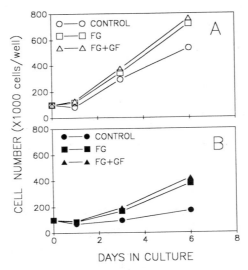

Figure 1: Fibrin glue promotes endothelial cell proliferation. Sheep aortic ECs (panel A) and human aortic ECs (panel B) were seeded into 6 well tissue culture plates, either left untreated (controls, tissue culture grade polystyrene) or covered with fibrin glue (FG). Cell proliferation was assessed using a non-invasive, fluorimetric assay, Alamar blue, as previously described in detail (42). Ctrl: ECs grown on polystyrene surface (gold standard); FG: ECs grown on fibrin glue; FG+GF: ECs grown on fibrin glue supplemented with growth factors (30 µg/ml endothelial cell growth supplement).

Most recently, we have applied all of the above principles and successfully tested Endothelialized ventricular assist devices both *in vitro* in a mock-loop configuration, as well as *in vivo* in calf models. Success is currently measured in terms of the retention of a functional EC monolayer up to 28 days after implantation (Lelkes *et al.*, manuscript in preparation). Thus, after two decades of great initial enthusiasm and equally great frustration (11,36,59), we have finally begun to prove that endothelialization of complex cardiovascular prostheses is possible an realizable. An arduous road lies ahead, which includes improvement of the procedures, and numerous *in vitro* tests of the functionality and, most importantly, of the nonthrombogenicity of the EC monolayer inside the blood sacs. Moreover, we will have to unequivocally prove the clinical benefits in preclinical animal trials, and then translate the procedures into the realm of an operating room.

SKELETAL MUSCLE VENTRICLES

Skeletal muscle, when appropriately trained by repetitive electrical stimulation, will acquire some of the biochemical properties of fatigue-resistant cardiac muscle (34,47). In its original concept, cardiomyoplasty is aimed at assisting the failing myocardium in its contractile function by transposing an electrically conditioned, fatigue-resistant skeletal muscle and wrapping it around the myocardial wall (6,7,35). One of the potential applications of such a "conditioned" skeletal muscle, in most instances the *musculus latissimus dorsi* (LDM), is to generate an endogenous cardiac assist device (54). By wrapping the LDM around a synthetic mandrel, a fibrous pouch is generated, which, when connected to the circulation via bioprosthetic valves and conduits, can serve as a skeletal muscle ventricle, SMV (50). This SMV can provide substantial assistance to an ailing heart, as has been proven in preclinical animal studies (41,48).

A recurrent problem associated with the long-term use of such endogenous cardiac assist devices is the thrombogenicity of the fibrous blood-contacting surface inside the SMV (1,49). In the past, we have chosen to enhance the hemocompatibility of SMVs by using a similar approach as described above for grafts, TAHs and VADs: we designed and optimized methodologies to endothelialize the luminal surfaces of SMVs *in situ* in an operating room setting (29,55,56). After successfully endothelializing both static and electrically conditioned but non-pumping SMVs, we recently subjected our hypothesis to the ultimate *in vivo* test by connecting several of the endothelialized SMVs to circulation. To the best of our knowledge, this is the first time that endothelialized cardiovascular protheses, other than tubular grafts, have been tested *in vivo*. Our initial results, obtained after 7 days in circulation, suggest a significant reduction, although not complete abolition, yet, of the incidents of thrombosis inside the SMVs (57). We found that most (ca. 80%), but not all, of the SMV surface was lined with a functional monolayer of ECs. Thrombus formation was noted in areas devoid of EC coverage. At this time, we attribute the failure to provide a complete EC lining to mere technical difficulties; it might be a problem of the seeding technique, or it might be related to surgical procedures at the time of SMV connection to the circulation. Future research will have to address these questions as well as devise new approaches for reinforcing the process of endothelialization from within by "therapeutic angiogenesis", i.e. by stimulating angiogenic mechanisms inside the skeletal muscle-bioprosthesis.

NOVEL APPLICATIONS OF "THERAPEUTIC ANGIOGENESIS" IN DYNAMIC CARDIOMYOPLASTY

More than half a century ago, Leriche (46) and Beck (33) used unconditioned muscle flaps to assist in the contractility of a failing heart. Today, this form of cardiac surgery is recognized as static cardiomyoplasty. In 1985, Carpentier, in France (3, 4, 28), and Magovern, in the USA (34), made the first clinical attempts to transplant a preconditioned skeletal muscle as a driver of a failing myocardium. This new approach was termed dynamic cardioplasty (8). By using the LDM, dynamic cardioplasty provides a surgical means to create autogenous cardiovascular assist devices without the need for introducing foreign bioprosthetic materials into the circulatory system (5). A major advantage of this approach is that the natural EC lining of the patient's own (but poorly pumping) heart remains in place. An essential disadvantage of dynamic cardiomyoplasty is its dependence on a long-term formation of a unique hybrid muscle comprised of both cardiac and skeletal muscle in the areas of contact between the two tissues.

Postoperatively, effective healing and functionality of the "repaired" heart relies on an efficient neovascularization of the wounded area. However, past experience has shown that such an angiogenic response does not readily occur. Both the traumatized, ischemic skeletal muscle and the mobile myocardial wall have compromised vasculature with low angiogenic potential, as the two muscles are not in a tight contact with each other, at least initially. Typically, neovascularization of the compromised tissues takes 6-8 months after conventional dynamic cardiomyoplasty. For some patients, such a long healing process might be too late.

A novel approach to resolving these problems involves therapeutic angiogenesis (21), which we proposed to apply to accelerate "wound healing" in dynamic cardiomyoplasty. Specifically, we suggested utilizing fibrin meshwork (see above) both as a physical glue between the myocardium and skeletal tissue and as an angiogenic interface. We considered fibrin glue an excellent candidate for this application, as it readily forms a neovascularized pseudolayer between the contacting tissues, and also has the capacity to serve as a depot for the local delivery of antibiotics, anabolic drugs and/or growth factors.

To test our approach, we performed dynamic cardiomyoplasty in four adult sheeps. Prior to the experimental trials, myocardial damage was artificially inflicted by limiting the blood supply through the intermediate branch of the circumflex coronary artery. At the start of the cardiomyoplasty procedure, the LDM was completely mobilized and introduced into the chest through the second intercostal space. The mobilized muscle was then wrapped around the heart and, surgically, two pockets were created by a line of sutures between the muscle and myocardium (Figure 2).

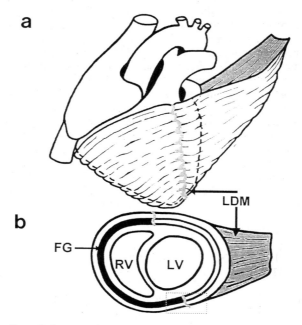

Figure 2: Schematics of the experimental setup for dynamic cardiomyoplasty: a. general view of a skeletal muscle flap wrapped around the failing heart, b. detailed cross-section outlining the location of the two experimentally introduced pockets between the ailing myocardium and fatigue resistant skeletal muscle (LDM). One pocket (FG) contained fibrin glue, the other served as control. RV and LV denote right and left ventricles, respectively. The broken line encloses the area from which specimens were obtained for histomicroscopic examination.

Approx 50 ml of fibrin glue was administered into one pocket, the other served as a control. Finally, muscle and myocardium leads were connected to a cardiosynchronized myostimulator. Electrical stimulation of the LDM was started 5 days after surgery. The stimulus was a single pulse every second beat of the native heart beat. The intensity and frequency of the stimulus were increased every ten days. After 8 weeks of continuous electrical stimulation, the animals were sacrificed and samples were excised. Visual examination and light microscopic studies of the tissues were performed to assess the results of the trials.

As seen in Figure 3, in the control pocket, the skeletal and cardiac muscles are separated, even after 8 weeks, by a visible gap, preventing any organic contact between the two tissues. In contrast, in the fibrin glue-filled pocket a new pseudolayer has been created, which physically bridges the gap between the myocardium and the skeletal muscle (Figure 4). Moreover, blood vessels of various diameters, including fully developed arteries and veins can be detected in the fibrin layer. At higher magnification, neovascularization of the fibrin glue layer is clearly visible in the form of numerous small capillaries, lined by (von Willebrand factor-positive) ECs (Figure 5).

234

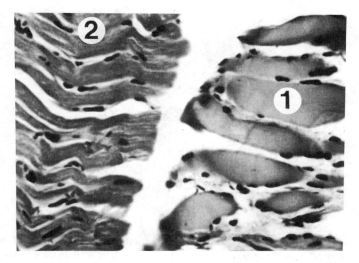

Figure 3: Gross appearance of the skeletal muscle/myocardium interface 8 weeks after conventional dynamic cardiomyoplasty. 1) skeletal muscle and 2) myocardium. Note that in the absence of fibrin glue there is no organic, physical contact between the two tissues. Original magnification: 250 x.

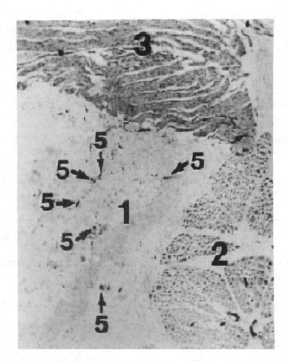

Figure 4: Histological section of the contact-area between cardiac and skeletal muscle. H&E stained histological section which was obtained from the macrosection shown in Fig. 2. 1) pseudolayer of fibrin glue; 2) Latissimus Dorsi muscle; 3) myocardium; 5) newly formed blood vessels of significant diameter within the fibrin glue pseudolayer. Original magnification: 37.5 x.

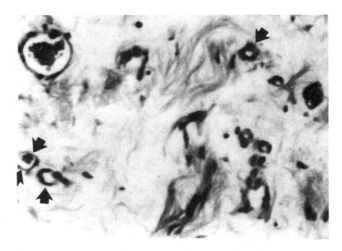

Figure 5 Neovascularization in the fibrin glue pseudolayer between skeletal and cardiac muscle. Individual EC lined capillaries (arrows) are visible, as well as capillaries containing red-blood cells trapped in the vessel lumen. Original magnification: 500 x.

At least some of the newly formed capillaries are functional, as inferred from the inclusion of erythrocytes. These preliminary studies prove the feasibility of our concept and provide evidence that "factitious angiogenesis" can be used for accelerating the healing process and providing for an organic bridge between the two tissues. Current experiments are underway to asses the origin of the newly formed capillaries in the fibrin glue monolayer (from the skeletal muscle or from the cardiac tissue), to improve the angiogenic response by including growth factors in the fibrin glue mixture, and to evaluate the clinical impact of factitious angiogenesis in dynamic cardiomyoplasty.

CONCLUSIONS

The most exciting aspect of these studies is the recognition that the generation of new blood vessels by factitious angiogenesis has become one of the central issues in successful approaches to tissue engineering. As demonstrated in this communication, the application of the basic principles of angiogenesis has significantly advanced the generation of biocompatible artificial blood conduits. More advanced work is needed, though. Some of it is already in progress, aimed at re-creating not only the intimal EC layer of a "neo-vessel", but also at re-constructing the entire vessel wall by including smooth muscle cells (cardiac myocytes), fibroblasts, neurons, etc. Interestingly, once all the components for this process are in place (co-cultures in permissible matrices) and exposed to the appropriate physical environment, such as pulsatile shear-stress (or microgravity), the cues contained in and secreted from such a tissue-specific mixture of cells is sufficient to orchestrate the correct three-dimensional arrangement in the nascent blood-conduits (or for that matter, organoids). It is the goal of current and future studies to understand these tissue-specific interactions and to apply the knowledge acquired to the rational design of biocompatible cardiovascular protheses.

ACKNOWLEDGMENTS: The original studies reported in this communication were supported, in part, by grants-in-aid from the Milwaukee Heart Research Foundation, Promeon/Medtronic Inc. and the American Heart Association, Wisconsin Chapter (GIA # 92-65-26C to PIL). We are grateful to Dr. Tom Hayman for critically reading this manuscript.

REFERENCES

1. **Anderson, D. R., A. Pochettino, R. L. Hammond, E. Hohenhaus, A. D. Spanta, C. R. Bridges, Jr., S. Lavine, R. D. Bhan, M. Colson, and L. W. Stephenson.** 1991. Autogenously lined skeletal muscle ventricles in circulation: up to nine months' experience. *J. Thorac. Cardiovasc. Surg.* 101:661-670.

2. **Bell, E.** 1991. Tissue engineering: a perspective. *J. Cell. Biochem.* 45:239-241.

3. **Carpentier, A. and J. C. Chachques.** 1985. Myocardial substitution with a stimulated skeletal muscle: first successful clinical case. *Lancet* 1:1267.

4. **Carpentier, A. and J. C. Chachques.** 1987. Latissimus dorsi cardiomyoplasty to increase cardiac output. In *Heart valve replacement: current status and future trends.* G. Rabago and D. A. Cooley, editors. Future Publishing Co., Inc. Mount Kisco, N.Y. 473-486.

5. **Carpentier, A. and J. C. Chachques.** 1991. Clinical dynamic cardiomyoplasty: method and outcome. *Semin. Thorac. Cardiovasc. Surg.* 3:136-139.

6. **Carpentier, A., J. C. Chachques, and P. A. Grandjean (eds.).** 1991. *Cardiomyoplasty.* The Bakken Research Center Series, Volume 3. Futura Publishing Company, Inc. Mount Kisco, NY.

7. **Chachques, J. C., P. A. Grandjean, T. A. Pfeffer, P. Perier, G. Dreyfus, V. Jebara, C. Acar, M. Levy, I. Bourgeois, J. N. Fabiani, A. Deloche, and A. Carpentier.** 1990. Cardiac assistance by atrial or ventricular cardiomyoplasty. *J Heart Transplant* 9:239-251.

8. **Chiu, R. C. J., ed.** 1986. *Biomechanical cardiac assist: cardiomyoplasty and muscle-powered devices.* Futura Pub. Co. Mount Kisco, NY.

9. **Courtney, J. M., N. M. K. Lamba, S. Sundaram, and C. D. Forbes.** 1994. Biomaterials for blood-contacting applications. *Biomaterials* 15:737-744.

10. **Dvorak, H. F., V. S. Harvey, P. Estrella, L. F. Brown, J. McDonagh, and A. M. Dvorak.** 1989. Fibrin containing gels induce angiogenesis: implications for tumor stroma generation and wound healing. *Lab. Invest.* 57:673-686.

11. **Eskin, S. G., H. D. Sybers, W. O'Bannon, and L. T. Navarro.** 1982. Performance of tissue cultured endothelial cells in a mock circulatory loop. *Artery* 10:159-171.

12. **Freed, L. E., G. Vunjak-Novakovic, and R. Langer.** 1993. Cultivation of cell-polymer cartilage implants in bioreactors. *J. Cell. Biochem.* 51:257-264.

13. **Freshney, R. I.** 1994. *Culture of animal cells. A manual of basic technique.* Wiley-Liss, Inc, New York.

14. **Galvan, D. L., B. R. Unsworth, T. J. Goodwin, J. Liu, and P. I. Lelkes.** 1995. Microgravity enhances tissue-specific neuroendocrine differentiation in cocultures of rat adrenal medullary parenchymal and endothelial cells. *In Vitro Cell. Dev. Biol.* 31:10A.

15. **Golden, M. A., S. R. Hanson, T. R. Kirkman, P. A. Schneider, and A. W. Clowes.** 1990. Healing of polytetrafluoroethylene arterial grafts is influenced by graft porosity. *J. Vasc. Surg.* 11:838-845.

16. **Goodwin, T. J., T. L. Prewett, D. A. Wolf, and G. F. Spaulding.** 1993. Reduced shear stress: a major component in the ability of mammalian tissues to form three-dimensional assemblies in simulated microgravity. *J. Cell. Biochem.* 51:301-311.

17. **Goodwin, T. J., W. F. Schroeder, D. A. Wolf, and M. P. Moyer.** 1993. Rotating-wall vessel coculture of small intestine as a prelude to tissue modeling: aspects of simulated microgravity. *Proc. Soc. Exp. Biol. Med.* 202:181-192.

18. **Greisler, H. P., D. J. Cziperle, D. U. Kim, J. D. Garfield, D. Petsikas, P. M. Murchan, E. O. Applegren, W. Drohan, and W. H. Burgess.** 1992. Enhanced endothelialization of expanded polytetrafluoroethylene grafts by fibroblast growth factor type 1 pretreatment. *Surgery* 112:244-255.

19. **Herring, M. B., R. Compton, D. R. LeGrand, and A. L. Gardner.** 1989. Endothelial cell seeding in the management of vascular thrombosis. *Semin. Thromb. Hemost.* 15:200-205.

20. **Hognes, J. R. and M. VanAntwerp (eds).** 1991. *The Artificial Heart: Prototypes, Policies, and Patients.* National Academy Press, Washington DC.

21. **Höckel, M., K. Schlenger, S. Doctrow, T. Kissel, and P. Vaupel.** 1993. Therapeutic angiogenesis. *Arch. Surg.* 128:423-429.

22. **Ishibashi, K. and T. Matsuda.** 1994. Reconstruction of a hybrid vascular graft hierarchically layered with three cell types. *ASAIO J.* 40:M284-M290.

23. **Ito, Y. and Y. Imanishi.** 1989. Blood compatibility of polyurethanes. *Crit. Rev. Biocompat.* 5:45-104.

24. **Kanda, K., T. Matsuda, and T. Oka.** 1993. In vitro reconstruction of hybrid vascular tissue: hierarchic and oriented cell layers. *ASAIO J.* 39:M561-M565.

25. **Klement, B. J. and B. S. Spooner.** 1993. Utilization of microgravity bioreactors for differentiation of mammalian skeletal tissue. *J. Cell. Biochem.* 51:252-256.

26. **Kohler, T. R., J. R. Stratton, T. R. Kirkman, K. H. Johansen, B. K. Zierler, and A. W. Clowes.** 1992. Conventional versus high-porosity polytetrafluoroethylene grafts: clinical evaluation. *Surgery* 112:901-907.

27. **Langer, R. and J. P. Vacanti.** 1993. Tissue engineering. *Science* 260:920-926.

28. **Langer, R. and J. P. Vacanti.** 1995. Artificial Organs. *Scientific American.* 273:130-133.

29. **Lelkes, P. I., H. Gao, J. R. Edgerton, and C. W. Christensen.** 1994. Endothelial cell seeding of latissimus dorsi muscle pouches. *J. Surg. Res.* 57:460-469.

30. **Lelkes, P. I. and M. M. Samet.** 1991. Endothelialization of the luminal sac in artificial cardiac prostheses: a challenge for both biologists and engineers. *J. Biomech. Eng.* 113:132-142.

31. **Lelkes, P. I., M. M. Samet, C. W. Christensen, and D. L. Amrani.** 1992. Factitious angiogenesis: endothelialization of artificial cardiovascular prostheses. In *Angiogenesis in health and disease.* M. E. Maragoudakis, P. Gullino, and P. I. Lelkes, editors. Plenum Press, New York, NY. 339-353.

32. **Lelkes, P. I., D. M. Chick, M. M. Samet, V. Nikolaychik, G. A. Thomas, and R. L. Hammond, and L. W. Stephenson.** 1994. Factitious angiogenesis: not so factitious anymore? The role of angiogenic processes in the endothelialization of artificial cardiovascular prostheses. In **Angiogenesis: Molecular Biology, Clinical Aspects.** M. E. Maragoudakis, P. Guillino, and P. I. Lelkes, editors. Plenum Press, New York, NY. 321-331.

33. **Lerich, R.** 1933. Essai experimentale de traitement de certains infarctus du myocarde et de l'aneurysme du coeur par une graffe de muscle strie. *Bull. Soc. Nat. Chir.* 59:229-234.

34. **Magovern, G. J.** 1991. Introduction to the history and development of skeletal muscle plasticity and its clinical application to cardiomyoplasty and skeletal muscle ventricle. *Semin. Thorac. Cardiovasc. Surg.* 3:95-97.

35. **Magovern, G. J., F. R. Heckler, S. B. Park, I. Y. Christlieb, G. A. Liebler, J. A. Burkholder, T. D. Maher, D. H. Benckart, G. J. Magovern, Jr., and R. L. Kao.** 1988. Paced skeletal muscle for dynamic cardiomyoplasty. *Ann. Thorac. Surg.* 45:614-619.

36. **Mansfield, P. B., A. R. Wechezak, and L. R. Sauvage.** 1975. Preventing thrombus on artificial vascular surfaces: true endothelial cell linings. *Trans. Am. Soc. Artif. Intern. Organs* 21:264-272.

37. **Massia, S. P. and J. A. Hubbell.** 1992. Tissue engineering in the vascular graft. *Cytotechnology* 10:189-204.

38. **Menger, M. D., F. Hammersen, P. Walter, and K. Messmer.** 1990. Neovascularization of prosthetic vascular grafts: quantitative analysis of angiogenesis and microhemodynamics by means of intravital microscopy. *Thorac. Cardiovasc. Surgeon* 38:139-145.

39. **Montesano, R., G. Schaller, and L. Orci.** 1991. Induction of epithelial tubular morphogenesis in vitro by fibroblast-derived soluble factors. *Cell* 66:697-711.

40. **Müller-Glauser, W., P. Zilla, M. Lachat, B. Bisang, F. Rieser, L. von Segesser, and M. Turina.** 1993. Immediate shear stress resistance of endothelial cell monolayers seeded in vitro on fibrin glue-coated ePTFE prostheses. *Eur. J. Vasc. Surg.* 7:324-328.

41. **Niinami, H., A. Pochettino, and L. W. Stephenson.** 1991. Use of skeletal muscle grafts for cardiac assist. *Trends Cardiovasc. Med.* 1:122-126.

42. **Nikolaychik, V. V., M. M. Samet, and P. I. Lelkes.** 1994. A new, cryoprecipitate-based coating for improved endothelial cell attachment and growth on medical grade artificial surfaces. *ASAIO J.* 40:M846-M852.

43. **Noishiki, Y., Y. Yamane, M. Furuse, and T. Miyata.** 1988. Development of a Growable Vascular Graft. *ASAIO J.* 34:308-313.

238

44. **Peppas, N. A. and R. Langer.** 1994. New challenges in biomaterials. *Science* 263:1715-1720.

45. **Pepper, M. D. and R. Montesano.** 1991. Proteolytic balance and capillary morphogenesis. *Cell Differentiation* 32:319-328.

46. **Petrosky, B. V.** 1966. Surgical treatment of cardiac aneurysms. *J. Cardiovasc. Surg.* 7:87-95.

47. **Pette, D.** 1991. Changes in phenotype expression of stimulated skeletal muscle. In *Cardiomyoplasty. The Bakken Reserach Center Series Volume 3.* A. Carpentier, J. C. Chachques, and P. A. Grandjean, editors. Futura Publishing Company, Inc. Mount Kisco, NY. 19-31.

48. **Pochettino, A., D. R. Anderson, R. L. Hammond, S. Salmons, and L. W. Stephenson.** 1991. Skeletal muscle ventricles. *Semin. Thorac. Cardiovasc. Surg.* 3:154-159.

49. **Pochettino, A., F. W. Mocek, H. Lu, R. L. Hammond, A. D. Spanta, T. L. Hooper, H. Niinami, R. Ruggiero, M. L. Colson, and L. W. Stephenson.** 1992. Skeletal muscle ventricles with improved thromboresistance: 28 weeks in circulation. *Ann. Thorac. Surg.* 53:1025-1032.

50. **Pochettino, A., A. D. Spanta, R. L. Hammond, D. R. Anderson, C. R. Bridges, Jr., P. Samet, H. Ninami, E. Hohenhaus, S. Salmons, and L. W. Stephenson.** 1990. Skeletal muscle ventricles for total heart replacement. *Ann. Surg.* 212:345-352.

51. **Samet, M. M., D. M. Wankowski, V. Nikolaychik, and P. I. Lelkes.** 1994. Endothelial cell seeding with rotation of a ventricular blood sac. *ASAIO J.* 40:M319-M324.

52. **Schoen, F. J.** 1991. Biomaterials science, medical devices, and artificial organs: synergistic interactions for the 1990s. *Trans. Am. Soc. Artif. Intern. Organs* 37:44-48.

53. **Schwarz, R. P., T. J. Goodwin, and D. A. Wolf.** 1992. Cell culture for three-dimensional modeling in rotating-wall vessels: an application of simulated microgravity. *J. Tiss. Cult. Meth.* 14:51-58.

54. **Stephenson, L. W., J. A. Macoviak, F. Armenti, T. Bitto, J. D. Mannion, and M. A. Acker.** 1986. Skeletal muscle for potential correction of congenital heart defects. In *Biomechanical cardiac assist: cardiomyoplasty and muscle-powered devices.* R. C.-J. Chiu, editor. Futura Pub. Co. Mount Kisco, N.Y. 129-139.

55. **Thomas, G. A., P. I. Lelkes, D. M. Chick, S. Isoda, H. Lu, H. Nakajima, R. L. Hammond, H. L. Walters III, and L. W. Stephenson.** 1995. Skeletal muscle ventricles seeded with autogenous endothelium. *ASAIO J.* 41:204-211.

56. **Thomas, G. A., P. I. Lelkes, D. M. Chick, H. Lu, T. A. Kowal, R. L. Hammond, H. Nakajima, H. O. Nakajima, A. D. Spanta, and L. W. Stephenson.** 1995. Endothelial lined skeletal muscle ventricles: open and percutaneous techniques. *J. Card. Surg.* In press.

57. **Thomas, G. A., P. I. Lelkes, S. Isoda, D. Chick, H. Lu, R. L. Hammond, H. Nakajima, H. L. Walters III, and L. W. Stephenson.** 1995. Endothelial cell-lined skeletal muscle ventricles in circulation. *J. Thorac. Cardiovasc. Surg.* 109:66-73.

58. **Watt, F. M.** 1991. Cell culture models of differentiation. *FASEB J.* 5:298-294.

59. **Wechezak, A. R., R. F. Viggers, L. R. Sauvage, and P. B. Mansfield.** 1984. Endothelial cell rounding associated with long-term implantations of left ventricular assist devices. *Scanning Electron Microscopy* 3:1353-1360.

60. **Wu, M. H.-D., Q. Shi, A. R. Wechezak, A. W. Clowes, I. L. Gordon, and L. R. Sauvage.** 1995. Definitive proof of endothelialization of a Dacron arterial prosthesis in a human being. *J. Vasc. Surg.* 21:862-867.

61. **Zilla, P., R. Fasol, M. Grimm, T. Fischlein, T. Eberl, P. Preiss, O. Krupicka, U. von Oppell, and M. Deutsch.** 1991. Growth properties of cultured human endothelial cells on differently coated artificial heart materials. *J. Thorac. Cardiovasc. Surg.* 101:671-680.

62. **Zilla, P. P., R. D. Fasol, and M. Deutsch (eds.).** 1987. *Endothelialization of vascular grafts.* Karger, Basel, Switzerland.

A METHOD FOR THE *IN VIVO* QUANTITATION OF

ANGIOGENESIS IN THE RABBIT CORNEAL MODEL

Robert W. Shaffer and Martin Friedlander

The Scripps Research Institute
Department of Cell Biology
La Jolla, CA 92037

ABSTRACT

Angiogenesis is stimulated by a variety of factors and underlies the pathology of many disease states, including cancer and blinding eye disease. It is often necessary to quantitate the angiogenic response to various stimuli and determine the efficacy of putative inhibitory agents. We describe a non-invasive method for measuring angiogenesis in a rabbit corneal model with precision and ease, regardless of geometrical irregularities in the vessel growth pattern.

INTRODUCTION

In the normal state, angiogenesis (defined as the growth of new blood vessels from pre-existing ones) is tightly regulated; it is limited to development and, in the adult, wound healing and ovulation. Our understanding of the molecular events involved in the angiogenic process has advanced significantly since the purification of the first angiogenic molecules only a decade ago (Shing et al., 1984). We know that this process, under physiological conditions, is highly regulated and may be induced by specific angiogenic molecules such as basic and acidic fibroblast growth factor (Folkman, 1992), vascular endothelial growth factor (Perrara et al., 1992), angiogenin (Fett et al., 1986), transforming growth factor (Antonelli-Orlidge et al., 1989), tumor necrosis factor-α (Beutler and Cerami, 1986), and platelet derived growth factor (Ishikawa et al., 1989). It is thought that angiogenesis can also be suppressed by inhibitory molecules such as interferon-α (Ezekowitz et al., 1992), thrombospondin (Rastinejad et al., 1989), or angiostatin (Derynck, 1990). It is the balance of these stimulators and inhibitors that is thought to tightly control the normally quiescent capillary vasculature (D'Amore, 1994). When this balance is upset, as in certain disease states, capillary endothelial cells are induced to proliferate, migrate and ultimately differentiate.

Angiogenesis plays a central role in a variety of diseases including cancer (Fidler and Ellis, 1994; Folkman, 1995) and ocular neovascularization (D'Amore, 1994). Sustained growth and metastasis of a variety of tumors has also been shown to be dependent on the growth of new host blood vessels into the tumor in response to tumor derived angiogenic factors (Folkman, 1992). Proliferation of new blood vessels in response to a variety of stimuli occurs as the dominant finding in the majority of eye diseases that blind; diabetic retinopathy, macular degeneration, rubeotic glaucoma, interstitial keratitis and retinopathy of prematurity are a few examples. Ordinarily, the proliferation of new blood vessels in the eye is very tightly regulated; tissue damage can stimulate release of angiogenic factors resulting in capillary proliferation (Folkman and Shing, 1992; Casey et al., 1994). Recent reports

Molecular, Cellular, and Clinical Aspects of Angiogenesis
Edited by Michael E. Maragoudakis, Plenum Press, New York, 1996

suggest that vascular endothelial growth factor (VEGF) may be the dominant stimulus in experimentally induced iris neovascularization (Miller et al., 1994) as well as endogenous neovascular retinopathies (Aiello et al., 1994). While these data seem to show a correlation between intraocular VEGF levels and ischemic retinopathic ocular neovascularization, a role for FGF can not be ruled out (Hanneken et al., 1991). In fact, both basic and acidic FGF are known to be present in the normal adult retina, even though detectable levels are not consistently correlated with neovascularization. This may be largely due to the fact that FGF binds very tightly to charged components of the extracellular matrix and may not be readily available in a freely diffusable form that would be detected by standard assays of intraocular fluids.

The few published reports of the use of angiogenic inhibitors to prevent ocular neovascularization in experimental animal models or human disease have met with limited success. For example, α-interferon was shown to have anti-angiogenic activity in treating pulmonary hemangiomatosis (White et al., 1987), but had no demonstrated effect compared to controls when tested in patients with chorioretinal neovascular membranes associated with age-related macular degeneration (Poliner et al., 1993; Chan et al., 1993). Recently, thalidomide was administered orally to rabbits with FGF-induced corneal neovascularization (D'Amato et al., 1994); the area of angiogenesis was reduced by 36% when a teratogenic dose was used. This inhibition could be increased to 52% if, in addition to the thalidomide, rabbits were pretreated with the maximally tolerated dose of total body irradiation. While the clinical promise of angiogenic inhibitors such as angiostatin, fumagillin-derivative AGM1470 and thalidomide is great, the mechanism whereby they inhibit angiogenesis remains unknown and their potential use largely untested.

A novel conceptual approach towards understanding the problem of unregulated angiogenesis has focused on the role of cell adhesion molecules in regulating the relationship between proliferating vascular cells and their environment (Brooks et al., 1994). This class of adhesion receptors, called integrins, are expressed as heterodimers consisting of an α and β subunit on all cells (Hynes, 1992) One such integrin, $\alpha_v\beta_3$, is the most promiscuous member of this family and allows endothelial cells to interact with a wide variety of extracellular matrix components (Cheresh, 1987). Our understanding of the molecular differences between quiescent and proliferating capillary endothelial cells, and the relationship of $\alpha_v\beta_3$ to this process, has been significantly advanced by recent observations from the laboratory of Dr. David Cheresh. Focusing on the adhesive and migratory properties of angiogenic blood vessels, his group has recently identified the integrin $\alpha_v\beta_3$ as a marker of proliferating blood vessels in human granulation tissue, but not in normal skin from the same individual (Clark et al., 1994). Reasoning that this integrin may be expressed on the surface of new blood vessels in response to angiogenic stimuli, the same group demonstrated, using the chick chorioallantoic membrane (CAM) model, that blood vessels induced to proliferate with cytokines had a four-fold increase in the integrin $\alpha_v\beta_3$ compared to preexisting quiescent vessels (Brooks et al., 1994a). In fact, a monoclonal antibody to $\alpha_v\beta_3$ (LM609) specifically inhibited this angiogenic response (Clark et al., 1994). When tumor explants were used instead of cytokines to stimulate neovascularization, similar results were observed. In the presence of integrin antagonists tumor angiogenesis and growth are disrupted and tumor regression occurs (Brooks et al., 1994b). The mechanism of inhibition of angiogenesis appears to be due to selective apoptosis of sprouting blood vessels. Taken together, these data strongly suggest that the adhesion between a proliferating endothelial cell and its substrate are necessary for angiogenesis to proceed; if the cell is prevented from receiving the appropriate signal that is transduced when $\alpha v\beta_3$ binds to specific components of the extracellular matrix, the cell is induced to undergo apoptosis. Since this response is observed in endothelial cells induced to undergo angiogenesis by a variety of stimuli, it would suggest that the underlying mechanism whereby LM609 inhibits angiogenesis is generic and occurs late in the angiogenic program without regard to the specific stimulus. As a potential therapeutic agent, this would make LM609 very attractive.

Our laboratory has been using the rabbit corneal micropocket model to test the efficacy of integrin antagonists in the inhibition of ocular angiogenesis (Friedlander et al., 1995a; Friedlander et al., 1995b). Since these studies require quantitative analysis, we have developed a relatively simple, highly reproducible method for evaluating the extent of corneal neovascularization in live rabbits.

Current Methods for the Quantitation of Angiogenesis

The rabbit cornea itself is avascular, but is surrounded by a network of perilimbal vessels (Ruben, 1981). Thus, the rabbit cornea is an excellent model for studying neovascularization since new blood vessels arising from the limbal vasculature can easily be identified. Several methods exist for the quantitation of angiogenesis in the rabbit corneal model. In detecting an angiogenic response, many investigators simply monitor the presence or absence (+/-) of new blood vessel growth (Ziche et al., 1989; Polverini et al., 1977). Graded scales have also been used to provide semi-quantitative evaluation of corneal angiogenesis. In a study of the angiogenic response to corneal implants of psoriatic tissue, vessel growth was graded on a scale of 1+ to 5+ (Malhotra, 1989). 1+ indicated mild congestion, yet no new vessel formation, 2+ represented some new vessel growth, 3+ indicated growth midway to the stimulus, 4+ represented vessels extending to the stimulus, and 5+ signified extensive vessel growth to the stimulus. A study to determine the neovascular effects of extended contact lens wear simply assessed blood vessel growth as severe (+++), moderate (++), mild (+), or no response (-) (Madigan et al., 1990). A third method ranked blood vessel growth on a scale from 0 (minimum) to +6 (maximum) based on the fractional vessel growth distance from the limbus to the stimulus (Mahoney et al., 1985), yet another study used a grading scale from 1 to 4 based on measured vessel growth from the limbus (Frucht et al., 1984). Many of these grading scales are subject to observer bias.

In an effort to more objectively quantitate angiogenic responses, several investigators have measured the stimulated vessel length from the limbus. While some measure the length of the most central vessel (Jensen et al., 1986), others have measured the length of the leading vessel (Duffin et al., 1982). In contrast, many investigators take multiple caliper readings from standardized or random positions along the limbus and determine a mean value from this data (Cooper et al., 1980; Sholley et al., 1984). Comparison of data between studies is difficult because of these variations in vessel length measurement protocols.

Image analysis techniques have also been used in measuring the angiogenic response. One such method involves morphometric analysis (Glatt and Klintworth, 1986). Rabbits are perfused with India ink under anesthesia to create contrast between vessels and the corneal background. The animals are then sacrificed and flat preparations of the corneas are made. A grid is superimposed on the cornea and the number of intersections between the grid and vessels are counted to obtain a reflection of vessel quantity. A second method involves computerized planimetry and corneas are prepared in a similar manner (Glatt and Klintworth, 1986). A stylus is used to trace all blood vessels to be quantitated and a total vessel length is obtained. This method is time consuming and one is prone to retracing errors while collecting data. A third technique for image analysis involves computerized image analysis to obtain vessel density (Proia et al., 1988). Animals are perfused with India ink, euthanized, and flat preparations of the corneas are made. A television camera mounted to an operating microscope is used to obtain images, and computerized pixel analysis determines the presence or absence of vessels at a given point based on pixel shade. Lighting variations and imperfections in the cornea may produce false identification of vessel localization. All three of these image analysis methods must be performed post-mortem.

An *in vivo* technique using an image analysis approach has recently been described (Conrad et al., 1994). Rabbits are positioned in a stereotactic holding device to ensure reproducible positioning. Corneas are illuminated with monochromatic light at a wavelength of peak hemoglobin absorption to maximize contrast of the blood vessels. Several digital images are taken at 15 degree intervals along the circumference of the cornea. A computer is used to process and create a single montage from the overlapping images. The montage is then enhanced and pixel counting based on density is used to quantitate the vessels. Although capable of obtaining several data points at differing times from a single cornea, this method is complex, costly, and time-consuming.

Several methods exist for determining the area of neovascularization for rabbit corneas. One such technique reported by BenEzra quantitates new blood vessels by creating a triangular region (BenEzra, 1978; BenEzra et al., 1990). Using measuring calipers, the distance of vessel growth from the limbus and half of the active base of neovascularization are measured. These values are multiplied to obtain a geometrical approximation. A second similar technique reported by D'Amato quantitated the anti-angiogenic effects of thalidomide using an alternate geometric approximation (D'Amato et al., 1994). In this model, calipers are used to measure the number of "clock hours" of stimulation along the limbus and the distance of vessel growth from the limbus. Using formula (1), where r is equal to the measured corneal radius, L is equal to vessel length from the limbus, and C is equal to the number of clock hours estimated, the area of a circular band segment representing the region of vessel growth is obtained.

$$\text{Area of neovascularization} = C/12 \times 3.1416[r^2 - (r-L)^2] \qquad (1)$$

It should be noted, however, that neovascularization does not always occur in an ordered geometric pattern, and neither of these methods fully account for this.

We have a new technique for quantitating the neovascular response which has distinct advantages over those previously used. With simple slide photography and a scanning imaging densitometer, one is able to obtain multiple data points from a single rabbit in a non-invasive manner quickly and with ease. This method is highly reproducible and sensitive to small changes in vessel area. In addition, multiple observers may obtain consistent results. We have also compared our method for quantitating neovascular area with those of BenEzra and D'Amato and suggest advantages inherent in our technique. Finally, we consider the use of this technique in calculating the relative vessel quantity in the cornea through optical density measurements.

EXPERIMENTAL DESIGN

Basic fibroblast growth factor (bFGF) was used as the angiogenic stimulus; it produces a strong, highly reproducible neovascular response in the rabbit cornea. bFGF (Genzyme Cambridge, MA) was stabilized with carafate (sucralfate, Marion Merrell Dow, Cincinnati, Ohio) and cast in Hydron (poly(hydroxyethyl methacrylate)) pellets as described by D'Amato et al. (1994).

Six white albino New Zealand rabbits (2.5-3.5 kg) were anesthetized by intramuscular injection of Ketamine HCl (50 mg/kg; Fort Dodge Laboratories Inc. Fort Dodge, Iowa) and subcutaneous injection with both Atropine Sulfate (0.13 mg/kg; Elkins-Sinn Inc. Cherry Hill, N.J.) and Xylazine HCl (2 mg/kg; Rugby Labs Inc. Rockville Centre, N.Y.). In addition, eyes were topically anesthetized with Opticaine Tetrachloride. Photographs were taken of each cornea at 16X magnification under consistent lighting conditions with Kodak Ektachrome 64T slide film prior to surgery using a camera mounted on a Wild operating microscope. Corneas were irrigated topically with 0.9% saline (BSS, Alcon Labs Ft. Worth, TX) throughout the surgical procedure. A single pellet was surgically implanted into a corneal micropocket such that the distance from the limbus to the pellet was 2 mm, and a drop of tobramycin (Tobrex 0.3%; Falcon Ophthalmics Inc., Fort Worth, TX) was topically applied. Three animals were given implants O.D. and three were given implants O.S. Surgical calipers were used to measure the diameter of each cornea for use in the quantitative conversion factors.

The corneas were examined using a Wild M490 operating microscope on days 0 (post-operative),2,4,6,8,10, and 12. Rabbits were sedated with an intramuscular injection of Ketamine HCl (33 mg/kg) and topical Opticaine Tetrachloride. Four photographs of each cornea with implant were taken . Rabbits were positioned for photography such that the point located half way between the limbus and the neovascular front, and directly between the pellet and the limbus, was oriented planar to the camera lens. When in focus, this yielded a maximum field of view of neovascularization for the camera. Between each photograph, the rabbit was repositioned so that the four photographs could be compared to assess the ability to position the rabbit in the same manner. Animals were monitored until alert and then returned to their cages. Rabbits were euthanized at day 13 by lethal injection of sodium pentobarbital euthanasia solution (Henry Schein Inc., Port Washington, N.J.) into the marginal ear vein.

Data Acquisition

For each day of observation the set of four slides was scanned by transmittance at a resolution of 64um and a gray scale of 256 shades using BIO-RAD's GS-670 imaging densitometer and were converted into a computer image by BIO-RAD's Molecular Analyst 1.1 software. Using the software's magnification tool, the images were expanded to 100%, and the histogram gray-scale plot was then adjusted to lighten the image until the greatest contrast of corneal surface to blood vessels was obtained. For density analysis, this procedure does not alter the underlying image data. Using the lasso tool, the area of neovascularization was outlined beginning at the limbus, and the resultant raw area was then obtained. Hemorrhage and non-corneal vessels were not included in the analysis. The diameter of the cornea was measured on a selected pre-operative slide along the axis of the pellet. A conversion factor of 1/5.44 for raw area cm^2 to actual area mm^2 was then calculated using equation (2):

$$\text{Conversion Factor} = [(\text{Actual Diameter})^2/(\text{Slide Diameter})^2] \times 100 \qquad (2)$$

The outline was cleared and retraced three times to determine the precision of the computer-assisted analysis. Bias was avoided for successive measurements because data were not viewed until after the outlining procedure was completed and no values were discarded. This procedure was repeated for the remaining three slides for the same rabbit on the same day of observation.

Ability of Independent Observers to Obtain Consistent Data

Rabbit #5 on post-operative day 8 was chosen at random for observer comparison analysis. An independent investigator with no clinical ophthalmological experience or familiarity with our protocol was asked to quantitate the neovascular response four times for each of the four slides for this rabbit and day. She was briefly taught the technique and was then left unaccompanied to obtain data sets. She did not have access to any previously acquired data and all of her readings were used in this comparison.

Comparison of Three Quantitative Techniques

Our described technique was compared with two other quantitative methods which also measure neovascularization in terms of area. Both D'Amato's (1994) circular band segment method and BenEzra's (1990; 1978) triangular approximation method were used as described to quantitate slide photographs previously obtained. The slide giving maximum neovascular area through the use of our quantitative method for each cornea at each time point was used in our comparison. A single measurement was determined for each cornea for each of the alternative techniques, and these were compared with the mean values of the four readings taken using our method.

Vessel Quantity Determination

Measures of vessel quantity were obtained for Rabbit #5 (randomly selected) at days 6, 8, 10, and 12. Days 2 and 4 were discarded due to massive corneal edema that obscured vessel detail. The same slides from the area quantitations were used. Following outlining, both area and "volume" (expressed as cm^2 x (optical density)) were recorded. The volume was divided by the area to obtain the optical density of the region. Using landmarks on the cornea, a similar region was outlined on each of the four post-operative slides, the same measurements were recorded, and the mean value of Volume/Area was determined to be the background optical density for the avascular cornea. The difference in background and neovascular optical density was multiplied by the neovascular spot area to obtain a relative value for blood vessel quantity (cm^2 x O.D.). This value was multiplied by the conversion factor to obtain actual relative vessel quantity. This procedure was performed a total of four times for each slide.

DATA ANALYSIS AND RESULTS

Demonstration of Quantitative Technique

Quantitation of the angiogenic response in the rabbit corneal model was highly reproducible. Figure 1 presents mean values and standard deviations calculated for the four data values quantitated from each slide from Rabbit #1 O.D., and is typical of the data obtained from the remaining five rabbits. Neovascular area increased with time, plateaued by day 10, and began to regress. Low standard deviation values suggest an ability to use the Molecular Analyst 1.1 software to consistently measure the neovascular area from a single slide. Upon inspection of data from all six rabbits, we noted that standard deviations for individual slides approached zero for very small measurements of neovascularization, thus we were able to discriminate changes as small as 0.1 mm^2 in new blood vessel area. As the neovascular area increased, standard deviation values for each slide increased and began to level off.

Slide	Mean Value mm^2	Standard Deviation mm^2
2A	0	
2B	0	
2C	0	
2D	0	
4A	3.27	+/- 0.250
4B	3.36	+/- 0.259
4C	4.01	+/- 0.790
4D	3.97	+/- 0.397
6A	9.74	+/- 0.831
6B	11.2	+/- 1.00
6C	12.5	+/- 0.329
6D	12.2	+/- 0.634
8A	15.3	+/- 0.401
8B	17.6	+/- 0.301
8C	17.5	+/- 0.700
8D	18.5	+/- 0.596
10A	31.3	+/- 1.56
10B	29.2	+/- 1.07
10C	32.7	+/- 0.858
10D	30.3	+/- 0.432
12A	30.6	+/- 0.494
12B	32.2	+/- 1.18
12C	31.6	+/- 0.853
12D	31.3	+/- 1.70

Figure 1. Mean values and standard deviations of the four data values generated from each slide of Rabbit #1 O.D. Slide numbers indicate post-operative days and letters distinguish the four slides taken at each time point. No neovascularization was observed on day two, although vasodilation of the limbal vessels was apparent. The area of neovascularization increased with time and began to level off at post-operative day 12. Standard deviation values increased slightly as the neovascular area increased.

From all data obtained from the six rabbits, we frequently noted that mean values among slides from the same rabbit and day of observation differed significantly, suggesting that a source of error was present in our photographic technique. Lens magnification was kept constant throughout the photographic procedure and all photographs used were well focused. It is important to position the rabbit cornea such that the greatest surface area of neovascularization is in the plane of the camera lens; failure to do so results in an underestimation of vessel growth. Since the rabbit eye is manually proptosed and held in position, there is variability between each photographic observation. Therefore, we presume that the photographic image with the largest mean value for the four measurements most closely approximated the true area of neovascularization. By measuring four slides, we statistically eliminated, to a degree, the variability created by slight changes in rabbit positioning. Figure 2 illustrates the neovascular area data generated from rabbit #1 and includes only data from the slide with the greatest mean value from each time point.

Independent Reader Comparison

In order to determine reader-associated variability, an independent reader was chosen to quantitate randomly selected slides for comparison with the data obtained by the primary reader. Rabbit #5 at day 8 was randomly selected and each of the four slides were quantitated four times. Mean values and standard deviations for a single investigator's reading of a slide were calculated.

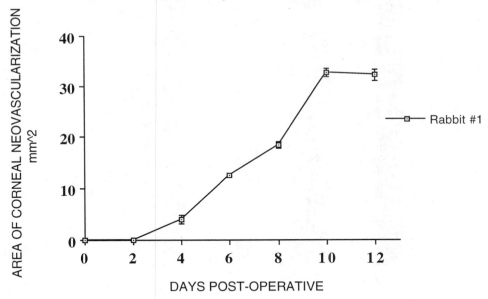

Figure 2. The time course of the bFGF-induced angiogenesis in Rabbit #1 O.D. By post-operative day #4, new vessels had invaded the corneal stroma. By post-operative day 10, the area of neovascularization plateaued. Error bars represent the standard deviations from the mean calculated for each set of four values generated from a single slide. Only the slide with the greatest calculated area of neovascularization for each time point is shown, as discussed in the text.

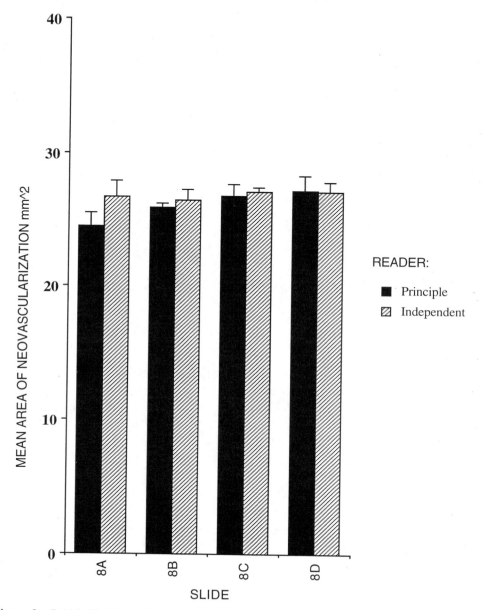

Figure 3. Rabbit #5 O.S. at day 8 was randomly chosen for observer comparison analysis. An independent observer with minimal prior exposure to our technique was briefly taught the method. She was left unaccompanied to calculate each of four slides four separate times. The mean values and standard deviation values she obtained are shown and are juxtaposed to similar data obtained by the principle observer from the same slides. In all cases, the difference between the two mean values for a given slide was less than the sum of the two standard deviations computed for each slide.

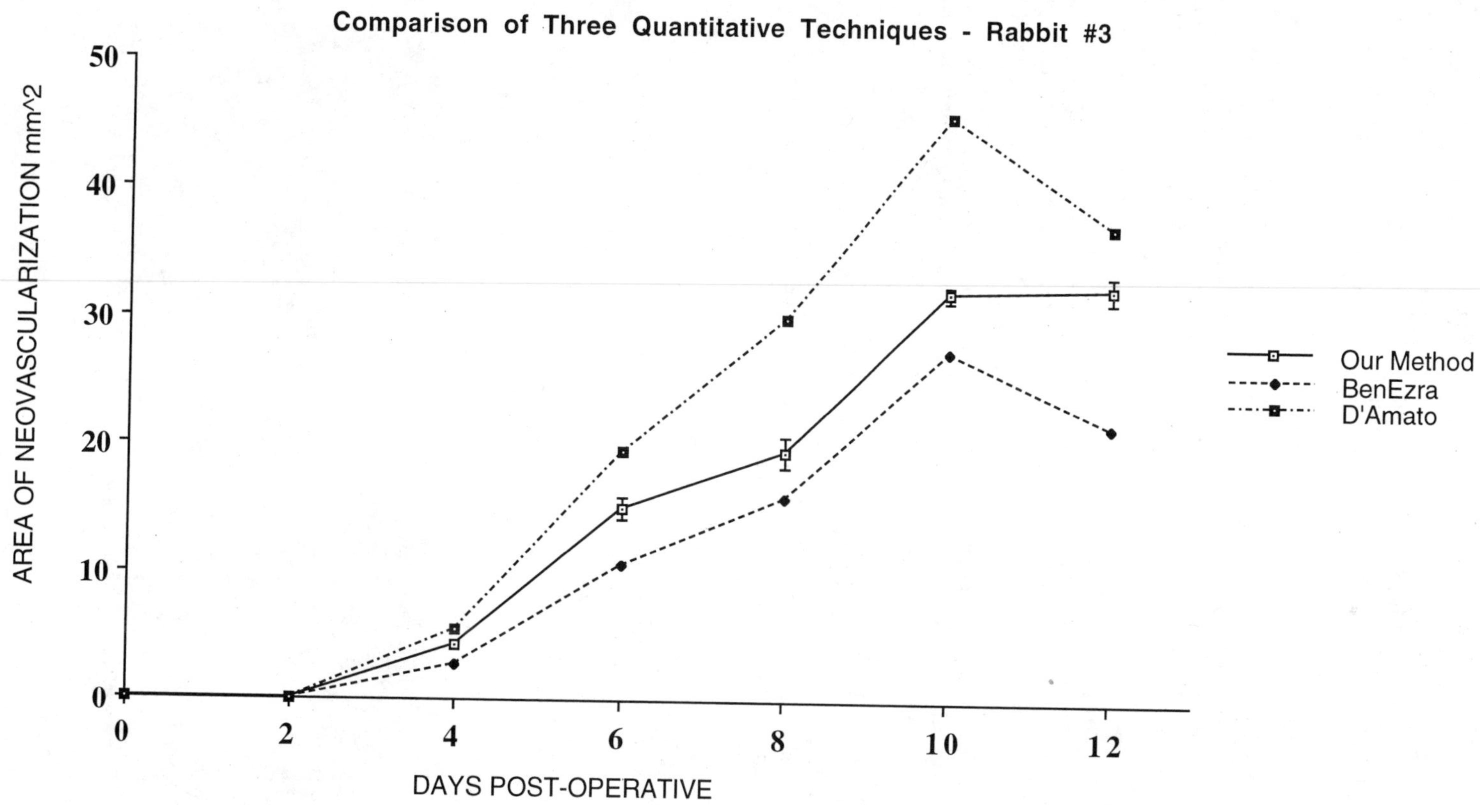

Figure 4. Slides from each time point for Rabbit #3 O.D. were quantitated using our described method. The slide with the greatest calculated mean value for each time point is graphically represented above, and standard deviations are represented by the error bars. The same slides were individually quantitated using both BenEzra's triangular approximation method and D'Amato's circular band segment approximation. In comparison to the data generated using our technique, values determined using BenEzra's method were significantly higher and those determined using D'Amato's method were significantly lower. All data points from the alternative two methods were outside the range of one standard deviation from our calculated means.

Figure 3 illustrates mean value and standard deviation data generated by each observer. The sum of the two observer's standard deviations for each slide was computed and was taken to represent the maximum acceptable difference between the two mean values determined for that given slide. The differences in mean values fell within the standard deviation sums for each of the four individual slides suggesting that the two independent readers were capable of generating similar neovascular area measurements from the same computerized image.

Comparison of Three Quantitative Techniques

Figure 4 illustrates neovascular area results for rabbit #3 using our quantitative method and techniques published by BenEzra and D'Amato. These results were similar to those obtained from the remaining five rabbits. BenEzra's method for quantitating angiogenesis estimates the geometrical growth of new blood vessels in the cornea as two right triangles sharing a common leg. Based on our observations of numerous neovascularized corneas, the overall shape of vessel growth in this model is complex (Figure 5). Even when growth patterns are typical a curved vessel front is observed, and superimposed triangles often exclude portions of this front. Therefore, we would anticipate that BenEzra's technique underestimates the area of neovascularization. As expected, data generated using BenEzra's technique was significantly lower than those values produced using our technique that accounted for the exact geometry of vessel growth.

D'Amato's method for the quantitation of angiogenesis is similar to BenEzra's in that it approximates the area of blood vessel growth according to a geometric algorithm, but assumes a circular band segment shape for this region. Points existing along the limbus further from the angiogenic stimulus typically exhibit less vessel elongation than the point closest to the stimulus, yet D'Amato's method assumes the vessel length is constant from all points along the limbus where an angiogenic response is present. Therefore, we anticipated a general overestimation of blood vessel area. In five of the six rabbits, D'Amato's method generated data that were significantly greater than data produced from the same slides using our technique, consistent with our expectations.

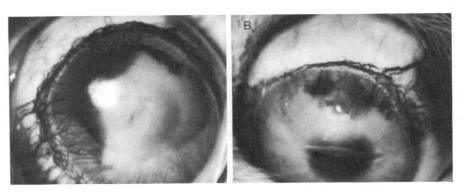

Figure 5. Neovascularization in the rabbit corneal model at post-operative day 12. The vessel growth can be irregular and complex.

Typically, investigators compare the neovascular response to a stimulus/inhibitor with a control cornea. Therefore, generated data need not be exact values assuming they deviate by the same factor of error, because this factor will be eliminated once the two values are compared. However, methods that do not account for geometrical inconsistencies decrease the likelihood that two values for different growth responses will differ by the same factor. Therefore, we suggest that our method for the quantitation of angiogenesis may be more accurate than others used for comparison because it does consider exact geometry of vessel growth, even in the most irregular of growth patterns.

Vessel Quantity Determination

Using the Molecular Analyst Software, we attempted to determine relative vessel quantity, defined as area x optical density, for neovascular corneas. Rabbit #2 was randomly selected for observation and the same slides used for vessel area determination were used in this analysis. Edema resulted in a clouding of the cornea, and the corneal shade of gray, as read by the densitometer, changed which made it impossible to subtract out a background optical density for the cornea using photographs taken post-operatively. Upon inspection of slide photographs from days two and four, we determined that the edema associated with the neovascularization was too great to effectively determine vessel quantity, and these slides were discarded. Edema was less severe in corneas at post-operative days 6,8,10 and 12, and these were used for analysis.

Blood vessels were typically a darker shade of gray than the cornea, and once background (error in density associated with the avascular cornea) was subtracted, the adjusted optical density represented new vessel formation. When multiplied by the actual surface area, a relative value for vessel quantity was produced.

From our calculated results (not shown), we noted that mean vessel quantities varied greatly among slides taken of the same cornea and time point, indicating that the photographic technique was highly prone to variables that may have influenced our results. Furthermore, our data suggested that the blood vessel quantity dropped significantly at days 8 and 10 and then increased once again at day 12, inconsistent with observations made by other investigators that the vessel quantity increases linearly with time (Conrad et al., 1994). Although differences in exposed neovascular area contributed to this error, other factors were clearly involved.

Edema associated with the neovascular response was apparent in all corneas quantitated and was, most likely, a significant source of error. In addition to changing the apparent density of the cornea, edema may have clouded the vessels, causing them to appear less dark and have a lower optical density. Furthermore, slight changes in the angle of the cornea during photography may have produced a different glaring effects captured by photography. In addition, any variation in film processing among slides may have contributed to variation in the apparent optical densities of those slides.

DISCUSSION

The rabbit corneal pocket assay for angiogenesis has proven to be a widely accepted model for the study of angiogenic stimulators and inhibitors. While numerous reports in the literature describe methods for evaluating new blood vessel growth in this model, most are descriptive using a subjective grading system. The studies that quantitate neovascularization are either expensive, tedious, imprecise or require the use of post-mortem specimens. We describe a simple, cost-effective method of quantitating corneal angiogenesis that is also accurate and highly reproducible.

The method reported has a number of advantages: (1) it can be performed on live animals, permitting multiple time points to be recorded from a single animal; (2) the area of neovascularization can be precisely quantitated by using Bio-Rad Molecular Analyst 1.1 program's lasso tool, thus eliminating a significant source of error inherent in other techniques due to geometric irregularities in corneal vessel growth; (3) it is rapid and simple with a high degree of reproducibility (a typical set of eight animals can be anesthetized and photographed in less than half an hour and the resulting set of 32 slides can be scanned and quantitated four separate times in another 30 minutes); (4) little or no experience is necessary for using the imaging densitometer and the software for quantitation; and (5) the technique is cost-effective, using the same operating microscope necessary for the surgeries to acquire the photographic data and using a common piece of laboratory core equipment that has many additional uses besides the slide scanning and quantitation (e.g., protein gel electrophoretic analysis). It is also possible to use this method to quantitate relative vessel quantity with optical density measurements. However, appropriate correction for background density makes this application difficult; variables such as lighting and corneal positioning, and underlying structures (e.g., iris) can contribute to artifactual values. Furthermore, the use of vessel quantity, as calculated by optical density, may also be misleading; since vessels do not grow into the cornea in a single plane, multiple vessels may superimpose on one another throughout the corneal stroma. Optical density measurements cannot accurately reflect this.

We have used this method of quantitating corneal angiogenesis to evaluate a number of antibody, peptide and non-peptide antagonists of VNR integrins. This technique not only provides quantitative information, but permits the efficient, accurate storage and retrieval of large amounts of data generated from these types of studies.

ACKNOWLEDGMENTS

We extend special thanks to Miyuki Sugita for her surgical skill, to Christine Kincaid for her role as independent investigator in the quantitative comparison, and to David Cheresh and Peter Brooks for numerous helpful discussions. This work was supported by grants to Martin Friedlander from The Department of Academic Affairs Clinical Research Program and the Lions Sight First Diabetic Retinopathy Research Program of the American Diabetes Association, the Ralwhbnbg Fund and the Robert Mealey Laboratory for the Study of Macular Degenerations.

REFERENCES

Aeillo, L.P., Avery, R.L., Arrigg, P.G., Keyt, B.A., Jampel, H.D., Shah, S.T., Pasquale, L.R., Theime, H., Iwamoto, M.A., Park, J.E., Nguyen, H.V., Aiello, L.M., Ferrara, N., and King, G.L., 1994, Vascular endothelial growth factor in ocular fluid of patients with diabetic retinopathy and other retinal disorders, *New Eng. J. Med.* 331: 1480-1487.

Antonelli-Orlidge, A., Saunders, K.B., Smith, S.R., and D'Amore, P.A., 1989, An activated form of TGF-beta is produced by co-cultures of endothelial cells and pericytes, *Proc. Natl. Acad. Sci. USA* 86: 4544.

BenEzra, D., Hemo, I., and Maftzir, G., 1990, In vivo angiogenic activity of interleukins, *Arch. Ophthalmology* 108: 573-576.

BenEzra, D., 1978, Neovasculogenic ability of prostaglandins, growth factors, and synthetic chemoattractants, *American Journal of Pathology* 86: 455-461.

Beutler, B. and Cerami, A., 1986, Cachectin and tumour necrosis factor as two sides of the same biological coin, *Nature* 320: 584.

Brooks, P., Clarke, R., and Cheresh, D., 1994, Requirement of vascular integrin alpha-v-beta-3 for angiogenesis, *Science* 264: 569.

Brooks, P., Montgomery, A., Rosenfeld, M., Reisfeld, R., Hu, T., Klier, G., and Cheresh, D., 1994, Integrin alpha-v-beta-3 antagonists promote tumor regression by inducing apoptosis of angiogenesis blood vessels, *Cell* 79: 1157-1104.

Chan, C.K., Kempin, S.J., Noble, S.K., and Palmer, G.A., 1993, The treatment of choroidal neovascular membranes by alpha interferon, *Ophthal.* 101: 289.

Cheresh, D.A., 1987, Human endothelial cells synthesize and express an Arg-Gly-Asp-directed adhesion receptor involved in attachment to fibrinogen and von Willebrand factor, *Proc. Natl. Acad. Sci. USA* 84: 6471-6475.

Clark, R.A.F., Tonnesen, M.G., Gailit, J., and Cheresh, D., 1994, Integrin alpha-v-beta-3 on capillary sprouts promotes granulation tissue formation during porcine wound healing, *J. Clin. Invest.*

Conrad, T. J., Chandler, D. B., Corless, J. M., and Klintworth, G. K., 1994, In vivo measurement of corneal angiogenesis with video data acquisition and computerized image analysis, *Laboratory Investigation* 70: 426-433.

Cooper, C. A., Bergamini, M. V. W., and Leopold, I. H., 1980, Use of flurbiprofin to inhibit corneal neovascularization, *Arch. Ophthalmology*, 98: 1102-1105.

D'Amato, R. J., Loughnan, M. S., Flynn, E., and Folkman, J., 1994, Thalidomide is an inhibitor of angiogenesis. *Proceedings of the National Academy of Sciences, USA.* 91: 4082-4085.

D'Amore, P.A., 1994, Mechanisms of retinal and choroidal neovascularization, *Invest. Ophthal. Vis. Sci.* 35: 3974-3979.

Derynck, R., 1990, Transforming growth factor-alpha, *Mol. Reprod. Dev.* 27: 3.

Deutsch, T. A., and Hughes, F. W., 1979, Suppressive effects of indomethacin on thermally induced neovascularization of rabbit corneas, *American Journal of Ophthalmology* 87: 536-540.

Duffin, R. M., Weissman, B. A., Glasser, D. B., and Pettit, T. H., 1982, Flurbiprofin in the treatment of corneal neovascularization induced by contact lenses, *American Journal of Pathology* 93: 607-614.

Ezekowitz, R.A.B., Mulliken, J.B., and Folkman, J., 1992, Interferon alpha-2a therapy for life threatening hemangiomas of infancy, *New Eng. J. Med.* 326: 1456-1463.

Fett, J. W., Strydom, D. J., and Lobb, R. F., 1986, Isolation and characterization of angiogenesis in angiogenic protein from human carcinoma cells, *Biochem.* 24: 5480.

Fidler, F.J. and Ellis, L.M., 1994, The implications of angiogenesis for the biology of and therapy of cancer metastasis, *Cell* 79: 185-188.

Folkman, J., 1995, Angiogenesis in cancer, vascular, rheumatoid and other disease, *Nature Medicine* 1: 27-31.

Folkman, J., and Shing, Y., 1992, Angiogenesis, *The Journal of Biological Chemistry* 267: 10931-10934.

Folkman, J., 1992, The role of angiogenesis in tumor growth, *Sem. Canc. Biol.* 3: 65.

Friedlander, M., Shaffer, R., Kincaid, C., Brooks, P., and Cheresh, D., 1995, An antibody to the integrin $\alpha v\beta 3$ inhibits ocular angiogenesis, *Invest. Ophthalm. Vis. Sci.* 36: S1047

Friedlander, M., Brooks, P., Shaffer, R., Kincaid, C., Varner, J., and Cheresh, D., 1995, Definition of two angiogenic pathways by distinct α_V integrins, *Science* 270:1500-1502.

Frucht, J., and Zauberman, H., 1984, Topical indomethacin effect on neovascularization of the cornea and on prostaglandin E2 levels, *British Journal of Ophthalmology* 68: 656-659.

Glatt, H. J., and Klintworth, J. K., 1985, Quantitation of neovascularization in flat preparations of the cornea, *Microvascular Research* 31:104-109.

Hanneken, A., de Juan, E., Cutty, O.A., Fox, O.M., Schiffer, S., and Hzelmerade, R., 1991, Altered distribution of basic fibroblast growth factor in diabetic retinopathy, *Arch. Ophthal.* 109: 1005-1011.

Hynes, R.O., 1992, Integrins: Versatility, modulation, and signaling in cell adhesion, *Cell* 69: 11-25.

Ishikawa, F., Miyazono, K., Hellman, U., Drexler, H., Wernstedt, C., Hagiwara, K., Usuki, K., Takaku, F., Risau, W., and Heldin, C.-H., 1989, *Nature* 338: 557.

Jensen, A. J., Hunt, T. K., Scheuenstuhl, H., and Banda, M. J., 1986, Effect of lactate, pyruvate, and pH on secretion of angiogenesis and mitogenesis factors by macrophages, *Laboratory Investigation* 54: 574-578.

Leavesley, P.I., Schwartz, M.A., Rosenfeld, M., and Cheresh, D.A., 1993, Integrin beta-1 and beta-3 mediated endothelial cell migration is triggered through distinct signaling mechanisms, *J. Cell Biol.* 121: 163-170.

Madigan, M. C., Penfold, P. L., Holden, B. A., and Billson, F. A., 1990, Ultrastructural features of contact lens-induced deep corneal neovascularization and associated stromal leukocytes, *Cornea* 9(2): 144-151.

Malhotra, R., Stenn, K. S., Fernandez, L. A., and Braverman, I. M., 1989, Angiogenic properties of normal and psoriatic skin associate with epidermis, not dermis, *Laboratory Investigation* 61: 162-165.

Miller, J.W., Adamis, A.P., Shima, D.T., D'Amore, P.A., Moulton, R.S., O'Reilly, M.S., Folkman, J., Dvorak, H.F., Brown, L.F., Berse, B., Yeo, T.-K., and Yeo, K.-T., 1994, Vascular endothelial growth factor / vascular permeability factor is temporally and spatially correlated with ocular angiogenesis in a primate model, *Am. J. Pathol.* 145: 574-584.

Perrara, N., Houck, K., Jakeman, L., and Leung, D. W., 1992, Molecular and biological properties of neovascular endothelial growth factor family of proteins, *Endocrin Dev.* 13: 18-92.

Poliner, L.S., Tornambe, P.E., Michelson, P.E., and Heitzmann, J.G., 1993, Interferon alpha-2a for subfoveal neovascularization in age-related macular degeneration, *Ophthal.* 100: 1417.

Polverini, P. J., Cotran, R. S., Gimbrone, M. AS., and Unanue, E. R.,1977, Activated macrophages induce vascular proliferation, *Nature* 269: 804-806.

Proia, A. D., Chandler, D. B., Haynes, W. L., Smith, C. F., Suvarnamani, C, Erkel, F. H., and Klintworth, G. K., 1988, Quantitation of corneal neovascularization using computerized image analysis, *Laboratory Investigation* 58: 473-479.

Rastinejad, F., Polverri, P.J., and Bouck, N.P., 1989, Regulation of the activity of a new inhibitor of angiogenesis, *cell* 56: 345-355.

Ruben, M., 1981, Corneal vascularization, *International Ophthalmology Clinics* 21: 27-38.

Shing, Y., Folkman, J., Sullivan, R., Butterfield, C., Murray, J. and Kagsburn, 1984, Heparin affinity: Purification of a tumor derived capillary endothelial cell growth factor, *Science* 223: 1296.

Sholley, M. M., Ferguson, G. P., Seibel, H. R., Hugo, R., Montour, J. L., and Wilson, J. D., 1984, Vascular sprouting can occur without proliferation of endothelial cells, *Laboratory Investigation* 51: 624-634.

White, C.W., Sondheimer, H.M., and Crouch, E.C., 1987, Treatment of pulmonary hemangiomatosis with recombinant interferon-alpha-2a, *N. Eng. J. Med.* 320: 1197.

Ziche, M., Alessandri, G., and Gullino, P.M., 1989, Gangliosides promote the angiogenic response, *Laboratory Investigation* 61: 629-634.

EVIDENCE THAT TENASCIN AND THROMBOSPONDIN-1 MODULATE SPROUTING OF ENDOTHELIAL CELLS

A.E. Canfield[1] and A.M. Schor[2]

[1]University of Manchester and [2]University of Dundee
Manchester M13 9PT and Dundee DD1 4HR, U.K.

Cultured endothelial cells undergo a reversible transition from a resting (cobblestone) to an angiogenic (sprouting) phenotype. This transition mimics the early events of angiogenesis. We have previously reported that the addition of exogenous xylosides inhibits endothelial cell sprouting and modifies the extracellular matrix (ECM) synthesized by the cells (1, 2). We have now investigated whether endothelial sprouting is mediated by the nature of the ECM in contact with the cells. Accordingly, cell-free matrices deposited by bovine aortic endothelial cells (BAEC) were isolated. These matrices were produced under conditions in which the formation of the sprouting phenotype was permitted (controls) or inhibited (by the addition of exogenous xylosides). BAEC were then plated on these matrices and grown under conditions which promote sprouting. Sprouting proceeded normally on control matrices, whereas it was inhibited when the cells were grown on matrices deposited in the presence of xylosides. The composition of the permissive and inhibitory matrices was then analysed. Inhibitory matrices contained reduced levels of tenascin and increased levels of thrombospondin-1 by comparison to the permissive matrices. In contrast, no differences were detected in the relative levels of laminin. The roles of tenascin and thrombospondin-1 in endothelial sprouting were confirmed using specific antibodies. Immunolocalisation studies revealed the presence of both proteins in sprouting cells. Antibodies to tenascin inhibited the formation of sprouting cells on permissive matrices and on gelatin-coated dishes without affecting cell growth. Tenascin synthesis was increased when sprouting cells were present in the cultures. Antibodies to thrombospondin-1 stimulated sprouting on inhibitory matrices. These results suggest that the transition from a resting to a sprouting phenotype is promoted by tenascin and inhibited by thrombospondin-1.

References

1.   Schor, A.M. and Schor, S.L. Inhibition of endothelial cell morphogenetic interactions *in vitro* by alpha- and beta-xylosides. *In Vitro* Cell Dev. Biol. 24, 659-668, 1988.
2.   Canfield, A.E., Sutton, A.B., Hiscock, D.R.R., Gallagher, J.T. & Schor, A.M. Alpha- and beta-xylosides modulate the synthesis of fibronectin and thrombospondin by endothelial cells. Biochim. Biophys. Acta, 1200, 249-258, 1994.

ANGIOSTATIC AND ANTI-INVASIVE THERAPY OF HUMAN MALIGNANT GLIOMAS USING BATIMASTAT (BB-94)

P.C. Costello[1], R.J. Sedran[1], T. Yazaki[2], C. Minniti[2], A. Chalivi[2], P. Brown[3], J. Rak[1] and R.L. Martuza[2]

[2]University of Toronto, [2]Georgetown University and [3]British Biotech Pharmaceuticals
Toronto, Ontario, Canada, Washington, DC, USA and Oxford, UK

Introduction: Gioblastome multiforme, the most malignant and common of human brain tumors are highly vascularized and are refractory to present treatment regimes. This class of tumors represent a likely target for angiostatic therapy. In this study, human cerebral microvessel endothelium derived from neurosurgical specimens and human tumor cells from glioblastoma multiforme were utilized for *in vitro* models of angiogenic and invasive properties. Xenografts of human glioblastoma multiforme cell lines were employed for *in vivo* analysis.

Methods: The following experiments were designed to assess the therapeutic potential of BB94 in the treatment of human malignant glial brain tumors. *In vitro*: 1) The effects of BB-94 on the growth rate of human glioma and cerebral microvascular endothelial cells in culture as assessed by cell counts and ^{3}H-thymidine incorporation.

2) The effects of BB-94 on the invasiveness (invasion chambers) and motility (Boyden chambers) of human glioma and cerebral microvascular endothelial cells in culture.

3) The inhibition of type IV collagenase activity released from glioma and endothelial cells as detected using zymography and ^{3}H-collagen activity assays.

In vivo: 1) The effects of BB-94 on the growth rate of a human malignant glioma implanted sub-cutaneously into nude mice.

2) The effect of BB-94 on the extent of vascularization of human malignant assessed by immunohistochemistry (Factor VIII).

Results: *In vitro:* BB-94 inhibited glioma and endothelial cell 3H-thymidine incorporation at 10 µg/ml concentrations but was ineffective at therapeutic doses of 1.0 and 0.1 µg/ml doses. Soft agar growth of glioblastoma derived cell lines was inhibited by up to 80% by 5 µg/ml BB-94. BB-94 significantly inhibited the motility and invasion of endothelium from 3 different patient derived 1° cultures and 3 glioma cell lines (U87, U251 and T98G). BB-94 inhibited the Type IV collagenase activity secreted from endothelium by 60% and from glioma cells by 35%.

In vitro: Most interestingly, the growth of sub-cutaneous implants glial tumors was inhibited by 60% upon daily i.p. injection of 50 mg/kg of BB-94. Histological exam of the tumors treated with BB-94 revealed a decrease in vascularity and increased tumor encapsulation when compared to the vehicle only controls.

Conclusions: These results indicate BB-94's potential as a therapeutic for human gliomas and that the mode of action is anti-invasive and angiostatic.

ANTI-ANGIOGENIC DIFFERENTIATION ACTION OF CURCUMIN IN HUMAN UMBILICAL VEIN ENDOTHELIAL CELLS

T. Deepa, P. Prasad, G.S. Sidhu and R.K. Maheshwari

Uniformed Services, University of Health Sciences
Bethesda, MD, USA

Curcumin, a dietary pigment isolated from *Curcuma longa* rhizomes, is a potent anti-oxidant and has been implicated as an inhibitor of tumor initiation and tumor promotion. We explored its anti-proliferative and anti-angiogenic properties in human umbilical vein endothelial cells (HUVEC). Cells were treated with curcumin (0.25-25 uM) for 3, 6, 12, 24 and 32 h and cellular proliferation was assayed by using tretrazolium salts (MTT), incorporation of 3H-thymidine and viable cell counting. Results showed that curcumin inhibited the proliferation of HUVEC in a dose dependent manner. HUVEC cells, when plated on a matrigel undergo angiogenic differentiation. When HUVEC were treated with curcumin either before, after or at time of plating on matrigel, there was a significant inhibition in the angiogenic response. This effect of curcumin was dependent on its concentration in the medium and 10uM curcumin completely inhibited this response. Studies from several laboratories have implicated the involvement of metalloproteases in extracellular matrix rearrangement and angiogenic differentiation. Gelatin zymography of the culture fluid showed that there was a marked inhibition in the activities of 67, 92 and 53 kDa metalloproteases after curcumin treatment. We are investigating whether this inhibition occurs at the transcriptional or posttranscriptional level. The altered levels of metalloproteases after curcumin treatment may play a role in the inhibition of angiogenic response of HUVEC by this compound.

MICROVESSEL COUNT (MC) AND p53 PROTEIN ACCUMULATION ARE STRICTLY CORRELATED AND AFFECT RECURRENCE AND DEATH IN NON SMALL CELL LUNG CARCINOMAS (NSCLC)

G. Fontanini, S. Vignati, D. Bigini, M. Lucchi, A. Mussi, C.A. Angeletti and C. Bevilacqua

University of Pisa, Institute of Pathology and Service of Thoracic Surgery
Pisa, Italy

Recent analysis in several types of human cancers has underlined the influence of new vessel formation (neoangiogenesis) and tumour suppressor gene alterations on tumour development and metastatic progression. The influence of MC (a measure of tumour angiogenesis) and the overexpression of p53 protein on metastatic onset and overall survival has also been demonstrated in Non Small Cell Lung Cancer (1-2). Neoangiogenesis is controlled by the local balance between factors that stimulate new vessel growth and factors that inhibit it. Since an up-regulation of an angiogenesis inhibitor by wild-type p53 protein has recently been proven (3) we analysed on the one hand the prognostic impact of MC and p53 protein accumulation in NSCLC progression, and on the other hand the interrelation between the microvasculature pattern and p53 protein accumulation. 73 patients (66 males and 7 females, mean age of 63 ± 6.2, range 46-79) resected for NSCLC between April 1991 and December 1992 (median follow-up 25 months) were analysed. Frozen and formalin-fixed tumour samples were collected to detect p53 cellular accumulation and to evaluate microvessel count respectively. Pab1801 (Oncogene Science, Manhasset, NY, USA) at 1:200 of diluition was used for p53 detection. Microvessels were highlighted by NCL-END Mab (anti-human haemopoietic progenitor cell antigen CD34) 1:100 of diluition. Avidin Biotin peroxidase methods was used in both cases, developing immunoreactions with diaminobenzidine. The most common histologic type was squamous carcinoma (61.6%). 49 out of 73 tumour (67.1%) were classified as T2; 22 out of 73 (30.1%) had metastatic involvement of hilar and/or mediastinal lymph nodes, and 32 patients developed distant metastases during follow up. 29 patients died for metastatic disease, while 44 were alive at the moment of analysis. P53 immunoreactivity was localized in the nuclei of neoplastic cells; 5% of positive cells was assumed as cut-off value to distinguish p53 negative from p53 positive tumour and in 48 out of 73 (65.7%) tumour samples (mean 45.6 ± 24.1; median 50) a p53 protein overexpression was detected. As we have previously reported, the MC was evaluated carefully counting the intratumoural vessels in the most intense areas of neovascularization at 25X of magnification. The median value of 15 distinguished the tumours with low ($\leq$15 vessels/HPF) or high (>15 vessels HPF) MC. In univariate analysis histotype, T-status, N-status, MC and p53 accumulation were shown to significantly affect recurrence and death. Stepwise logistic regression showed that MC, N-status, and T-status maintained their independent prognostic role on overall survival ($p<0.005$; $p<0.006$; $p<0.006$), whereas MC alone was predictive of recurrence ($p<0.002$). Interestingly, a strong statistical association was observed between p53 nuclear accumulation and MC (linear regression: r=0.41; p=0.0003). Indeed, the mean MC in p53 positive tumours was significantly higher (25.2±14.7) than in negative ones (16.9±10.7) (p=0.01 Unpaired t-test). These results underline that in NSCLC:1) MC has an independent prognostic significance on recurrence and overall survival; 2) alterations of p53 tumour suppressor gene with consequent cellular protein accumulation correlate with an increased tumoral vascularization, supporting the hypothesis that wild-type p53 protein could regulate the angiogenetic process through a positive regulation of some of its inhibiting factors.

Supported by Italian Association for Cancer Research (AIRC).

References

1. Macchiarini, P., Fontanini, G., Hardin, M. et al. Relation of neovascularization to metastasis of non small cell lung cancer. Lancet, 348, 145-146, 1992.
2. Harpole, D., Herndon, J., Wolf, W. et al. A prognostic model of recurrence and death in Stage I Non small cell lung cancer utilizing presentation, histopathology, and oncoprotein expression. Cancer Res., 55, 51-56, 1995.
3. Dameron, K.M., Volpert O., Tainsky, M., Bouck, N. Control of angiogenesis in fibroblasts by p53 regulation of Thrombospondin 1. Science 265, 1582-1584, 1994.

HYPOXIA ENHANCES IN VITRO ANGIOGENESIS OF HUMAN MICROVASCULAR ENDOTHELIAL CELLS CULTURED ON MATRIGEL

Kenneth A. Hahn, Donald H. Schmidt, and Peter I. Lelkes

Univ. Wisc. Med. School, Milwaukee Clin. Campus
Milwaukee, WI, U.S.A.

Ischemia is known to stimulate myocardial collateral vessel formation. The process of neovascularization is presumably initiated by angiogenic growth factors, such as Vascular Endothelial Cell Growth Factor (VEGF) which may be released from hypoxic cardiac myocytes (1). On the other hand, other more ubiquitous growth factors, such as bFGF are also reportedly upregulated in ischemic tissues (2). Recently, hypoxia was shown, at least in some types, to directly activate protein kinase C (PKC), which in endothelial cells is central step in the signal transduction mechanism involved in angiogenesis (3). We hypothesized that in addition to upregulating VEGF in non-endothelial cells, hypoxia might also directly affect endothelial cell angiogenesis by activating PKC in endothelial cells and/or causing the autocrine release of angiogenic growth factors. We tested this hypothesis by culturing human dermal microvascular endothelial cells (HMVEC) on a complex artificial basement membrane, Matrigel. We and others have shown in the past that endothelial cells, cultured on Matrigel will rapidly form lumen-containing capillary-like tubes and that this system is a in vitro model for assessing cellular and molecular mechanisms of angiogenesis (4).

To test our hypothesis we exposed, at the time of plating, HMVEC for various amounts of time (5 -60 min) to hypoxia generated by oxygen deprivation. Subsequently normoxia was restored and the cells were maintained under standard tissue culture conditions. Eight (8) hours later the effect of hypoxia was evaluated by quantitating the length and complexity of the ensuing "capillary network". Optimal enhancement of in vitro angiogenesis was obtained following 40 min exposure to hypoxia (approx. 1% O_2). Under these conditions the capillary length was significantly increased from 237.5 ± 31.25 µm (ctrl, normoxia) to 562.5 ± 78.75 µm (hypoxia) (n=6, p<0.001). In agreement with our hypothesis, tube formation under both hypoxic and normoxic conditions, was enhanced by the PKC activator phorbol-myrystoyl acetate (PMA), and inhibited by the PKC inhibitor, H-7. PKC activation and hypoxia appeared to have an additive effect. Furthermore, the inhibitory effect of an oxygen radical scavenger, superoxide dismutase (SOD), on the formation of "capillaries" under both under normoxic and hypoxic conditions suggests that the generation of free radicals may play a hitherto unrecognized function in angiogenesis. Current experiments are under way to further evaluate the underlying cellular and molecular mechanisms in the context of our hypothesis.

REFERENCES
1. Banai, S., Shweiki, D., Pinson, A., Chandra, M., Lazarovici, G., and Keshet, E. Upregulation of vascular endothelial growth factor expression induced by myocardial ischaemia: implications for coronary angiogenesis. Cardiovasc Res 28:1176-1179, 1994.
2. Schaper, W., and Schaper, J. (eds) Collateral Circulation, KLuwer Academic Publishers, Boston, 1993.

3. Tsopanoglou, N.E., Pipili-Synetos, EW., and Maragoudakis, M. Protein kinase C involvement in the regulation of angiogenesis. J. Vasc. Res 30:202-208, 1993.

4. Grant, D.S., Lelkes, P.I., Fukuda, K., and Kleinman, H.K. Intracellular mechanisms involved in basement membrane induced blood vessel differentiation in vitro. In Vitro Cell Develop. Biol. 27A: 327-336, 1991.

SIMULATED MICROGRAVITY FACILITATES IN VITRO ANGIOGENESIS IN ORGAN-SPECIFIC CO-CULTURES OF ADRENAL PARENCHYMAL AND ENDOTHELIAL CELLS

Daniel L. Galvan, Jim Liu, Brian R. Unsworth, and Peter I. Lelkes

Marquette Univ. and Univ. Wisc. Med. School
Milwaukee, WI, U.S.A.

Angiogenesis is presumed to be primarily mediated by angiogenic growth factors, such as FGF and VEGF (1). Vascular cell interactions are also important regulators, as demonstrated by the inhibition of endothelial cell proliferation in co-cultures with pericytes or parenchymal cells (2). In vitro studies of angiogenesis, using traditional cell culture, are limited by the confinement to two dimensions or employment of exogenous substrates. Simulated microgravity in the NASA rotating wall vessels (RWVs) offers an advantage for three-dimensional tissue culture, by combining reduced shear stress and randomized gravitational vectors, thus allowing for protracted cell-cell interactions and high local concentrations of potentially important paracrine growth factors. Thus, the NASA vessels offer a novel cell culture environment to explore the role of heterotypic cellular interactions in tissue specific differentiation, as already demonstrated in several cellular systems (3).

In extension of our previous studies (4), we cultured rat adrenal medullary endothelial cells (RAME) and chromaffin cell-derived pheochromocytoma cells (PC12) under simulated microgravity conditions. After 20 days in the RWV, PC12 cells in monoculture acquired an enhanced neuroendocrine phenotype, as assessed by enhanced steady-state expression of mRNA for many of the catecholamine synthesizing enzymes as well as for the extracellular matrix proteins collagen IV and fibronectin. However, while large aggregates were formed, no organization reminiscent of adrenal acini was present. Interestingly, RAME cells when cultured as above, formed large aggregates of monolayer covered microcarrier beads, without discernable effects on gene-expression, with the exception of genes for select basement membrane proteins, such as laminin A.

When the two cell types were co-cultured as above, the ensuing aggregates contained structures resembling adrenal acini. Nests of PC12 cells were surrounded on all sides by a flattened endothelial lining. The endothelial cells were polarized, with basal localization of the elongated nuclei. In some instances endothelial-lined vascular spaces were discernable surrounding the acini-like structures, suggestive of neovascularization.

These results suggest that the formation of blood vessels can be mimicked and studied in vitro, given the right culture conditions, such as the unique environment of the NASA vessels and the concomitant presence of organotypic parenchymal cells.

REFERENCES:
1. Folkman, J. and Klagsbrun, M, Angiogenic Factors, Science, 235:442-447, 1987.
2. Orlidge, A. and D'Amore, P., Inhibition of Capillary Endothelial Cell Growth by Pericytes and Smooth Muscle Cells, Journal of Cell Biology, 105:1455-1462, 1987.
3. Goodwin, T., Prewett, T. et al., Reduced Shear Stress: A Major Component in the Ability of Mammalian Tissues to Form Three-Dimensional Assemblies in Simulated Microgravity, Journal of Cellular Biochemistry, 51:301-311, 1993.
4. Mizrachi, Y., Narranjo, J. et al., PC12 Cells Differentiate into Chromaffin Cell Like Phenotype in Co-Culture with Adrenal Medullary Endothelial Cells, Proceedings of the National Academy of Science, USA, 87:6161-6165, 1990.

ANGIOGENIN RECEPTORS ON BOVINE AORTIC SMOOTH MUSCLE CELLS

E. Hatzi, M. Moenner and J. Badet

Lab. CRRET, CNRS URA 1813, INSERM- University of Paris
94010 Créteil, France

Angiogenin (ANG), a 14 kDa polypeptide, has been shown to be a potent inducer of blood vessel formation in experimental models, in vivo, and to exhibit a ribonucleolytic activity (1). A functional enzymatic active site is required to express its angiogenic property. Its presence in normal plasma suggests that it might be involved in endothelium homeostasis. ANG specific receptors have been first demonstrated on endothelial cells (2). In smooth muscle cells (SMCs), ANG has been shown to induce activation of phospholipase C and cholesterol esterification (3). Since interactions between endothelium and vascular SMCs play an important role in the biology of the blood vessel, we have studied the interactions of angiogenin with bovine aortic SMCs.

Human recombinant ANG was shown to affect proliferation and migration of SMCs. Radioligand binding studies using ^{125}I-ANG were performed under equilibrium conditions, at 4°C, on cells at the density of 30,000 cells/cm^2. The saturable interactions were shown to be specific since a large excess of the unlabelled molecule reduced ^{125}I-ANG binding by more than 80%. Analyses of binding data using the LIGAND program suggested two apparent families of interactions on SMCs. Angiogenin (17,000 molecules/cell) bound to high affinity sites with an apparent dissociation constant of 0.2 nM. The second component, with an apparent dissociation constant of 65 nM, involved 3.5 10^6 molecules/cell. The number of cell-binding sites decreased as the cell density increased. These high and low affinity binding sites were cell specific since similar results were obtained on SMCs grown in serum-free conditions. Iodinated ANG was covalently linked to SMC surfaces using bi(sulfosucci-nimidyl)suberate. Several complexes were visualized by autoradiography after SDS-gel electrophoresis of the solubilized cells. A molar excess of 500-fold of the unlabelled molecule abolished the labelling of the complexes that migrated at an apparent molecular mass of 167, 131, 100, 72, 62, 47, 31 kDa; the major ones being at 72, 47, 31 kDa. The multiplicity of the bands obtained suggest complex interactions of ANG with cell membranes.

These results strongly support the existence of specific receptors for angiogenin on bovine aortic smooth muscle cells.

References

1.   Riordan, J.F., Vallee, B.L. Human angiogenin, an organogenic protein. Br. Cancer, 57:587, 1988.
2.   Badet, J., Soncin, F., Guitton, J-D., Lamare, O., Cartwright, T., Barritault, D.

Specific binding of angiogenin to calf pulmonary artery endothelial cells. Proc. Natl. Acad. Sci., 86: 8427, 1989.

3. Moore, F., Riordan, J.F. Angiogenin activates phospholipase C and elicits a rapid incorporation of fatty acid into cholesterol esters in vascular smooth muscle cells. Biochemistry, 29: 228, 1990.

A HIGHLY SULPHATED LAMINARIN SULPHATE INHIBITS ANGIOGENESIS *IN VITRO* AND TUMOUR GROWTH *IN VIVO*

R. Hoffman and D. Paper

MRC Clinical Oncology & Radiotherapeutics Unit and
University of Regensburg
Cambridge, UK and Regensburg, Germany

Angiogenesis is necessary for the growth of solid tumours and so inhibitors of angiogenic molecules may have anti-tumour activity. A number of angiogenic growth factors and cell adhesion molecules only become functional when bound to heparan sulphate co-receptors. Based on these observations, we have synthesized a series of laminarin sulphates with different degrees of sulphation as heparan sulphate analogues and evaluated the effects of these agents on angiogenesis and tumour growth.

Increasing the degree of sulphation increased the activity of the laminarin sulphates as inhibitors of basic fibroblast growth factor (bFGF) binding and as inhibitors of the proliferation of bFGF-dependent foetal bovine heart endothelial (FBHE) cells. The most sulphated laminarin sulphate, LAM S5, (degree of suphation 2.31) inhibited binding of bFGF to low and high affinity sites on BHK cells with IC50 values of 12 ± 8 µg/ml and 69 ± 66 µg/ml respectively, and inhibited bFGF-stimulated DNA synthesis and cell proliferation of FBHE cells with IC 50 values of approx. 1µg/ml. Inhibition of DNA synthesis was reversed by excess bFGF suggesting that bFGF antagonism is involved in the inhibition of FBHE cells.

LAM S5 inhibited tubule formation by human microvessel endothelial cells (HMEC-1) cultured on Matrigel indicating that LAM S5 may have anti-angiogenic activity. This was confirmed in the chick chorioallantoic membrane (CAM) assay. 10µg pellets of LAM S5 produced zones of inhibition on about 70% of the CAMs.

The anti-coagulant activity of LAM S5 was determined since LAM S5 is structurally similar to heparin. Anti-coagulant activity of LAM S5 in the APTT test was 26.2 USP-U/mg. This indicates that plasma levels of about 9.5 µg/ml should be achievable before there is a significant effect on blood coagulation. (Corresponding values in the APTT test for heparin and pentosan polysulphate were 147 and 32.7 USP-U/mg). When administered *in vivo*, maximum tolerated doses of LAM S5 in mice before haemorrhagic effects were observed were about 12 mg/kg given iv daily 5x/week. Using this regimen, LAM S5 produced a tumour growth delay of the murine fibrosarcoma RIF-1 of about 4d. Tumour growth delay was increased to 6d by combining LAM S5 with the corticosteroid tetrahydrocortisol. LAM S5 was only a weak inhibitor of the proliferation of RIF-1 cells *in vitro* (IC50 about 50µg/ml), suggesting that inhibition of angiogenesis may be involved in the inhibition of RIF-1 tumour growth *in vivo*.

MICROVASCULAR NETWORK REMODELING IN THE NEONATAL RAT BRAIN

A.G. Hudetz, G. Fehır, M.L. Schulte

The Departments of Anesthesiology and Physiology, Medical College of Wisconsin
Milwaukee, WI 53226, U.S.A.

An intriguing characteristic of the ontogenic development of cerebral vasculature is the rapid differentiation of cerebral surface microvessels (leptomeningeal vascular plexus) during the first three postnatal weeks (1). The physiological and cellular mechanisms of this cerebrovascular remodeling process are unclear. We previously hypothesized that blood flow or wall shear stress in microvessels played a role in the regulation of network differentiation (2). The objective of our present work was to determine and correlate changes in microvascular density, network pattern and flow velocity in the leptomeningeal vascular plexus of the rat during postnatal development *in vivo*. To this end, microvascular diameter, segment length, and vascular density of the leptomeningeal vascular network were measured from video-recordings of the microcirculation visualized through a cranial window in 0-15 day old Sprague-Dawley rats. Velocity of erythrocytes in the microvessels was measured by frame tracking of fluorescently labeled red blood cells.

We found that the density of surface microvessels decreased significantly by the second week after birth. Anastomosing vascular polygons, characteristic of newborn networks, became less numerous and larger in diameter during the postnatal two week period indicating progressive rarefaction of the networks. Vessel diameter and red cell velocity showed transient increases at 1.5 weeks. The velocity/diameter ratio, however, increased by the age of 1.5 weeks and remained unchanged afterwards.

We conclude that postnatal remodeling of the leptomeningeal vascular network is associated with rarefaction and increased flow velocity/vessel diameter ratio, suggesting an increase in microvascular wall shear rate. These changes may contribute to further vascular differentiation and redistribution of blood flow from superficial to intracortical vasculature in the developing brain.

REFERENCES:
1. Bar, T.H.., Miodonski, A., Budi Santoso, A.W.: Postnatal development of the vascular pattern in the rat telencephalic pia-arachnoid. Anat. Embryol. 174:215-223, 1986.
2. Hudetz, A.G., Kiani, M.F.: The role of wall shear stress in microvascular network adaptation. Adv. Exp. Med. Biol. 316:31-39, 1992.

This work was supported by the National Science Foundation Grants BCS-9001425 and BES-9411631.

COMPUTER SIMULATION OF THE IN VITRO TUBE FORMATION BY ENDOTHELIAL CELLS CULTURED ON MATRIGEL

Antal G. Hudetz[*], Vangelis G. Manolopoulos and Peter I. Lelkes

(*) Departments of Anesthesiology and Physiology, Medical College of Wisconsin and Laboratory of Cell Biology, Department of Medicine, University of Wisconsin Medical School
Milwaukee, WI USA

Matrigel, the complex basement membrane isolated from EHS tumors, is an established system for the study of cellular and molecular mechanisms of *in vitro* angiogenesis (1). Visually, this *in vitro* angiogenic process bears a remarkable similarity to vascular development *in vivo*, for example during embryonic development in the brain (2). Based on the previous, successful modelling (3) of the angiogenic process, we hypothesized that we could use a similar algorithm to model *in vitro* tube formation of endothelial cells seeded onto Matrigel. In the process of this modelling we anticipate that we will learn details about some of the biophysical/biochemical parameters which govern the formation of complex capillary-like networks. Based on video-microscopic observations of the time-course, extent, and shape of "*in vitro* angiogenesis", we developed a first-order simulation with a set of boundary conditions which correspond to the following physiological parameters:
_ cell density
_ average size of round and elongated cells
_ plating at random
_ cell migration at constant speed and against a gradient of an "angiogenic-factor" generated   by surrounding cells
_ gradual elongation of migrating cells

The results of our initial modelling indicate:

The modelling critically depends on realistic values for cell density and velocity of cell migration and elongation.

Random migration, based on nearest neighbor-interaction, without taking into consideration the existence of a field, corresponding to soluble, "angiogenic factors"/receptor, does not result in formation of any tubes, but in cell clumping. This result is reminiscent of the reality of fibroblasts, plated on Matrigel.

A purely diffusion-based angiogenic factor field (_ r^{-2}), results in some form of incipient tube formation. However, the patterns of migration and cell accumulation are unrealistic.

By including a second factor, assigned to the sensitivity of the cells to the "angiogenic factor" probably via a receptor, we modified the field to be _r^{-4}. Under this conditions, we obtained a more realistic simulation, including the formation of complex, capillary-like networks and branching.

267

Our initial results suggest that this may be a useful approach to test some of the mechanistic hypotheses of endothelial cell tube formation on Matrigel.

REFERENCES:
1. Grant DS, Lelkes, PI, Fukuda K, Kleinman HK. Intracellular mechanisms involved in basement membrane induced blood vessel differentiation in vitro. In Vitro Cell._Dev. Biol. 27A:327-336, 1991.
2. Bar TH, Molodonski A, Budi Santoso AW. Postnatal development of the vascular pattern in the rat telencephalic pia-arachnoid. Anat. Embryol. 174:215-223, 1986
3. Kiani MF, Hudetz AG. Computer simulation of growth of anastomosing vascular networks. J. Theor. Biol. 150:547-560, 1991

ANGIOGENESIS IN CHRONIC PANCREATITIS AND PANCREATIC CARCINOMA

R. Kuehn, C. Bloechle, A. Niendorf[#], J.R. Izbicki, P.I. Lelkes[*]

Dept. Surg., Univ. , ([#]) Dept. Pathol., Univ. Hamburg and ([*]) Lab.
Cell. Biol., Dept. Med., Univ. Wisc. Med. School
Hamburg, Germany and Milwaukee, WI, USA

Progression from chronic inflammatory alterations in chronic pancreatitis (CP) to malignant transformations of pancreatic carcinomas (PCA) has not been proven yet, although ample evidence
is found for histological and molecular similarities between both conditions. A previously reported reduction of the vascular supply contrasts with recent findings, that CP is accompanied by an overexpression of angiogenic growth factors (1). Based on previous studies on tumorangiogenesis (2) we hypothesize that angiogenesis might play a pivotal role in the etiology and histopathology of CP and PCA. **Aim of the study:** Angiogenic activity is compared in CP vs. PCA by assessing microvascular density and characterizing the expression of Vascular Endothelial Growth Factor (VEGF) (3). Furthermore activation antigens like ICAM and VCAM are examined regarding their diagnostic or prognostic significance in CP and PCA. **Materials and methods:** 5µm paraffin sections of surgical specimens from 18 patients with CP and 10 with PCA were immunostained for the specific endothelial markers von Willebrand Factor (vWF) and PECAM-1, as well as for cell adhesion molecules ICAM-1, VCAM-1. The microvascular density was evaluated by quantifying the number of positively stained vessels. Per microscopic field ($0.679mm^2$ at 380x) the absolute number and relative density of vessels in a "hot spot" area were determined both by manual counting and by computer aided digitized color image analysis. All values were normalized and statistically evaluated using ANOVA (significance at $p<0.001$). **Results:** The mean number of vessels per mm^2 is significantly increased in cancerous tissue (132±53.3) and in CP (97.1±32.1) compared to controls (n=14; 14.0±4.9). The number of blood vessels in PCA is significantly higher than that in CP. The relative vessel density is significantly increased in both PCA (41.3±11.2%) and CP (30.6±11.1%) vs. controls (12.0±5.6%) and also in PCA vs. CP. Enhanced expression of VEGF is mainly detected in ductal and centroacinar cells as well as in tumor cells. Strong expression of ICAM-1 is predominantly seen in ductal cells whereas VCAM is mostly expressed in acinar cells. **Conclusions:** In support of our hypothesis, our results indicate that the vascular supply in both CP and PCA is increased, although the mass of functional exocrine parenchyma is decreased. The expression of cell adhesion molecules is enhanced under both pathological conditions.

REFERENCES:
1. Friess H et al. Increased expression of acidic and basic fibroblast growth factors in

chronic pancreatitis. Am J Pathol 1994, 144: 117-128.

2. Folkman J. What is the evidence that tumors are angiogenesis dependent. J Natl Cancer Inst 1989, 82 (1):4-6.

3. Ferrara N et al. The Vascular Endothelial Growth Factor family of polypeptides. J Cell Biochem 1991, 47: 211-218.

FACTITIOUS ANGIOGENESIS III: CAPITALIZING ON ANGIOGENIC MECHANISMS TO CREATE BIOCOMPATIBLE CARDIOVASCULAR PROSTHESES AND FUNCTIONAL ORGANOIDS

Peter I. Lelkes, Mark M. Samet, Dawn M Wankowski, Viktor V. Nikolaychik, Valery S. Chekanov, Daniel L. Galvan (*) and Brian R. Unsworth (*)

University of Wisconsin Medical School, and (*) Marquette University, Milwaukee, WI, USA

We recently coined the term "factitious angiogenesis" (1,2) to denote the deliberate use of angiogenic processes and mechanisms for biotechnological purposes, e.g., for creating biocompatible cardiovascular prostheses, such as artificial vascular grafts, cardiac assist devices and total artificial hearts. In this presentation, we will review novel developments in the field and exemplify the successful application of "factitious angiogenesis" with recent results from the literature as well as from ongoing studies in our laboratory. Specifically, we will discuss:

• use of angiogenesis-promoting experimental conditions (e.g., addition of angiogenic growth factors) to facilitate endothelial cell coverage of novel porous small diameter and large-bore vascular grafts.

• induction of angiogenesis in the process of endothelialization of skeletal muscle ventricles

• promotion of angiogenesis in enhancing the efficacy of "classical" cardiomyoplasty.

All these examples can be classified as belonging to the exciting new field of "tissue engineering", .i.e. the regeneration of biological tissues *in vitro* by using innovative cell-culture modalities. Indeed, recent heterotypic co-culture studies have shown that the re-formation of functional tissues (organoids) from isolated constituent cells, critically depends on the generation of new endothelial cell-lined blood vessel-like structures within the "organoids". Implantation experiments are under way to test whether these "neo-vessels" will anastomose to the vasculature of a host organ.

REFERENCES :

1. Lelkes, P.I., Samet, M.M., Christensen, C.W. and Amrani, D.L.. (1992) Factitious angiogenesis: endothelialization of artificial cardiovascular prostheses. In: Angiogenesis in health and diseases (Maragoudakis, M.E., Gullino, P. and Lelkes, P.I., eds.), Plenum Press, New York and London, pp. 339-351.

2.. Lelkes, P.I., Chick, D.M., Samet, M.M., Nikolaychik, V., Thomas, G.A., Hammond, R.L., Isoda, S. and Stephenson, L.W. (1994) Factitious Angiogenesis: Not so factitious anymore? The role of angiogenic processes in the endothelialization of artificial cardiovascular prostheses. In: Angiogenesis: Molecular Biology, Clinical Aspects (M.E. Maragoudakis, P. Gullino, and P.I. Lelkes eds.), Plenum Press, New York and London, pp. 321-331.

ENDOTHELIAL CELL HETEROGENEITY AND ITS IMPORTANCE IN ANGIOGENESIS

Peter I. Lelkes, Vangelis G. Manolopoulos, Kenneth A. Hahn, Soverin Karmiol(*), Mark M. Samet, Matthew D. Silverman, Dawn M. Wankowski, Shaosong Zhang, and Brian R. Unsworth(#)

Univ. Wisc. Med. School, (#) Marquette Univ., and (*) Clonetics Corp. Milwaukee, WI and San Diego, CA, U.S.A.

An increasing body of experimental evidence suggests that endothelial cells are as heterogeneous as the blood vessels and lymphatics from which they are derived. Angiogenesis, the formation of new blood vessels, is closely related to the cellular and molecular biology of the "endothelium". In this presentation we will further advance our contention of the importance of endothelial cell heterogeneity, as reflected by distinct phenotypic and genotypic differences between EC derived from various vascular beds. We will exemplify functional EC heterogeneity with recent results from the literature as well as from ongoing studies in our laboratory. These studies underline the differential regulation of a number of angiogenesis-related phenomena in cultured endothelial cells derived from large vessels and from the microvasculature, for example:
• "in vitro angiogenesis", e.g., on Matrigel
• production of extracellular matrix proteins under static and dynamic   conditions
• production of basement membrane protein- degrading enzymes (collagenases)
• signal transduction mechanisms, such as cAMP, PKC and their regulation by angiogenic agonists
• effects of mechanical forces (cyclic strain, flow-induced shear stress) on cell adhesion molecules (ICAM -1, VCAM -1, PECAM -1)
• effects of mechanical forces (cyclic strain, hydrostatic pressure) on the regulation of cell surface protein expression (e.g. tissue factor)
Understanding the cellular and molecular basis of endothelial cell heterogeneity will be an important step towards developing selective pharmacological tools for organ-specific control (promotion or inhibition) of angiogenesis in health and disease.

REFERENCES
1. Lelkes, P.I., and Unsworth, B.R. (1992) The role of heterotypic interactions between endothelial cells and parenchymal cells in organspecific differentiation: a possible trigger for vasculogenesis. In: Angiogenesis in health and diseases (Maragoudakis, M.E., Gullino, P. and Lelkes, P.I., eds.) pp. 27-41.
2. Lelkes, P.I., Manolopoulos, V.G., Chick, D. and Unsworth, B.R. (1994) Endothelial cell heterogeneity and organ-specificity. In: Angiogenesis: Molecular Biology, Clinical Aspects (M.E. Maragoudakis, P. Gullino, and P.I. Lelkes eds.), <u>Plenum Press</u>, New York and London, pp. 15-28.

THERAPEUTICALLY ENHANCED ANGIOGENESIS AFTER CARDIOMYOPLASTY

V.V. Nikolaychik, P.I. Lelkes and M.M. Samet and G.V. Tchekanov and V.N. Chekanov

Sinai Samaritan Medical Center, University of Wisconsin Medical School Milwaukee, Wisconsin 53233, USA

Cardiomyoplasty is aimed at assisting failing myocardium in its contractile function by transposing a stimulated, fatigue-resistant skeletal muscle and wrapping it around the myocardial wall. Postoperatively, effective healing and functionality of the "repaired" heart relies on efficient neovascularization of the wounded area. However, experience shows that such angiogenic response does not occur because the traumatized, ischemic skeletal muscle and mobile myocardial wall have compromised vasculature, low angiogenic potential, and are not in a tight contact with each other.

A novel approach to enhance wound healing through induction of neovascularization, termed *"therapeutic angiogenesis,"* relies on use of pharmacological and biological factors. Using autologous fibrin glue as a bioactive interface, we adopted this approach to cardiomyoplasty. To this end, the following steps were included in the surgical procedure: **a.** reduction of tissue lesions (resulting from local ischemia reperfusion) by delivering iron-chelating compounds (deferoxamin) and protease inhibitors (aprotinin) into the skeletal muscle, **b.** activation of angiogenesis in donor skeletal muscle and on recipient myocardial site before intervention, **c.** reinforcement of topical muscle-to-muscle contact at the site of the desired neovascularization through modulation of the fibrin interface which, with time, become organized into immature vascularized tissue.

To date we have concluded 4 procedures which, within 2 month period, have demonstrated the prospect and effectiveness of this strategy. At present, experiments are underway in which further integration of the bioactive interface is accomplished by incorporating in fibrin glue meshwork viable cells derived from vascular wall.

Supported by the Milwaukee Heart Research Foundation

DIFFERENTIAL REGULATION OF EXTRACELLULAR MATRIX PROTEIN SECRETION AND DEPOSITION BY CULTURED ENDOTHELIAL CELLS

E. Papadimitriou[#@], M.E. Maragoudakis[#], V.G. Manolopoulos[@], B.R. Unsworth[*] and P.I. Lelkes[@]

[#]Dept of Pharmacology, Univ of Patras Medical School, [@]Dept of Cell Biology, Univ of Wisconsin Medical School, [*]Dept of Biology, Marquette Univ.
Patras, Greece; Milwaukee, Wisconsin, USA

Endothelial cells in culture synthesize numerous extracellular matrix proteins that are either deposited into the extracellular matrix, or released into the culture medium. In the present work we studied, by either direct or indirect enzyme-linked immunoassays, which of the well known intracellular signal transduction pathways, namely those of adenylate cyclase, protein kinase C and calcium mobilization, might be involved in the secretion and deposition of the extracellular matrix proteins, fibronectin, laminin, collagen-IV and collagen-I, by cultured endothelial cells. Increase of the intracellular cAMP levels by either forskolin or dib-cAMP caused an increase in the deposition of all the ECM proteins studied, with a concomitant decrease in the corresponding amounts released into the culture medium. Activation of protein kinase C (PKC) by PMA, on the other hand, resulted in a decrease in the ECM protein deposition, without any effect on the amounts of the same molecules released into the culture medium. Finally, an increased ECM deposition was also the result of intracellular calcium chelation, which was followed by a decrease in the corresponding amounts released into the culture medium of RAME cells.

REFERENCES:

1. Papadimitriou, E., Unsworth, B.R., Maragoudakis, M.E. and Lelkes, P.I. "Time-course and quantitation of extracellular matrix maturation in the chick chorioallantoic membrane and cultured endothelial cells." Endothelium 1(3): 207-219,1993.

2. Papadimitriou, E., Manolopoulos, V.G., Maragoudakis, M.E., Unsworth, B.R and Lelkes, P.I. "Thrombin receptor activation increases the deposition of extracellular matrix proteins in cultured endothelial cells." Submitted for publication.

Supported by a collaborative grant from NATO to M.E. Maragoudakis and P.I. Lelkes.

A LOW MOLECULAR WEIGHT FRACTION FROM SHARK CARTILAGE INHIBITS ANGIOGENESIS AND COLLAGENASE TYPE IV PRODUCTION

Platon Peristeris, Paraskevi Andriopoulou, Eleftheria Missirlis and Michael E. Maragoudakis

University of Patras Medical School, Department of Pharmacology, 26110 Rio, Patras, Greece

In view of the avascular nature of cartilage this tissue was investigated as potential source of angiogenesis inhibitors. Different fractions from shark cartilage (SC) have been shown to inhibit tumour neovacularisation (1) and angiogenesis in the rabbit cornea and the chick chorioallantoic membrane (CAM) (1,2). The antiangiogenic and antitumor activity has been proposed to be the result of direct inhibition of interstitial collagenase (collagenase I) (2).

In the present study a 2-10 kDa fraction of SC was isolated after 4M HCl-Guanidine extraction and fractionation by ultrafiltration. Analysis by SDS-PAGE of the 2-10 kDa fraction showed a single proteic band of approximately 9 kDa.

In the CAM assay in vivo this fraction produced a dose-dependent inhibition of angiogenesis ranging from -28.5 ± 6.8 % to -78 ± 3.9 % at 2.5 µg/disc and at 50 µg/disc respectively as measured by the biosynthesis of collagenous proteins as an index of angiogenesis. This was confirmed by morphological evaluation. The antiangiogenic activity was unaffected by heat, acid, alcali, trypsin, heparinase and chondroitinase ABC treatment while papain digestion caused a 70% reduction. The SC fraction, at 25 µg/ml, also prevented tube formation by human umbilical vein endothelial cells in the Matrigel system.

Antiangiogenic action of SC appears to be unrelated to direct inhibition of type IV collagenase, purified from a Walker 256 carcinoma extract. On the contrary, collagenase type IV production from isolated chick fibroblasts was completely blocked by addition of 25 µg/ml of SC in the culture medium. The 2-10 kDa SC fraction also reversed the proangiogenic effect of phorbol esters (PMA) and this may explain the mechanism of action by inhibiting protein kinase C, which is known to regulate angiogenesis (3).

References

1. Oikawa, T., H. Ashino-Fuse, M. Shinamura, U. Koide and T. Iwagushi, A novel1 angiogenic inhibitor derived from Japanese shark cartilage. Cancer Letters 51, 181-186, 1990.
2. Lee, A. and R. Langer, Shark cartilage contains inhibitors of tumor angiogenesis. Science 221, 1185-1187, 1983.
3. N.E. Tsopanoglou, G.C. Haralabopoulos and M.E. Maragoudakis. Opposing effect on modulation of angiogenesis by protein kinase C and cyclic AMP-mediated pathways. J. Vasc. Res. 31, 195-204, 1994.

DIFFERENTIALLY EXPRESSED GENES DURING ANGIOGENESIS

E. Pröls, S. Voit, and M. Marx

University of Erlanger-Nürnberg, Dept. of Nephrology

Erlangen 91054, Germany

Neovascularization of the tissue is induced in several pathological processes, such as the progression of tumors or in inflammatory events. Thus, the specific inhibition of angiogenesis is in focus of clinical interest.

The goal of this work is to inhibit pathologically occuring angiogenesis by either overexpression of angiogenesis repressing genes or by inhibiting the expression of angiogenesis promoting genes. Prerequisite for this is the identification of genes specifically involved in the process of angiogenesis. This has been successfully achieved by comparing the gene expression pattern of endothelial cells cultured in two different culture systems: the conventional two-dimensional (2D) and the three-dimensional (3D) culture system, where the cells are embedded in collagen type I gels. In the 2D-system, the cells proliferate and grow as a monolayer until confluency whereas the cells in the 3D-system stop proliferating and differentiate into capillary tubes (1). The gene expression pattern of the endothelial cells derived from both culture systems were compared using the differential display method published by Liang and Pardee (2).

By means of this powerful technique, three differentially expressed genes were identified and their expression pattern verified by northern blot analysis. Two of the genes are expressed during *in vitro* angiogenesis (i.e. in the 3D-system) but not in the proliferating, nondifferentiating cells of the 2D system. The third gene is highly expressed in the 2D systme and less expressed in the initial phase of the 3D system with increasing intensity over the time course up to 72 hours.

Mesangial cells, which were studied in parallel, showed no modulation in gene expression with respect to the three genes of interest suggesting the observed gene expression pattern being specific for endothelial cells and angiogenetic processes.

A cDNA-library was constructed to get hand on the complete cDNA sequences and to be enabled to characterize the isolated genes by functional assays.

References

1. Merwin, J.R., Anderson, J.M., Kocher, O., van Itallie, C.M., and J.A. Madri. Transforming growth factor β1 modulates extracellular matrix organization and cell-cell junctional complex formation during in vitro angiogenesis. J. Cell Physiol. 142:117-128, 1990.
2. Liang, P. and A.B., Pardee. Differential display of eukaryotic messenger RNA by means of polymerase chain reaction. Science 257:967-971, 1992.

ASSESSMENT OF ANGIOGENESIS IN HUMAN CERVICAL LESIONS

Ravazoula P.[1], Zolota V.[1], Gerokosta A.[1], Hanjiconti O.[2] and M.E. Maragoudakis[3]

University of Patras Medical School, Department of Pathology[1], Department of Radiology[2] and Department of Pharmacology[3] 261 10 RIO, Patras, Greece

Angiogenesis has been extensively studied in invasive carcinomas and has been independently correlated with tumor growth and metastasis. Previous studies however, have not revealed the correlation between microvessel density and conditions that take place much before the onset of tumor formation (i.e. dysplastic lesions).

To determine when neovascularization is initiated, tumor vascularity was quantified in a series of cervical lesions : 92 dysplasias (31 mild, 24 moderate and 36 severe), 11 infiltrating squamous cell carcinomas and 3 HPV infections. Microvessels were visualized by monoclonal antibody against factor VIII-related antigen (DAKO), using streptavidin - peroxidase immunohistochemical method. Vessel density was quantified in 3 high power fields(hpf) of the most vascular areas, by two independent observers and the mean was recorded for these 3 counts.

Mean vessel counts for each category of cervical lesions were as follows:

Lesion category	Number	Mean vessel count per hpf	
Normal (adjascent to dysplasia)	36	9 (4-17)	
			$p<0,001$
CIN I	31	13 (4-25)	
			$P<0.01$
CIN II	25	17 (12-25)	
			$p<0,05$
CIN III	36	20(10-44)	
Inf. carcinomas	11	17 (12-28)	
HPV lesions	3	16 (15-19)	

As it is shown there was a statistically significant increase of vascularity in the dysplastic lesions compared with normal cervical epithelium as well as with progressively greater atypia. No significant differences in vascularity were noted between severe cervical dysplasias and infiltrating carcinomas.

Our findings suggest that angiogenesis may be an important event in tumor initiation and conversion of normal epithelium into a cancer. This hypothesis can be explained by the knowledge that several angiogenic molecules are products of protooncogenes that may be activated at the initial steps of carcinogenesis.

THYMOSIN-β4 ENHANCES ANGIOGENESIS IN MOUSE LYMPHOID ENDOTHELIAL CELL LINE

G.S. Sidhu, A.K. Singh, T. Deepa, P. Prasad, V. Karthik*, D.S. Grant and R.K. Maheshwari

Uniformed Services, University of Health Sciences
Jefferson Medical College*
Bethesda, MD and Philadelphia, PA, USA

We have studied the potential role of thymosin-β4 gene (Thyβ4), which is over expressed in human umbilical vein endothelial cells (HUVEC) on matrigel. SVEC4-10 cells, a mouse lympoid endothelial cell line were co-transfected with pThyβ4 expressing Thyβ4 and pcDNA3 containing neomycin resistance marker by electroporation. Transiently transfected cells express Thyβ4 in the cytoplasm and produced tube networks within 2-4 h on matrigel. The tube formation was 2-3 fold higher in transfected cells as compared to controls. Our lab has previously shown that human IFNα_{2a} enhanced the tube formation in a dose dependent manner (Maheshwari et al, 1991), therefore, we examined the expression of Thyβ4 in IFNα_{2a} treated cells. Northern analysis revealed a significant increase (3-4 fold) in mRNA transcripts in human IFNα_{2a} treated HUVEC cells as compared to control, suggesting a possible mechanism by which IFNα may enhance angiogenesis. These studies suggest a possible role for Thyβ4 in vessel formation or angiogenic differentation and would facilitate the development of agonists and antagonists for effective treatments of pathogenic neovascularization such as in Kaposi's sarcoma and malignant tumor growth. We have also developed a stably transduced SVEC4-10 cell line in which the expression of Thyβ4 is inducible by isopropyl-β-D-thiogalactoside (IPTG). Thyβ4 gene was cloned in the vector pOPRSV1CAT which contains lac operator and was co-transfected with lac repressor expressing vector p3'SS by calcium phosphate mediated transfection with glycerol shock after transfection. In this cell line, the expression of Thyβ4 can be reversibly turn on and off in order to investigate its physiological role in angiogenic differentation.

References

1. Maheshwari, R.K., Srikantan, V., Bhartiya, D., Kleinman, H.K., Grant, D.S. Differential effects of interferon gamma and alpha on in vitro model of angiogenesis. Journal of Cellular Physiology 146, 164-169, 1991.

LIPOPROTEIN (α) AS AN ACUTE-PHASE PROTEIN IN REPAIR OF TISSUE INJURY AND ANGIOGENESIS

D. Sgoutas, D. Apanay, Y-C Wang and O. Lattouf

Emory University, School of Medicine
Atlanta, GA 30322, USA

This study investigates changes in plasma lipoproteins and compares them to changes in proteins like C-reactive protein, and α_1-antitrypsin in patients undergoing coronary artery bypass grafting (CABG). CABG involves use of extracorporeal circulation by use of a cardiopulmonary bypass pump, which necessitates hemodilution due to the infusion liquids. Throughout the procedure and in the postoperative period a significant decrease is observed in all plasma lipoprotein concentrations, except lipoprotein (α) [Lp(α)]. Lp(α) shows a significant rise during CPB, peaking at about 80% of basal levels at 2 hrs after the start of CPB and it remains high throughout the postoperative period. The higher-density Lp(α) particles which contain the high molecular-weight apolipoprotein (α) [apo(α)] isoforms increased more than the lower-density and consequently the lower-molecular-weight containing isoforms of apo(α). The results suggest that Lp(α) may be considered an acute-phase protein, because it responded to CPB in the acute-phase mode resembling well known acute-phase proteins, such as C-reactive protein and α_1-antitrypsin. This property of Lp(α) is attributable to the apo(α) component which shows an aminoacid sequence homology with corresponding domains of plasminogen. Because of this apo(α) component, Lp(α) is an inhibitor of plasminogen activation and it is associated with thrombogenic systems. This suggest a regulatory function of Lp(α) in thrombogenesis especially where repair of injured tissues is needed. Immunohistochemical studies have already shown that Lp(α) may play an important role in the repair of injured tissues, especially during angiogenesis.

References

1. Sgoutas, D.S., Lattouf, O.M., Finlayson, D.C. and Clark, R.V. Paradoxical response of plasma lipoprotein(a) in patients undergoing cardiopulmonary bypass, Atherosclerosis, 97: 29-36, 1992.
2. Maeda, S., Abe, A., Seishima, M., Makino, K., Noma, A. and Kawade, M. Transient changes of serum lipoprotein(a) as an acute phase protein, Atherosclerosis, 10:672, 1989.
3. Loscaisio, J. Lipoprotein(a). A unique risk factor for atherothrombotic disease. Arteriosclerosis, 10:672, 1990.

ANALYSIS OF MICROVASCULAR DENSITY IN HUMAN BREAST AND LUNG CARCINOMA: METHODOLOGICAL STUDIES

R.L. Smither, N. Pendleton, K. Lessan, E. Heerkens, J. Morris, and A.M. Shor

Paterson Institute for Cancer Research, Univ. Dept. of Geriatric Medicine, Dept. of Dental Surgery & Periodont., Univ. of Dundee Manchester, U.K.

High microvascular density (MVD) in the area of most intense vascularisation has been shown to be associated with poor prognosis in many tumour types (1). However, some studies have found that tumour MVD did not have prognostic value, and it has also been suggested that such conflicting results may have resulted from differences in the methodology used, or in heterogeneity in the tumour vasculature.

In this study we have examined two methodological aspects which are relevant to the assessment of MVD in histological sections of breast and lung carcinoma, namely: histochemical procedures and tumour heterogeneity.

Our data demonstrated no significant differences between staining with two distinct endothelial markers (CD31 and VWF) or between three different methods of assessing vascular density. Tumour vascular heterogeneity was analysed in a group of 4 lung carcinoma (A, B, C, and D) using FVIII as vascular marker. Blocks were taken from three regions of the tumour: the periphery, the centre of the tumour, and midway between these. There was microvascular heterogeneity between multiple blocks taken from each of the three regions $(F(2,117)=62: p<0.001$ and $F(2,417)=116: p<0.001$ respectively). However, the variation between tumours was significantly greater than this (Tumours B, C > A, D: $F(3,43) = 9,1, p<0.001$). We then assessed whether determining the microvascular density in a single section is sufficient to provide an estimate of the overall microvascular density in a given tumour. The highest MVD found for each tumour was taken to rank the tumours as an "approximate ranking" (B=C>A=D). Selecting random sections from each tumour resulted in the correct approximate ranking in 68% of the cases.

Our results highlight the need to determine the level of confidence for different indices of angiogenesis before these may be successfully applied to the clinic.

Reference

1. Gasparini, G. et al., Tumour microvessel density, p53 expression, tumour size, and peritumoral lympatic vessel invasion are relevant prognostic markers in node-negative breast carcinoma. J. Clin. Oncol. 12(3), 454-466, 1994.

INHIBITION OF METALLOPROTEINASES ALTERS PROTEIN SYNTHESIS AND MORPHOGENESIS BY ENDOTHELIAL CELLS: IMPLICATIONS FOR MATRIX MODULATION OF ANGIOGENESIS

A.B. Sutton, A.E. Canfield, B. McLaughlin, J.B. Weiss, A. Howell, and A.M. Shor

Christie Hospital, University of Manchester, Hope Hospital and Dundee University
Manchester, Dundee, U.K.

BB94 (Batimastat) is a synthetic, low molecular weight, inhibitor of matrix metalloproteinases (MMPs) that has been shown to convert ascites tumours into avascular solid tumours (1). This observation suggests that BB94 may be anti-angiogenic. We have therefore examined the effects of BB94 on various aspects of endothelial cell behaviour in vitro which are germane to angiogenesis. Collagenase, gelatinase and stromelysin activities were detected in the conditioned medium of endothelial cell cultures. The activities of these enzymes were effectively inhibited by BB94 at concentrations of between 1 and 10µM. At these concentrations, BB94 did not effect endothelial cell proliferation or migration; it did, however, inhibit endothelial cell sprouting and morphogenesis induced by high concentrations of serum, vascular endothelial cell growth factor (VEGF) or basic fibroblast growth factor (bFGF). That is, BB94 inhibited the transition of endothelial cells from a cobblestone to a sprouting phenotype and the subsequent association of these sprouting cells into complex networks and tubule-like structures. This drug was also able to disrupt an established sprouting cell network. These findings indicate that BB94 may inhibit both the early and late stages of angiogenesis. The inhibition of endothelial morphogenesis by BB94 was reversible. This drug was also found to have effects that were apparently unrelated to its anti-enzymatic activity. Namely, BB94 specifically decreased the relative levels of thrombospondin-1, fibronectin and two high molecular weight proteins (Mr>220KDa) secreted by the cells. These effects occurred in the absence of any changes in total protein synthesis or the steady state mRNA levels of thrombospondin-1 and fibronectin. Our data indicate that BB94 may be useful in anti-angiogenic and/or anti-tumour vasculature therapies. It is unclear, however, whether this therapeutic potential is due to the drug's anti-MMP activity or to its ability to modulate protein synthesis.

Reference

1. Davies, B., Brown P.D., Crimmin, M.J. & Balkwill, F.R. A synthetic matrix metalloproteinase inhibitor decreases tumour burden and prolongs survival of mice bearing human ovarian carcinoma xenografts. Cancer Research, 53, 2087-2091, 1993.

ANGIOGENIC ACTIVITY OF EDMF (ENDOTHELIOMA DERIVED MOTILITY FACTOR)

G. Taraboletti, V. Vergani, D. Belotti, A. Garofalo, P. Borsotti, A. Bastone and R. Giavazzi

Institute of Pharmacological Research "Mario Negri"
24125 Bergamo, Italy

As a model to study molecular mediators of endothelial cell recruitment which occurs in the formation of vascular tumors, we have used an endothelioma cell line (eEnd.1), transformed by polyoma virus middle T oncogene, which *in vivo* forms hemangioma-like tumors. We found that endothelioma cells *in vitro* released an angiogenic factor (named EDMF- endothelioma derived motility factor), a 40-65 kDa heparin-binding protein (1). EDMF induced endothelial cell chemotactic and haptotactic motility (migration in response to soluble and substrate-bound attractant, respectively) (1), increased matrix metalloproteinases production by endothelial cells and augmented their ability to migrate through a layer of reconstituted basement membrane (Matrigel), a process that requires matrix digestion by the invading cells. *In vivo*, EDMF stimulated angiogenesis, measured as hemoglobin content of a pellet of Matrigel containing the angiogenic factor, implanted s.c. (2).

In order to clarify the role of EDMF-stimulated metalloproteinases production in vascular tumors and endothelial cell recruitment, we have used batimastat (BB94, British Biotech, Ltd., Oxford, UK), a synthetic matrix metalloproteinase inhibitor. The effect of batimastat was studied on tumor formation by endothelioma cells injected subcutaneously in nude mice. Daily topic treatment with batimastat (30-0.3 mg/kg) inhibited tumor growth, with a significant increase in doubling time and animal survival (2). *In vitro* batimastat did not affect endothelioma or endothelial cell proliferation, nor endothelial cell invasion through Matrigel (2). *In vivo*, batimastat inhibited EDMF-induced angiogenesis (2). These findings indicate that the production of metalloproteinases is a key step in the process of endothelial cell recruitment, and that treatments which affect the activity of these enzymes might inhibit angiogenesis and the growth of vascular tumors by blocking endothelial cell recruitment.

References

1. Taraboletti, G., Belotti, D., Dejana, E., Mantovani A., and Giavazzi, R. Endothelial cell migration and invasiveness are induced by a soluble factor produced by murine endothelioma cells transformed by polyoma virus middle T oncogene. Cancer Res. 53:3812-3816, 1993.
2. Taraboletti, G., Garofalo, A., Belotti, D., Drudis, T., Borsotti, P., Scanzianni, E., Brown, P.D., and Giavazzi, R. Inhibition of angiogenesis and murine hemangioma growth by batimastat, a synthetic inhibitor of matrix metalloproteinases. J. Natl. Cancer Inst. 87: 293-298, 1995.

DETECTION AND QUANTITATION OF BRAIN ENDOTHELIAL THROMBOMODULIN mRNA

N. D. Tran, V. Ly Wong, S. Schreiber, J. Bready* and M. Fisher

University of Southern California, School of Medicine and
Amgen Corporation*
1333 San Pablo St., MCH 246 Los Angeles, California 90033, USA

Thrombomodulin (TM), an endothelial cell (EC) membrane protein, plays a regulatory role in the protein C anticoagulant pathway. TM is decreased in brain regions where infarction is common in humans (Brain Res. 556:1-5). The ability to detect TM mRNA and quantitate small changes in TM mRNA concentration may provide insight into mechanisms that predispose the brain to ischemic events. We analyze TM mRNA in an angiogenesis model in which bovine brain EC, grown between two collagen matrices, can form capillary-like structures. In situ hybridization shows TM mRNA localized to the EC. We also use a quantitative-competitive polymerase chain reaction assay for the quantitation of TM mRNA. This assay uses an internal control cDNA which co-amplifies with the target cDNA using the same primers, thus increasing accuracy by significantly reducing the differences in primer annealing efficiency. This quantitative competitive PCR method is capable of detecting less than one picogram of TM mRNA from relatively small numbers of cells, and can be used to study TM mRNA regulation.

PARTICIPANT PHOTO

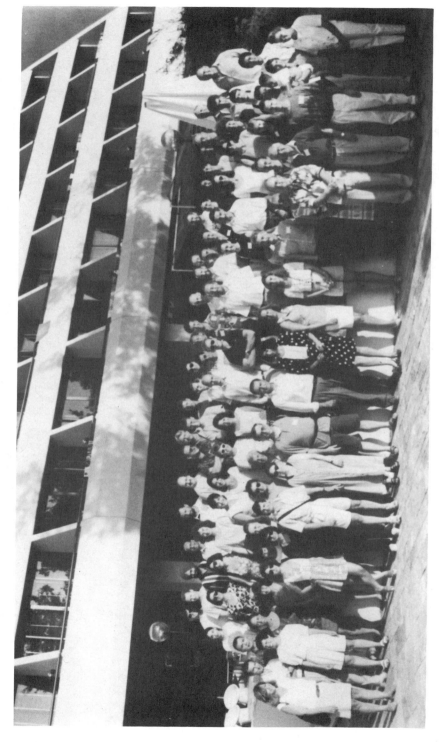

Participants of the NATO Advanced Studies Institute "Molecular, Cellular and Clinical Aspects of Angiogenesis" held at Porto Carras, Sithonia Hotel, Halkidiki, Greece, during 16-27 June, 1995

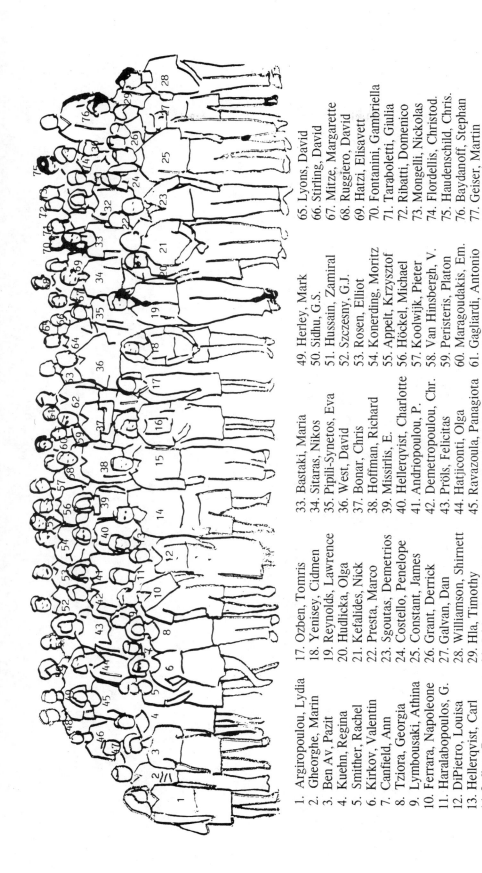

1. Argiropoulou, Lydia
2. Gheorghe, Marin
3. Ben Av, Pazit
4. Kuehn, Regina
5. Smither, Rachel
6. Kirkov, Valentin
7. Canfield, Ann
8. Tziora, Georgia
9. Lymbousaki, Athina
10. Ferrara, Napoleone
11. Haralabopoulos, G.
12. DiPietro, Louisa
13. Hellerqvist, Carl
14. Lelkes, Peter
15. Maragoudakis, Mich.
16. Marmara, Anna

17. Ozben, Tomris
18. Yenisey, Cidmen
19. Reynolds, Lawrence
20. Hudlicka, Olga
21. Kefalides, Nick
22. Presta, Marco
23. Sgoutas, Demetrios
24. Costello, Penelope
25. Constant, James
26. Grant, Derrick
27. Galvan, Dan
28. Williamson, Shirnett
29. Hla, Timothy
30. Schlenger, Karlheinz
31. Ciomei, Marina
32. Papadimitriou, Lily

33. Bastaki, Maria
34. Sitaras, Nikos
35. Pipili-Synetos, Eva
36. West, David
37. Bonar, Chris
38. Hoffman, Richard
39. Missirlis, E.
40. Hellerqvist, Charlotte
41. Andriopoulou, P.
42. Demetropoulou, Chr.
43. Pröls, Felicitas
44. Hatjiconti, Olga
45. Ravazoula, Panagiota
46. Nor, Anne-Adina
47. BenEzra, David
48. Rose, Wesley

49. Herley, Mark
50. Sidhu, G.S.
51. Hussain, Zamiral
52. Szczesny, G.J.
53. Rosen, Elliot
54. Konerding, Moritz
55. Appelt, Krzysztof
56. Höckel, Michael
57. Koolwijk, Pieter
58. Van Hinsbergh, V.
59. Peristeris, Platon
60. Maragoudakis, Em.
61. Gagliardi, Antonio
62. Zacharakis, George
63. Thompson, Douglas
64. Winkler, James

65. Lyons, David
66. Stirling, David
67. Mitze, Margarette
68. Ruggiero, David
69. Hatzi, Elisavett
70. Fontanini, Gambriella
71. Taraboletti, Giulia
72. Ribatti, Domenico
73. Mongelli, Nickolas
74. Flordellis, Christod.
75. Haudenschild, Chris.
76. Baydanoff, Stephan
77. Geiser, Martin

287

PARTICIPANTS

ANDRIOPOULOU, P.

Instituto di Richerche Pharmacolog. "Mario Negri", Via Eritrea 62, 20157 Milano, ITALY

APPELT, K.

Agouron Pharmac. Inc., 3565 General Atomics Court San Diego, California 92121 1121, USA

BASTAKI, M.

University of Brescia, Department of Science, Biomedicine and Biotechnology, Via Valsabbine 19, I-25123 Brescia, ITALY

BAYDANOFF, S.

Dept. of Biology & Immunology, University School of Medicine, "ST. KLIMENT OHRIDSKI" str1, 5800 Pleven, BULGARIA

BENEZRA, D.

Department of Ophthalmology, Hadassah Medical Organization, Kiryat Hadassah, PO Box 12000, II-91120 Jerasulem, ISRAEL

CANFIELD, A.

School of Biological Sciences, University of Manchester, 2205, Stopford Building, Oxford Road, Manchester M13 9PT, UNITED KINGDOM

CATRAVAS, J.

Department of Pharmacology and Toxicology, Medical College of Georgia, Augusta, GA 30912-2300, USA

CIOMEI, M.

Pharmacia Research Center, R & D Experimental Oncology Department, Via per Pogliano, 20014 Nerviano (MI), ITALY

CONSTANT, J.

UCSF Wound Healing Laboratory, Department of Surgery, 513 Parnassus HSW 1652, San Francisco, CA 94143-0522, USA

COSTELLO, P.	Health Science Center, Cancer Biology Division, 2075 Bayview Avenue, Toronto, Ontario M4N 3M5, CANADA
DEMETROPOULOU, C.	Anatomisches Institut, Univ. of Mainz, Johannes Guteberg, Beckerweg 13, 55099 Mainz, GERMANY
DIPIETRO, L.	Loyola University Medical School, Burn & Shock Trauma Institute, Bldg. 110, Room 4251, 2160 South First Avenue, Maywood, Illinois 60153, USA
FERRARA, N.	Genentech Inc., 460 Point San Bruno Bldg., South San Francisco, CA 94080, USA
FISHER, M.	University of Southern California, School of Medicine, Dept. of Neurology: Stroke Research, 1333 San Pablo str., MCH 246 Los Angeles, California 90033, USA
FLORDELLIS, C.	University of Patras, Medical School, Dept. of Pharmacology, 265 00 Patras, GREECE
FONTANINI, G.	Institute of Pathology, University of Pisa, via Roma 57, 56126 Pisa, ITALY
FRIEDLANDER, M.	The Scripps Research Institute, Dept. of Cell Biology-Imm-11, 10666 North Torrey Pines Road, La Jolla, California 92037, USA
FOTSIS, T.	Univ. of Heidelberg INF 326, Room 222, 69120 Heidelberg 1, GERMANY
GAGLIARDI, A.	University of Kentucky, Dept. of Obstetrics & Gynecology, Room MN 318, 800 Rose Street, Lexington, Kentucky 40536-0084, USA
GALVAN, D.	University of Wisconsin, Department of Medicine, Mount Sinai Campus, 950 North Twelfth St., PO Box 342, Milwaukee WI53201-0342, USA
GEISER, M.	CIBA-GEIGY Ltd., CH-4002 Basle, SWITZERLAND
GHEORGHE, M.	Str. Panduri No. 3, Bloc P32, Et. 8, Ap. 30, Sector 5, 76229 Bucharest, ROMANIA
GOLDBERG, I.	Dept. of Radiation of Oncology, Long Island Jewish Med. Center, 270 05 76th Ave., New Hyde Park, NY 11040, USA

GRANT, D.	Thomas Jefferson Univ. Hospital, Department of Medicine, Cardeza Foundation for Hematol. Res., 1015 Walnut Str., Curtis Bldg., Room 703 Philadelphia, PA 19107-5099, USA
HARALABOPOULOS, G.	University of Patras Medical School, School of Health Sciences, Department of Pharmacology, 265 00 Rio, Patras, GREECE
HATZI, E.	Universite Paris val de Marné, Lab. de Recherche sur la Croissance Cellulaire, CRNS URA 1813 - Pr. Denis Barritault, Avenue du Général de Gaulle, 94010 Creteil, FRANCE
HATZICONTI, O.	University of Patras Medical School, School of Health Sciences, Department of Pharmacology, 265 00 Rio, Patras, GREECE
HATZIOANNOU, A.	Patras Endocrinology Clinic, Pantanassis 45 str., Patras, GREECE
HAUDENCHILD, C.	American Red Cross, The Jerome Holland Laboratory, 15601 Crabbs Branch Way, Rockville, MD 20855, USA
HELLERQVIST, C.	Vanderbilt University Sch. of Medicine, Department of Biochemisty, Nashville, Tennessee 37232-0146, USA
HELLERQVIST, CH.	Carbomed Inc., 5115 Maryland Way, Brentwood, TN 37027, USA
HERBST, C.	Anatomisches Institut, Univ. of Mainz, Johannes Guteberg, Beckerweg 13, 55099 Mainz, GERMANY
HERLEY, M.	The W. Alton Jones, Cell Science Centre, 10 Old Barn Road, Lake Placid, NY 12946, USA
HLA, T.	American Red Cross, Dept. of Mol. Biology, Biomed. Res. & Development, The Jerome Holland Laboratory, 15601 Crabbs Branch Way, Rockville, MD 20855, USA
HOCKEL, M.	University of Mainz, Dept. of Gynecology and Obstetrics, Langenbeck str. 1, 55131 Mainz, GERMANY

HOFFMAN, R.

MRC Clinical Oncology and Radiotherapeutics Unit, Medical Research Council Center, Hills Road, Cambridge CB2 2QH, UNITED KINGDOM

HUDLICKA, O.

Department of Pathology, The University of Birmingham, Vincent Drive, Birmingham B15 2TJ, UNITED KINGDOM

HUSSAIN, H.

Department of Surgery, University of California, Parnassus Ave., Room HSE 839, San Francisco, USA

KEFALIDES, N.

University of Pennsylvania, University City Science Center, 3624 Market Street, Philadelphia, Pennsylvania 19104, USA

KIRKOV, V.

Dept. of Clinical Pharmacology, Faculty of Medicine, University Hospital "Queen Giovanna", S Beio more St., 1527 Sofia, BULGARIA

KONERDING, M.

Anatomisches Institut, Univ. of Mainz, Johannes Guteberg, Beckerweg 13, 55099 Mainz, GERMANY

KOOLWIJK, P.

TNO Prevention and Health, P.O. Box 2215, 2301 CE Lieden, THE NETHERLANDS

KRITIKOU, S.

University of Patras Medical School, School of Health Sciences, Department of Pharmacology, 265 00 Rio, Patras, GREECE

KUEHN, R.

University of Hamburg, Department of Surgery, Hamburg, GERMANY

LAKKA, L.

Patras Endocrinology Clinic, Pantanassis 45, Patras, GREECE

LELKES, P.I.

University of Wisconsin, Department of Medicine, Mount Sinai Campus, 950 North Twelfth St., PO Box 342, Milwaukee WI53201-0342, USA

LYMBOUSSAKI, A.

A. Papagou 116, 157 72 Zografou, Athens, GREECE

MARAGOUDAKIS, M.E.

University of Patras Medical School, School of Health Sciences, Department of Pharmacology, 265 00 Rio, Patras, GREECE

MISSIRLIS, E.	University of Patras Medical School, School of Health Sciences, Department of Pharmacology, 265 00 Rio, Patras, GREECE
MITZE, M.	University of Mainz, Institute of Pathology, Langenbeck str. 1, 55101 Mainz, GERMANY
MONGELLI, N.	Pharmacia SpA, Department of Chemistry, Via Bisceglie 96, 20152 Milano, ITALY
MURPHY, C.	University of Heidelberg, Clinical Research Units, INF 326, Room 222, 69120 Heidelberg 1, GERMANY
NOR, A.A.	Str. Cetatea de Balta, Nr. 112-114, Bl. 7, SC. D., Ap. 33 Sector 6, Bucuresti, ROMANIA
OZBEN, T.	Department of Biochemistry, School of Medicine, Akdeniz University, 07070 Kampus, Antalya, TURKEY
PAPADIMITRIOU, E.	University of Patras Medical School, School of Health Sciences, Department of Microbiology, 265 00 Rio, Patras, GREECE
BEN AV, P.	American Red Cross, Molecular Biology Dept., Holland Laboratory, 15601 Crabbs Branch Way, Rockville, MD 20852, USA
PERISTERIS, P.	University of Patras Medical School, School of Health Sciences, Department of Pharmacology, 265 00 Rio, Patras, GREECE
PIPILI-SYNETOS, E.	University of Patras Medical School, School of Health Sciences, Department of Pharmacology, 265 00 Rio, Patras, GREECE
PRESTA, M.	University of Brescia, Department of Science, Biomedicine and Biotechnology, Via Valsabina 19, I-25123 Brescia, ITALY
PROLS, F.	Department of Nephrology, University of Erlangen-Nurnberg, Loschgestrabe 8 1/2, 91054 Erlangen, GERMANY
RAVAZOULA, P.	University of Patras Medical School, Department of Pathology, 265 00 Rio, Patras, GREECE

REYNOLDS, L.	North Dakota State University, Dept. of Animal & Range Service, Hultz Hall, PO Box 5727, Fargo, ND 58105-5727, USA
RIBATTI, D.	Institute of Human Anatomy, Histology and Embryology, Policlinico Piazza Giulio Cesare, 11, I-70124 Bari, ITALY
ROSE, R.W.	Thomas Jefferson Univ. Hospital, Department of Medicine, Cardeza Foundation for Hematol. Res., 1015 Walnut Str., Curtis Bldg., Room 703 Philadelphia, PA 19107-5099, USA
ROSEN, E.	Dept. of Radiation of Oncology, Long Island Jewish Med. Center, 270 05 76th Ave., New Hyde Park, NY 11040, USA
RUGGIERO, D.	Diabetic Microangiopathy Research Unit, LIPHA/INSERM U 352 INSA Lyon, Bldg 406, 20 avenue Albert Einstein, 69261 Villeurbanne Cedex, FRANCE
SCHLENGER, K.	University of Mainz, Dept. of Gynecology and Obstetrics, Langenbeckerstrabe 1, 55131 Mainz, GERMANY
SIDHU, G.S.	Department of Pathology, Uniformed Services, University of the Health Sciences, 4301 Jones Bridge Road, Bethesda, MD 20814-4799, USA
SGOUTAS, D.	Clinical Chemistry Laboratory, Emory University Hospital, Room F-153C, Atlanta, Georgia 30322, USA
SMITHER, R.	Christie CRC Research Center, Paterson Institute for Cancer Res., Christie Hospital NHS Trust, Wilmslow Road, Manchester M20 9BX, UNITED KINGDOM
SKOTSIMARA, P.	University of Patras Medical School, School of Health Sciences, Department of Toxicology, General Hospital of Patras "Agios Andreas", Patras, GREECE
STIRLING, D.	Celgene Corporation, 7 Powder Horn Drive, P.O. Box 4914, Warren, New Jersey 07059, USA
SZCZESNY, G.J.	Dept. of Transplantation Surgery Polish Academy of Sciences, Warsaw Chatubinskiego 5, POLAND

TARABOLETTI, G.

Instituto di Richerche Farmacologiche"Mario Negri", Laboratori Negribergamo, Via Gabazzeni 11, 24100 Bergamo, ITALY

THOMPSON, D.

University of Aberdeen, Department of Pathology, University Medical Bldgs, Foresterhill Aberdeen, AB9 2ZD, UNITED KINGDOM

TRESSLER, R.

Cancer Project, Department of Pharmacology, GLYCOMED Incoroporated, 860 Atlantic Avenue, Alameda, CA 94501, USA

TSOPANOGLOU, N.

University of Patras Medical School, School of Health Sciences, Department of Pharmacology, 265 00 Rio, Patras, GREECE

TZIORA, G.

University of Patras Medical School, School of Health Sciences, Department of Pharmacology, 265 00 Rio, Patras, GREECE

VAN HINSBERGH, V.W.M.

TNO Gaubius Institute, PO Box 430, 2300 AK Leiden, NETHERLANDS

WEST, D.

Department of Immunology, University of Liverpool, PO Box 147, Liverpool L69 3BX, UNITED KINGDOM

WILLIAMSON, S.

Thomas Jefferson Univ. Hospital, Department of Medicine, Cardeza Foundation for Hematol. Res., 1015 Walnut Str., Curtis Bldg., Room 703 Philadelphia, PA 19107-5099, USA

WINKLER, J.

Department of Inflammation Pharmacology, SmithKline Beecham Pharmac., 709 Swedeland Road, King of Prussia, PA 19406, USA

YENISEY, C.

Adnan Menderes University, Faculty of Medicine, Department of Biochemistry, 09100 Aydin, TURKEY

ZACHARAKIS, G.

26 East Lane, Wembley Park, London HA9 7NR, UNITED KINGDOM

INDEX